Eighth International Conference on Dissociative Recombination: Theory, Experiments and Applications 2010

AA002800

(DR 2010)

Journal of Physics: Conference Series Volume 300

Lake Tahoe, California, USA
16-20 August 2010

Editors:

Steven L. Guberman **Ann E. Orel**

ISBN: 978-1-61839-045-5
ISSN: 1742-6588

Printed from e-media with permission by:

Curran Associates, Inc.
57 Morehouse Lane
Red Hook, NY 12571

Some format issues inherent in the e-media version may also appear in this print version.

Printed by Curran Associates, Inc. (2011)

For permission requests, please contact the Institute of Physics at the address below.

Institute of Physics
Dirac House, Temple Back
Bristol BS1 6BE UK

Phone: 44 1 17 929 7481
Fax: 44 1 17 920 0979

techtracking@iop.org

Additional copies of this publication are available from:

Curran Associates, Inc.
57 Morehouse Lane
Red Hook, NY 12571 USA
Phone: 845-758-0400
Fax: 845-758-2634
Email: curran@proceedings.com
Web: www.proceedings.com

TABLE OF CONTENTS

REVIEWS

EXPERIMENT

NEW RINGS

THEORY

PLASMAS

ASTROCHEMISTRY AND AERONOMY

Journal of Physics: Conference Series **300** (2011) 011001 doi:10.1088/1742-6596/300/1/011001

Dedication – Susan L Greenblatt

Figure 1. Susan in May, 1994.

This volume is dedicated to the memory of Susan L Greenblatt, the wife of Steven L Guberman. Susan attended 6 of the 8 dissociative recombination (DR) meetings. Her advice and wise counsel played a vital role in the organization of several of these meetings. The fifth meeting in Chicago in 2001 was her idea and it would not have occurred without her encouragement.

Susan was always amused by the memory of the first group dinner at the second DR meeting at St Jacut in 1992. As we went around the dinner table identifying ourselves, it soon became her turn. Susan was a sociologist and after introducing herself she said: "I am not a chemist". A spontaneous chorus of attendees proclaimed "Neither are we!". Her husband and a few other chemists abstained.

In 1983, Susan and I established the Institute for Scientific Research (ISR). The name was chosen so as to span sociology and chemical physics. Four years prior, an ophthalmologist had diagnosed a rare retinal condition of unknown origin and advised her to change her profession to one that did not involve reading. (She was able to read for the rest of her life.) Twenty years later we learned that the cause of the retinal and all her other health problems was a recently discovered rare mitochondrial mutation. Her experience with ophthalmologists and her life-long keen sense of injustice, led her to write a grant proposal to the US Department of Education to survey all ophthalmologists in the US to determine whether they were aware of and whether they told their patients about resources and aids that could help them to continue reading and participating in everyday activities. As part of the grant and based upon the survey results, she proposed to set up low-vision training programs for ophthalmology residents. We knew that the competition for funding was intense and included several well-known and more established organizations. Nevertheless, the proposal was funded for three years and ISR had its first grant. After the completion of the survey, Susan put together a team of dedicated ophthalmologists who conducted the training at medical schools at Stanford, the University of Chicago and at a joint program of the Washington, DC area medical schools. The program was also presented at several national ophthalmology conferences. One result of the grant was the establishment, with a colleague, of another non-profit organization, Resources for Rehabilitation, which published books and recordings on low-vision and other disabilities for patients and professionals. Susan always felt that a measure of the value of one's life is the degree to which one can improve the lives of others. By this measure alone, she was remarkably successful.

Susan was able to attend the seventh DR meeting in Ameland, The Netherlands, but only because we were able to travel with a peritoneal dialysis machine to which she was connected each evening and for about 30 minutes during the afternoons. I shall always remember with sincere gratitude, the generosity and kindness of other attendees who helped us carry the machine on and off the ferry boat, the bus and the train on our way to and from the two meeting hotels. After a kidney transplant in June 2009, it was our hope that she would be able to attend the 2010 meeting. Susan would have enjoyed renewing many old friendships and making new ones.

Steven L Guberman

Journal of Physics: Conference Series **300** (2011) 011002 doi:10.1088/1742-6596/300/1/011002

Preface – Eighth International Conference on Dissociative Recombination (DR2010)

In dissociative recombination (DR), a molecular ion captures an electron and falls apart. One might initially conclude that this process is relatively simple. However, a review of the DR literature over the past 60 years, since the pioneering papers of Sir David Bates [1], shows that the study of DR has proved to be highly challenging for both theorists and experimentalists. Both this challenge and the ubiquitous importance of the process have made its study rewarding, endlessly interesting, and full of surprises.

DR is fundamentally the interaction of an electronic continuum with a vibrational continuum often connected by quasibound intermediate states. Autoionization in the first continuum competes with dissociation in the second. DR occurs at the total energy of an ion plus a free electron and is in the same region as vibrationally excited Rydberg states and other neutral superexcited states. All these states interact with each other prior to dissociation, forming a complex that controls the quantum yields of products. The complexity of the process has required both innovation and ingenuity on the part of both theorists and experimentalists. Over the years, it has become apparent that in order to compare theoretical and experimental results, the experiments must identify the electronic, vibrational and for low mass ions, the rotational state. The traditional disagreement between theoretical and experimental results has been due, in part, to the sensitivity of DR cross sections and rate constants to the state of the recombining ion. Great progress has been made in recent years in both flowing afterglow and storage ring experiments aimed at reducing this uncertainty.

The latest advances and insights in theory and experiment were the topics of the Eighth International Conference on Dissociative Recombination held at the Granlibakken Lodge at Lake, Tahoe, California, August 16–20, 2010. The conference was part of a series of conferences on DR that started at Chateau Lake Louise, Alberta, Canada in May, 1988 [2] and was followed in May 1992 [3] at L'Abbaye de Saint Jacut de la Mer, Brittany, France, in May, 1995 [4] at Ein Gedi, Israel, in June 1999 [5] on the island of Nässlingen in the Stockholm archipelago, Sweden, in August, 2001 [6] at Chicago, USA, in July, 2004 [7] at the Alte Mälzerei, Mosbach, Germany and in July, 2007 [8] at the Resort d'Amelander Kaap on the island of Ameland, The Netherlands. All papers from the last two conferences and this conference are freely available at http://iopscience.iop.org/1742-6596.

In keeping with the tradition of prior DR conferences, all papers in this volume have been refereed. Our thanks go to the referees for their efforts.

Travel support for conference participants was provided by NSF grant ATM-0838061 and NASA grant NNX09AQ73G to SLG. We thank Priscilla Kujawski for proofreading the Dedication.

Steven L Guberman
Institute for Scientific Research, 22 Bonad Road, Winchester, MA 01890, USA
Ann E Orel
Department of Applied Science, University of California, Davis, One Shields Ave., Davis, CA 95616, USA
Editors

Journal of Physics: Conference Series **300** (2011) 011002 doi:10.1088/1742-6596/300/1/011002

Participants of the 8th International Conference on Dissociative Recombination: Theory, Experiments and Applications.

1. Stephen Pratt
2. Chris Greene
3. Bastiaan Braams
4. Ed Grant
5. Christian Nordhorn
6. Steen Brønsted Nielsen
7. Dermot Madden
8. Radek Plašil
9. Daniel Savin
10. Jonathan Tennyson
11. Peet Hickman
12. Michael Fogle
13. Waffeu Tamo Francois Oliver
14. Christian Jungen
15. Ilya Fabrikant
16. Wolf Geppert
17. Oldřich Novotný
18. Randy Vane
19. Claude Krantz
20. Xavier Urbain
21. Hidekazu Takagi
22. Brian Mitchell
23. Andreas Wolf
24. Daren Stotler
25. Slava Kokoouline
26. David Schultz
27. Mourad Telmini
28. Ruth Malenda
29. Slim Chourou
30. Petr Dohnal
31. Julia Stützel
32. Ioan Schneider
33. Nicholas Shuman
34. Holger Kreckel
35. Robert Continetti
36. Henrik Buhr
37. Mats Larsson
38. Dirk Schwalm
39. Evelyne Roueff
40. Pascal Pernot
41. Stefan Rosén
42. Rainer Johnsen
43. Xiaohong Cai
44. Dan Haxton
45. Åsa Larson
46. Dahbia Talbi
47. Ann Orel
48. Steven Guberman
49. Jane Fox
50. Richard Thomas
51. Fangfang Ruan

References

[1] Bates D R 1950 Electron recombination in Helium *Phys. Rev.* **77** 718
Bates D R 1950 Dissociative recombination *Phys. Rev.* **78** 492

[2] Mitchell J B A and Guberman S L (ed) 1989 *Dissociative Recombination: Theory, Experiment and Applications* (World Scientific: Singapore)

[3] Rowe B R, Mitchell J B A and Canosa A (ed) 1992 *Dissociative Recombination: Theory, Experiment and Applications* (Plenum Press: New York)

[4] Zajfman D, Mitchell J B A, Schwalm D and Rowe B R (ed) 1996 *Dissociative Recombination: Theory, Experiment and Applications III* (World Scientific: Singapore)

[5] Larsson M, Mitchell J B A and Schneider I F (ed) 2000 *Dissociative Recombination: Theory, Experiment and Applications IV* (World Scientific: Singapore)

[6] Guberman S L (ed) 2003 *Dissociative Recombination of Molecular Ions with Electrons* (Kluwer/Academic Plenum: New York)

[7] Wolf A, Lammich L and Schmelcher P (ed) 2005 Sixth International Conference on Dissociative Recombination: Theory, Experiment and Applications *J. Phys.: Conf. Ser.* **4**

[8] Zande W J van der 2009 Seventh International Conference on Dissociative Recombination: Theory, Experiment and Applications *J. Phys.: Conf. Ser.* **192**

All papers published in this volume of *Journal of Physics: Conference Series* have been peer reviewed through processes administered by the proceedings Editors. Reviews were conducted by expert referees to the professional and scientific standards expected of a proceedings journal published by IOP Publishing.

Comments on developments in the theory of dissociative recombination over the past 60 years

Steven L. Guberman

Institute for Scientific Research, 22 Bonad Road, Winchester, MA 01890, USA

slg@sci.org

Abstract. Since the proposal, sixty years ago, of the direct dissociative recombination (DR) mechanism by Bates, there have been many important theoretical advances, especially within the past 30 years. Space limitations preclude the review of all significant theoretical contributions. Instead, I comment upon a few selected developments. Highlights of the early scattering theory of DR and of the powerful Multichannel Quantum Defect Theory (MQDT) are discussed. Enhancements to the latter approach have included the addition of second and all order K matrices, ion rotational motion, Rydberg states with excited cores and spin-orbit coupling between Rydberg states. Also discussed are DR by Born-Oppenheimer breakdown, angular product distributions in diatomic DR and oscillations in DR cross sections.

1. The beginning

It was a mild evening on December 31, 1949 in Times Square, New York City as 750,000 [1] gathered to celebrate the arrival of the New Year. But this was no ordinary crowd. Struggling to make his way to a hotel, through the throngs of celebrants, was a British physicist of 33 years who held a one year U. S. Navy fellowship that started at 12:01 AM. [2] David Bates had just arrived by air from London and with no room to carry his suitcase at his side, he had to hug it. Already an important contributor to our understanding of atmospheric physics, he would soon be recognized as one of the most talented molecular physicists of his generation. In the coming year, he would propose a mechanism for a fundamental but poorly understood process in molecular physics, dissociative recombination. This process would eventually be found to be omnipresent in the ionospheres of the Earth, other planets and their moons, in interstellar clouds and necessary to the interpretation of many new laboratory experiments. After a brief stay in Pasadena, "I spent a month at Princeton University at the invitation of Lyman Spitzer.... In that stimulating environment, I also perceived the mechanism of dissociative recombination. This remains my most original contribution in atomic physics...Before 1950 ended, I fully appreciated what a marvellous country the United States is for scientific research ... I sometimes wonder if I should have stayed here. Perhaps I would have stayed except for one factor, one vital factor: Nobody offered me a job [2]."

The inspiration for the proposed mechanism came from the 1949 paper of Biondi and Brown [3]. Using a microwave afterglow in He, they reported measurements of the ambipolar diffusion coefficient. There is no mention of dissociative recombination (DR) in the title or abstract. In what can only be described as an exceedingly bold step, they also reported the measurement of the dissociative recombination rate coefficient, 1.7 x 10^{-8} cm^3/sec, i.e. the coefficient for $He_2^+ + e^- \rightarrow He + He$. At that time, no firm theoretical foundation for such a process existed and in fact both the noncrossing rule and the great mass mismatch between the incoming electron and the nuclei made it appear unlikely. DR had been mentioned previously in attempts to understand the green glow of the ionosphere [4] and the magnitude of the electron concentration in the ionosphere [5]. In the experiment, the rate coefficient was determined by plotting the inverse electron density versus time. For a binary process, as for DR, the slope of the straight line should give the recombination coefficient. In fact, although Biondi and Brown had raised the pressure in order to suppress diffusion, we now know that the measurement was for ambipolar diffusion and not recombination. Had they extended the plot to longer times, it is possible that they may not have obtained a straight line. Bates argued [6], in a one page paper, that the recombination observed by Biondi and Brown could not be atomic because that process was known to have a much smaller rate coefficient than that reported [3]. Instead, the recombination had to be molecular ion recombination [6]. From the vantage point of 60 years, it is clear that the prediction of Bates was based upon an incorrectly interpreted experiment. Another microwave discharge afterglow experiment [7] in Ne gas obtained a DR rate coefficient of 1.1 x 10^{-7} cm^3/sec. In this case, the inverse electron density was plotted over a much shorter time than in the He afterglow. One needs to remember that in these first pioneering experiments many other factors, including impurities, may have contributed to the magnitude of the reported rate constants [8]. However, the rate constant for Ne_2^+ is in reasonable agreement with more modern measurements [9]. Moved by the experimental results, another one page paper by Bates [10] gave an expression for the direct DR rate coefficient by writing it as the product of a capture rate constant and a probability for dissociation competing with autoionization. He claimed that the expression gave rate constants consistent with the experimental measurements although the necessary theoretical input data was only poorly known at that time. In the direct mechanism, the electron is captured into a dissociative state in which the atoms separate before autoionization can play a significant role. This seminal paper has received only 128 citations according to the American Physical Society and 144 citations according to Google. The author is tempted to conclude, but cannot prove, that these citation counts are not accurate and are much too low.

By the 1960s, there was enough experimental and theoretical evidence to conclude that the DR of He_2^+ was not observed in the original He afterglow experiments [11]. However, careful experiments on other rare gases, where the rate constants are higher than for He_2^+, indicated that DR was indeed occurring and the mechanism of Bates seemed plausible.

Fifty nine years after the first experiment on the DR of $^4He_2^+$, a storage ring study [12] found a rate coefficient of 1.7±1.0 x 10^{-9} cm^3/sec at 300K for 95% v=0 [12]. The rate coefficients for excited v are much higher than for v=0 and the ions must be relaxed since a small admixture of v>0 can significantly change the reported v=0 rate constant. For $^4He^3He^+$, another storage ring study [13] found a rate constant of 6.0±3.0 x $10^{-10}$$cm^3$/sec at 300K. The heavier isotopomer should have the lower rate constant because the v=0 level will lie lower in the potential well and the wave function will be more contracted than that for the lighter isotopomer. However, the error bars on the experimental values are too large for an accurate comparison. A theoretical study [14] found a rate constant of 6.1 x 10^{-11} cm^3/sec at 300 K from v=0.

The first ab initio DR cross section calculation was reported by Bauer and Wu in 1955 [15] for H_2^+. Because of the primitive nature of computational facilities at the time, many compromises were necessary. They ignored electron-electron repulsion in the calculation of the potential curves and used plane waves instead of coulomb waves for the free electron. Also an incorrect DR cross section expression was derived.

The role of capture into Rydberg states that are not dissociative channels was first explored by Stabler [16]. This was not a study of DR but of rotational capture and served as the inspiration for Bardsley's later proposal of the indirect DR mechanism. Surprisingly, Stabler found that rotational capture rate coefficients were high, i.e. 10^{-7}- 10^{-8} cm^3/sec, even though energetic considerations require capture into high n Rydberg states and $\alpha_{cap} \propto 1/n^3$. He also found that $\alpha_{cap} \propto T_e^{-3/2}$ (where T_e is the electron temperature) exactly as would be found later by Bardsley for DR through rovibrationally excited intermediate states.

2. Scattering theory of DR

The first detailed application of scattering theory to DR, appropriate for low energy electrons, was given by Bardsley [17,18] in pioneering papers in which he showed that the direct recombination cross section from v=0 could be written as the product of a capture term and a survival factor, S_f, i.e. $\sigma_{DR} = \sigma_{cap} * S_f$ where $S_f = 1/|\beta_{0E}G_{00}|^2$. The evaluation of β involves a principal part integral and G is an integral over a Green function. These terms were not evaluated explicitly and it was assumed that the survival factor was unity in many later applications. The direct cross section for v=0 is given by [17]

$$\sigma_{\mathrm{direct},v_0} = \frac{\pi^2 r}{k^2 2}\Gamma_c \frac{|\zeta_0(R_c)|^2}{|E_d'(R_c)|} S_f \quad (1)$$

where r is the ratio of the multiplicity of the repulsive state to the ion state, k is the electron wave number ($\varepsilon = k^2/2m_e$), R_c is the internuclear distance at which the separation of the ion and neutral curves is ε, Γ_c is the electronic width at R_c, $\zeta_0(R_c)$ is the amplitude of the ground state ion wave function at R_c and E_d' is the dissociative curve slope at R_c. Bardsley showed that the rate constant for direct recombination varies as $T_e^{-1/2}$.

In a second paper [18], indirect DR by vibrational and rotational capture into bound Rydberg states was introduced. The mechanism was based upon Stabler's [16] earlier discussion of indirect capture but now coupled with the predissociation of the Rydberg state by the dissociative state of the direct mechanism. Bardsley showed that the indirect cross section for an electron of energy ε takes the Breit-Wigner form:

$$\sigma_{\mathrm{indirect}} = \sum_s \frac{\pi\hbar^2}{2m\varepsilon}\frac{r}{2}\frac{\Gamma_{sa}\Gamma_{sd}}{(E-E_s)^2 + \frac{1}{4}\Gamma_s^2} \quad (2)$$

where E is the total energy, s is the index denoting a resonance (a vibrationally excited Rydberg level), Γ_{sa} and Γ_{sd} are the widths for autoionization and dissociation respectively, $\Gamma_s = \Gamma_{sa}+\Gamma_{sd}$ and E_s is the resonance energy. Note that there is no coupling term between two resonances. This expression is often referred to as the isolated resonance approximation because it assumes that the individual resonances do not interact with each other, i.e. they are narrow and nonoverlapping. The indirect rate constant corresponding to (2) has a $T_e^{-3/2}$ temperature dependence in agreement with Stabler's result.

3. Multichannel quantum defect theory

Because direct and indirect recombination provide for two different pathways to the product atoms, there is interference between these paths. However, Bardsley's approach, providing two independent expressions for direct and indirect recombination does not account for this interference. This problem was solved with the application of Multichannel Quantum Defect Theory (MQDT) to DR first by Lee [19] and later by Giusti [20]. These pioneering contributions were in part based upon the earlier development of MQDT due to Seaton, Fano, Jungen, Greene and coworkers. The reader is referred to the volume of Jungen [21] where many of these papers are collected. The details of this approach will not be discussed here other than to say that the theory is formulated so that all electronic interactions between channels are included in a K matrix which is based upon the Lippmann-Schwinger equation and calculated perturbatively. Each channel is generally defined as a state of the ion coupled to a free or bound electron. The innovation of Lee and Giusti was the inclusion of the dissociative state as one

Journal of Physics: Conference Series **300** (2011) 012001 doi:10.1088/1742-6596/300/1/012001

of the channels. With this change, direct and indirect DR could be treated on an equal basis and interference between these mechanisms could be described accurately. In addition, this approach took advantage of the capability of MQDT to describe entire Rydberg series, each characterized by a quantum defect, μ, where $\mu = n - (2E)^{-1/2}$. Here n is the principal quantum number and E is the energy of the Rydberg level relative to the ionization limit. This advantage is of major importance since a scattering theory approach requires the separate treatment of each individual Rydberg state.

The papers of Lee and Giusti did not lead to the same results. Indeed Lee did not treat the dissociative channels and the ionization channels equivalently. This was corrected in the approach of Giusti who showed that Lee's direct cross section expression was incorrect. Giusti showed that the direct cross section expression did not require evaluation of the complex integrals found in the survival factor of equation (1). Instead, she found that the total cross section for v=0 reduced to the following direct form if the quantum defects are set to be independent of the internuclear distance, R, and only a single dissociative channel is included:

$$\sigma_{\mathrm{direct},v_0} = \frac{\pi r}{k^2 2} \frac{4\xi_{v_0}^2}{(1+\sum_v \xi_v^2)^2} \,. \tag{3}$$

In Eq (3), r and k have the same definitions as for Eq (1) and $\xi_v = \pi \int \chi_v(R) V_{el}(\varepsilon, R) F_d(R) dR$ where $\chi_v(R)$ is the ion vibrational wave function for level v, $F_d(R)$ is the dissociative wave function and $V_{el}(\varepsilon, R) = \langle \{\Phi_i(R)\phi(\varepsilon,R)\}|H|\Phi_d(R)\rangle$. In V_{el}, the curly brackets indicate that the product of the ion wave function, Φ_i, and the coulomb wave function, ϕ, is antisymmetrized, H is the electronic Hamiltonian, Φ_d is the wave function of the dissociative state and the integration is over the electronic coordinates. Remarkably, the same expression has been derived by Flannery [22] by simply requiring unitarity for the capture cross section. In equation (3), the summation in the denominator is over all open vibrational levels at the electron energy corresponding to k^2. The direct cross section including the survival factor is now easily calculated.

Using the MQDT approach, it was shown later [23] that the interference between direct and indirect DR could be described in the same way that Fano [24] originally described the interference of a photoionization continuum with a resonance.

Near a resonance, the DR cross section can be written as $\sigma = \sigma_D (q_v^d + e)^2/(1+e^2)$ where σ_D is the background (direct) cross section and e is the energy, taken to be zero at the resonance center. q_v^d is the Fano profile index for a Rydberg resonance in vibrational level v which is predissociated by state d. q_v^d can be calculated from the K matrix elements and quantum defects that enter the MQDT approach. As in Fano's original treatment [24], if q_v^d is zero, a dip occurs in the cross section at the resonance. A peak occurs if $|q_v^d|>>0$. Constructive and destructive interference occur for $q_v^d = -1$. For $q_v^d = 1$, the cross section is a reflection of the shape for $q_v^d = -1$. Figure 1 shows two dissociative routes of N_2, 2 and $4^3\Pi_u$. The $2^3\Pi_u$ potential curve crosses the ion between the turning points of v=0 and $4^3\Pi_u$ crosses between v=1 and v=2. Figure 2 illustrates the cross section shape for $q_1^d \approx 0$ and for $|q_1^d|>>0$ [25]. Both cross sections are for DR from v=0 and only the v=1

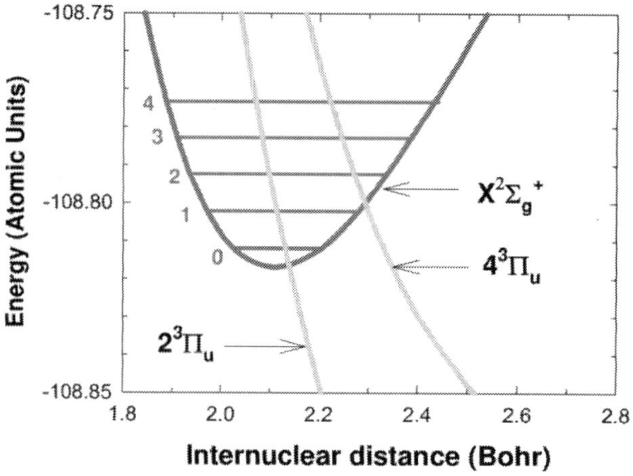

Figure 1. The 2 and $4^3\Pi_u$ potential curves of N_2 shown with the N_2^+ ground state.

resonances are included. The Fano profile index is given by [23] $q_1^d = (1/\pi)(V_1^{capt}/V_{0d}V_{1d})$ where $V_1^{capt} = -\langle\chi_0|\mu(R)|\chi_1\rangle$ and d=2 or 4 for the two $^3\Pi_u$ states. V_1^{capt} does not contain the dissociative state since it describes capture into the $^3\Pi_u$ Rydberg states (not shown figure 1) by Born-Oppenheimer break down. V_{vd} is the same as the matrix element for ξ_v (without π) shown below equation 3 for dissociative state d. V_{02} is much greater than V_{04}. The result is that $q_1^2 \approx 0$ leading to dips in the cross section along $2^3\Pi_u$ and $q_1^4 >> 0$ giving peaks along $4^3\Pi_u$. This is a general rule for DR: when direct DR is negligible, the cross section will have only peaks due to the resonances unless interference can occur between different Rydberg series. When direct DR is strong, the cross section will have only dips due to the resonances unless interference can occur between different Rydberg series, i.e. series having different ion vibrational levels as the series limit.

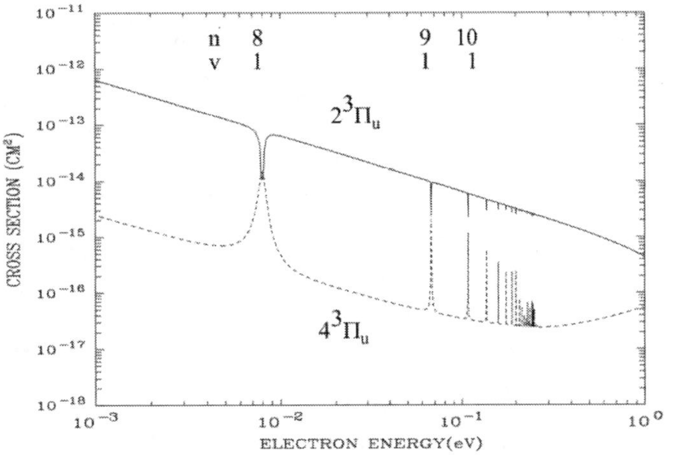

Figure 2. DR cross sections for N_2^+ [25].

When there is more than one dissociative curve of the same electronic symmetry, as in the case of N_2^+ DR, dissociation can become complicated. [26] For example, in figure 1, capture by v=0 can occur into the $2^3\Pi_u$ state which has a non-zero electronic matrix element with an intermediate Rydberg state. The Rydberg state is in turn connected to the $4^3\Pi_u$ state by a similar matrix element. Dissociation along $4^3\Pi_u$ completes DR. So DR can occur along a dissociative route that does not have a large direct matrix element for DR from the v=0 level. The same result ensues from capture by the v=0 ion into a v=1 Rydberg state (not shown in figure 1) by Born-Oppenheimer breakdown. The v=1 Rydberg state is predissociated by the $4^3\Pi_u$ state. The MQDT approach allows both mechanisms to interfere with each other.

4. MQDT extensions

The MQDT approach [20] is now the most commonly used approach for calculating low electron energy DR cross sections and rate constants. Since its introduction, many improvements and new capabilities have been added. The scattering matrix, S, is now calculated using Seaton's channel elimination method [27], an approach that ensures the unitarity of the S matrix.

The MQDT approach originally [20] used only a first order K matrix, i.e. the matrix elements were identical to the ξ_v (without π) given under equation 3 above (for the case of a single dissociative state). The method was later revised to include second order matrix elements [23]. Electronic matrix elements between ionization channels, which had been set to zero in the first order approach, were now calculated. These new matrix elements describe the electronic interaction of vibrationally excited Rydberg levels through their mutual interaction with an intermediate dissociative state. This interaction was shown to be quite important for O_2^+ [23]. In addition, the second order interaction was described with a Fano profile index and played an essential role in determining the cross section shape near a resonance [23]. Similar interactions had been explored previously by O'Malley [28] and Hickman [29]. The calculation of the K matrix was later revised so that it could be determined to all orders [30]. These full K matrix calculations have been applied to the DR of H_2^+, HeH^+ and isotopomers.

Rotational motion was first included in the MQDT approach by Takagi [31] and later by Schneider et al. [32,33]. Valcu et al. [33] found that rotation is important for light ions and in cases where indirect recombination is important and it is not important where the dominant process is direct recombination. This is consistent with the comments at the end of section 3.

Another improvement to the MQDT approach involved the inclusion of intermediate Rydberg states having an excited ion core. [34] These states play a similar role to the ground state core Rydberg states first studied by Stabler and Bardsley. However, these states are also fundamentally different from those with the ground state core. First, an incoming electron can be captured into an excited core state by an electronic matrix element and not by Born-Oppenheimer breakdown as is required for the ground core states. Second, the v=0 level of a core excited Rydberg state may exist at an energy above the ion v=0 level. On the contrary, all ground core Rydberg v=0 levels lie below the ion v=0 level. Since the first mention of this process (in the DR of N_2^+) [34], additional studies [35,36] of N_2^+ DR have found that for dissociative states that have large direct DR rate coefficients, the effect of excited core states on the rate coefficient is small, i.e. about 10%. However, in cases where the direct DR rate coefficient is small, excited core states have been shown to increase the DR rate coefficient by a factor of 4 [35] and to decrease it by a factor of 20 [36]. Additional studies of excited core states in DR have been reported for CD^+ [37, 38, 39, 40], CH^+ [38, 39, 40], and OH^+ [41].

Spin-orbit coupling between Rydberg states was incorporated into the MQDT approach in order to account for an important atomic emission in the Earth's ionosphere [42].The green line of atomic oxygen at 5577 Å is known to arise from the DR of O_2^+ but the only dissociative route that leads to the upper state, O(^{1}S), does not intersect the ground vibrational level of O_2^+. Because of the lack of a favorable intersection, the calculated [23] DR quantum yield of O(^{1}S) was much lower than the measured yield. This mystery continued for several years until it was shown [42] that the mixing of Rydberg states of different symmetries via spin-orbit coupling could account for the observed yield. The impetus for the study of spin-orbit mixing in DR grew out of the spectroscopic observations of the interaction between the $f^1\Sigma_u^+$ and $E^3\Sigma_u^-$ Rydberg states of O_2 [43]. However, before the theoretical calculations [42] were completed, it was not obvious that spin-orbit coupling could indeed be important for DR. The mechanism is shown in figure 3. The only dissociative route that leads to O(^{1}S) from the lowest vibrational levels is the $^1\Sigma_u^+$ valence state shown in the figure. This state has a negligible DR rate coefficient from v=0. Spin-orbit coupling mixes the Rydberg states of $^1\Sigma_u^+$ and $^3\Sigma_u^-$ symmetries and each of these Rydberg states is connected to the repulsive states of the same symmetry by an electronic coupling. The net result is that electron capture from v=0 can occur into the $^3\Sigma_u^-$ state which has a favorable crossing with v=0. Rather than dissociating entirely along the $^3\Sigma_u^-$ dissociative state, some of the flux exits through the $^1\Sigma_u^+$ state via the intermediate mixed Rydberg state. The quantum yield for O(^{1}S) from DR of v=0 is in agreement with laboratory and atmospheric measurements. The mechanism is similar to that illustrated by figure 2 except that in that case all curves are of the same symmetry and spin-orbit coupling does not play a role.

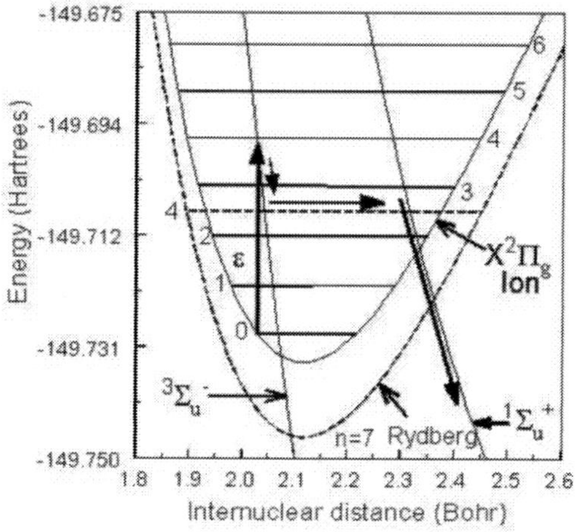

Figure 3. The potentials curves for the ground state of O_2^+ with the lowest 6 vibrational levels (solid line), an n=7 Rydberg state (dashed line) and the dissociative $^1\Sigma_u^+$ and $^3\Sigma_u^-$ states (solid lines).

5. DR by Born-Oppenheimer breakdown

Since the proposal of the direct mechanism [10] of DR, it was assumed for over 40 years that a favorable crossing of a dissociative potential curve within the turning points of an ion vibrational level was required for a high rate coefficient. In 1994, a calculation [44] on $^3HeH^+$ was reported which showed that a high DR rate coefficient was possible even in a case where there is no crossing of the ion potential by a neutral dissociative curve. The calculation was done with a revision of the first order MQDT approach incorporating derivative couplings. The rate constant was determined to be $2.6 \times 10^{-8} \times (T_e/300)^{-0.47}$ cm^3/sec. The main dissociative route at low electron energies was identified as the C state leading to He + H(n=2). The identity of the dissociation products was later confirmed in a storage ring experiment [45]. These results had implications for the DR of H_3^+ where it was also known [46,47] that no neutral surface crossed the ion at low electron energies. Both studies [46,47] found that the potential curves were too far apart for the indirect mechanism to play a role. In the absence of a crossing, it was assumed for many years that the DR of H_3^+ had to have a low rate coefficient. However, the ^3HeH results indicated that if the DR of H_3^+ proceeded by Born-Oppenheimer breakdown, the rate coefficient could indeed be high. Later calculations showed that this indeed was the case [48].

6. Angular distributions

The angular distributions of the products of DR are needed for the determination of product quantum yields in storage ring experiments. The products of DR are detected by the flash that they make when impacting a phosphor plate. In diatomic DR, the distribution of the distances between the two flashes for each DR event is dependent upon a number of factors of which one is the angular distribution of the two products relative to the space fixed electron beam. An excellent discussion of the experimentally derived angular distributions is given in a paper [49] on CH^+ DR. It has been shown [50] that the angular distribution of products is dependent upon the symmetry of the Coulomb partial wave that is captured in DR. Indeed, if only a single partial wave is dominant and if it is assumed that DR takes place before the molecule has a chance to rotate, the angular distribution of the products will be identical to the angular symmetry of the incoming electron partial wave. The allowed capture symmetries for the electron are determined by selection rules that are based upon the electronic symmetry of the ion ground state and that of the state that governs the motion of the product atoms. Surprisingly, it has only been recently [51] that angular distributions for $\ell > 1$ electrons (where ℓ is the angular momentum quantum number for the electron partial wave) have been used in the derivation of quantum yields. This is especially surprising since $\ell = 2$ is the dominant partial wave in H_2^+ DR. The recent analysis [51] for HD^+ DR has indeed shown the importance of the $\ell = 2$ partial wave. For HD^+, it was found that the predissociation time of the Rydberg resonances ($\approx 0.2 \times 10^{-12}$ sec) can be comparable to the rotational period, $\approx 0.8 \times 10^{-12} \times 1/J$ sec, where J is the capture state rotational quantum number. In this case, indirect DR is important and the slow rotation approximation is not entirely valid. (For HD^+, the dissociative curve does not cross within the turning points of the v=0 vibrational level). However, at energies where direct DR is important, the slow rotation approximation should be adequate since dissociation is much faster ($\approx 10^{-15}$ sec) than for indirect recombination. Also, in heavier molecules made up of first row atoms the rotational periods are longer than for HD^+ and the slow rotation approximation may be valid. For O_2^+, the rotational period is $\approx 10^{-11} \times 1/J$ sec. For molecular ions where direct DR is dominant, the slow rotation approximation will be valid unless the ion is highly rotationally excited or the dissociative route has an unusually low slope. Clearly, the validity of the slow rotation approximation should be assessed separately for each molecular ion.

7. Cross section oscillations

For cases where there are multiple dissociation routes that lead to the same products, interference between these routes can cause oscillations in the cross sections. These oscillations were found in a DR experiment [52] on HD^+ and have been confirmed with theoretical calculations [53]. The products of DR in this case are $H^+ + D^-$ and $H^- + D^+$. The only dissociative state included in the calculation is the lowest doubly excited $^1\Sigma^+$ state which dissociates directly to the ion pairs. This state has two crossings each with Rydberg states of the same symmetry arising from the H(n=2, 3, 4) +D(n=1) and D(n=2, 3, 4) +H(n=1) asymptotes. The dissociating flux can enter the Rydberg states at the first crossing and return to the doubly excited $^1\Sigma^+$ state at the second crossing. Wave packet propagation was used to calculate the cross sections. The multiple pathways generate the oscillations in the cross section. Oscillations have also been predicted for the H(1s) + H(2s) $\rightarrow H_2^+ + e^-$ associative ionization reaction [54]. The latter calculation was done with an MQDT approach using coupled equations for the nuclear motion. The interference in the cross section is due to the same process described for DR except that here only the H(1s) + H(2s) channel and the doubly excited $^1\Sigma_g^+$ channel are included at low energies.

8. Polyatomic DR

Significant advances have been made in the calculation of the DR cross section of the simplest polyatomic ion, H_3^+ [48, 55]. For further discussion of these advances, the reader is referred to a recent critical review [56].

Acknowledgments

This work is supported by U.S. National Science Foundation grant ATM-0838061 and by NASA grants NNX09AQ73G and NNX08AE67G.

References

[1] Faber H 1950 *New York Times* Jan. 1, p.1
[2] Bates D R 1987 *EOS* July 28 p. 659
[3] Biondi M A and Brown S C 1949 *Phys. Rev.* **75**, 1700
[4] Kaplan J 1931 *Phys. Rev.* **38**, 1048
[5] Bates D R and Massey H S W 1947 *Proc. Roy. Soc.* A **192** 1; Bates D R, Buckingham R A, Massey H S W and Unwin J J 1939 *Proc. Roy. Soc.* A **170** 322
[6] Bates D R 1950 *Phys. Rev.* **77** 718
[7] Holt R B, Richardson J M, Howland B and McClure B T 1950 *Phys. Rev.* **77** 239
[8] Johnsen, R private communication.
[9] Philbrick J, Mehr F J, Biondi M A 1969 *Phys. Rev. A* **181** 271
[10] Bates D R 1950 *Phys. Rev.* **78** 492
[11] Ferguson E E, Fehsenfeld F C and Schmeltekopf A L 1965 *Phys. Rev.* **138** 381
[12] Buhr H et al. 2008 *Phys. Rev.* A **77**, 032719
[13] Urbain X et al. 2005 *J. Phys. B* **38**, 43
[14] Carata L, Orel A E and Suzor-Weiner A 1999 *Phys. Rev.* A **59** 2804
[15] Bauer E and Wu T Y 1956 *Can. J. Phys.* **34** 1436
[16] Stabler R C 1963 *Phys. Rev.* **131** 1578
[17] Bardsley J N 1968 *J. Phys. B* **1** 349
[18] Bardsley J N 1968 *J. Phys. B* **1** 365
[19] Lee C M 1977 *Phys. Rev.* A **16** 109
[20] Giusti A 1980 *J. Phys.* B **13** 3867
[21] Jungen Ch 1996 *Molecular Applications of Quantum Defect Theory* (Bristol: Institute of Physics Publishing)
[22] Flannery M R 1995, Semiclassical theory of direct dissociative recombination, in *Atomic*

Collisions: A Symposium in honor of Christopher Bottcher, ed D R Schultz, M R Strayer, and J H Macek (New York: AIP Press) p 53
[23] Guberman S L and Giusti-Suzor A 1991 *J. Chem. Phys.* **95**, 2602
[24] Fano U 1961 *Phys. Rev.* **124** 1866
[25] Guberman S L 1993, Electron-Ion Continuum-Continuum Mixing in Dissociative Recombination, in *Dissociative Recombination: Theory, Experiment, and Applications*, ed B R Rowe, J B A Mitchell, and A Canosa (New York: Plenum) p 47
[26] Guberman S L 2003, Dissociative recombination mechanisms, in *Dissociative Recombination of Molecular Ions with Electrons*, ed S. L. Guberman (New York: Kluwer Academic /Plenum) p 1
[27] Nakashima K, Takagi H, and Nakamura H 1987 *J. Chem. Phys.* **86** 726
[28] O'Malley T F 1981 *J. Phys. B* ***14*** 1229
[29] Hickman A P 1987 *J. Phys. B* **20** 2091
[30] Takagi H, Hara S and Sato H 2009 *Phys. Rev.* A **79**, 012715; Takagi H 2000 Higher order contributions to the dissociative processes of diatomic molecules, in *Dissociative Recombination: Theory, Experiment and Applications IV*, ed M Larsson, J B A Mitchell and I F Schneider (Singapore: World Scientific), p. 180; Pichl L, Hiyama M and Nakamura H 2003 Analytical treatment of the K matrix integral equation in the dynamics of superexcited molecules, in *Dissociative Recombination of Molecular Ions with Electrons*, ed S L Guberman (New York: Kluwer/Plenum), p. 167; Pichl L, Nakamura H and Horáček J 2000 Comp. Phys. Comm. **124**, 1
[31] Takagi H, Kosugi N and Le Dourneuf, M 1991 *J. Phys.* B **24** 711; Takagi H 1993 *J. Phys.* B **26** 4815
[32] Schneider I F, Strömholm C, Carata L, Urbain X, Larsson M, and Suzor-Weiner A. 1997 *J. Phys.* B **30** 2687
[33] Vâlcu B, Schneider I F, Raoult M, Strömholm C, Larsson M and Suzor-Weiner A 1998 *Eur. Phys. J.* D **1** 71
[34] Guberman S L 1989 Ab initio studies of dissociative recombination, in *Dissociative Recombination: Theory, Experiment and Applications*, ed J B A Mitchell and S L Guberman (Singapore: World Scientific), p. 45
[35] Guberman S L 2007 *J. Phys. Chem.* A **111** 11254
[36] Guberman S L 2009 *Journal of Physics: Conference Series* **192** 012001
[37] Forck P et al. 1994 *Phys. Rev. Lett.* **72** 2002
[38] Amitay Z et al. 1996 *Phys. Rev.* A **54** 4032
[39] Carata L, Orel A E, Schneider I F and Suzor-Weiner A 2000 Core Excited Resonances in the Dissociative Recombination of CH^+ and CD^+, in *Dissociative Recombination: Theory, Experiment and Applications* ed Larsson M, Mitchell J B A and Schneider I F(Singapore: World Scientific), p 291
[40] Carata L, Orel A E, Raoult M, Schneider I F, Suzor-Weiner A 2000 *Phys. Rev. A* **62**, 052711
[41] Amitay Z et al. 1996 *Phys. Rev. A* **53** R644
[42] Guberman S L 1997 *Science* **278** 1276
[43] Lewis, B. R. et al. 1995 Phys. Rev. A 52 2717
[44] Guberman S L 1994 *Phys. Rev. A* **49** R4277
[45] Semaniak J et al. 1996 Phys. Rev. A 54 R4617
[46] Kulander K C and Guest M F 1979 *Journal of Physics B: Atomic and Molecular Physics* **12** L501
[47] Michels H H and Hobbs R H 1984 *Astrophysical Journal* **286** L27
[48] Kokoouline V and Greene C H 2003 *Phys. Rev. A* **68** 012703-1
[49] Amitay Z. et al. 1996 *Phys. Rev. A* **54** 4032
[50] Guberman S L 2004 *J. Chem. Phys.* **120** 9509

[51] Novotny S et al. 2008 *Phys. Rev. Lett.***100** 193201; Novotny S et al. 2009 *J. Phys. Conf. Ser.* **192** 012024
[52] Zong W et al. 1999 *Phys. Rev. Lett.* **83** 951
[53] Larson Å and Orel A E 2001 *Phys. Rev. A* **64** 062701
[54] Urbain X, Cornet A, Brouillard F and Giusti-Suzor A 1991 *Phys. Rev. Lett.* **66**, 1685
[55] Jungen C and Pratt S C 2009 *Phys. Rev. Lett.* **102** 023201-1
[56] Johnsen R and Guberman S L 2010 *Adv. At. Molec. Opt. Phys.* **57** 75

Journal of Physics: Conference Series **300** (2011) 012002 doi:10.1088/1742-6596/300/1/012002

DR Experiments in the 60 years since the Bates paper

Rainer Johnsen

University of Pittsburgh, Department of Physics and Astronomy,

Pittsburgh, Pa, 15260, USA

E-mail: rj@pitt.edu

Abstract. About 60 years have passed since David Bates in a 1950 Physical Review Letter to the Editor first clarified the dissociative recombination mechanism (DR) of molecular ions and provided the conceptual basis for the process that plays an important role in the ionospheres of Earth and other planets, the interstellar medium, and electrical discharges. His visionary (and characteristically "terse") thumbnail sketch of the DR mechanism only hinted at the subtleties that have kept atomic theorists and experimentalists busy for many years. In my talk, I will focus on experiments on N_2^+ recombination, beginning with plasma afterglows, and then move on to the "after"-afterglow period that is now dominated by the powerful storage-ring methods.

1. Introduction

A highly readable personal account of the early (pre-1950s) attempts to explain the fast recombination in the terrestrial ionosphere can be found in a paper written by Bates in 1988 [1]. It became gradually clear to him and others that explaining the ionization loss rates by radiative recombination or formation of negative ions was fraught with insurmountable difficulties and that something radically new was called for.

In the last paragraph of that paper Bates summarizes some of the "bizarre" sequence of events that led to his breakthrough work. (n (e) denotes the electron density)

"The contribution that scientists using radio-soundings have made to our knowledge of recombination in the E-layer is bizarre. If they had correctly deduced that n (e) is greater in the morning that in the afternoon they could not have made any estimate of the recombination coefficient. As it was they erroneously deduced that n (e) is greater in the afternoon than in the morning and deduced a bogus recombination coefficient from the apparent sluggishness of the diurnal variation. This bogus recombination coefficient is about 30 times smaller than the true recombination coefficient but by good fortune is approximately equal to the bogus recombination coefficient from the first microwave measurements of the decay of the electron density in a helium afterglow. Happily it was large enough to alert theorists to the occurrence of a hitherto unknown recombination process (both in the E layer and in the afterglow). "

The *"hitherto unknown recombination mechanism"*, of course, was the "Dissociative Recombination" that Bates introduced in a brief 1950 Letter to the Editor [2]. His Letter starts off with the sentence: *"By the successful exploitation of microwave techniques several groups of workers have recently derived interesting information on the decay of ionization"*. A few paragraphs later he says: *"The reaction can be thought of as taking place in two stages. As a result of a radiationless transition an excited molecule, XY' is formed. If this is unstable its constituent may move apart under the influence of their mutual repulsion thereby preventing auto-ionization".*

With these few words, Bates prepared the stage for extensive future theoretical refinements of the DR process, but he also implicitly admits that progress in understanding DR was often guided by, and sometimes forced by experimental evidence, as faulty, ambiguous, and contradictory as it may have been. The discovery of the mechanism of DR was clearly not a product of pure thought! Few theorists paid as close attention to experiments as Bates did, and those who had the good fortune to talk to him in person will recall his often critical and occasionally relentless questions regarding experimental procedures and conclusions.

It is not possible to review 60 years of all experiments on DR in a brief paper. Instead, I will retrace the history of measurements on one particular ion species, N_2^+, and while doing so digress occasionally to point out the strengths and limitations of different experimental techniques. I focus on N_2^+ not because it is a particularly abundant ionospheric species (which it is not because it is lost also by charge transfer with atomic and molecular oxygen) but because it has been studied extensively by theory and many experimental techniques.

The dissociative recombination of N_2+ in the vibrational ground state has four exoergic product branches:

$$\begin{aligned} N_2^+ + e^- &\rightarrow N(^4S) + N\,^4S) + 5.82eV \\ &\rightarrow N(^4S) + N\,^2D) + 3.44eV \\ &\rightarrow N(^4S) + N\,^2P) + 2.25eV \\ &\rightarrow N(^2D) + N^2D) + 1.06eV \end{aligned} \qquad (1)$$

Experiments and theory are expected to provide the recombination coefficient or cross section, as a function of either temperature or collision energy, and as a function of the N_2^+ vibrational state. Knowledge of the overall recombination coefficient often suffices to incorporate recombination in models of planetary ionospheres or other plasmas but one would also like to know the relative branching fractions and the kinetic energies given to the product atoms, which affect the rate of non-thermal escape of *N* atoms from some planetary atmospheres (e.g. from Mars).

2 Afterglow experiments

In the earliest methods to determine the recombination coefficient of N_2^+ a brief pulse of microwaves (f~3GHz) was admitted to a microwave cavity containing a sealed flask filled with pure nitrogen at pressures of a few Torr and at room temperature. If the microwave pulse is sufficiently intense, gaseous breakdown generates a weakly ionized plasma with initial electron densities around $n_e = 10^9$ cm^3/s. Since the Debye length is much smaller than the dimensions of the container the plasma is close to electrically neutral in most of the plasma volume. After the end of the breakdown pulse, the plasma decays by a both recombination and ambipolar diffusion to the walls. The first term on the RHS of the electron continuity equation

$$\frac{\partial n_e(t,\vec{r})}{\partial t} = -\alpha n_e^2(t,\vec{r}) + D_a \nabla^2 n_e(t,\vec{r}) \qquad (2)$$

describes the local electron loss due to recombination with recombination coefficient α, here assumed to be due to a single ion species, while the second term describes ambipolar diffusion losses, again assuming that only one ion species with ambipolar diffusion coefficient D_a is present. If the gas density is sufficiently high, D_a, being inversely proportional to density, becomes small and diffusion plays only a minor role during the initial afterglow phase (the "early afterglow). A commonly used, but not entirely satisfactory, approximation "corrects" for the diffusion loss of electrons by fitting the observed electron density decays to an equation of the form

$$\frac{dn_e(t)}{dt} = -\alpha n_e^2(t) - \frac{D_a n_e}{\Lambda^2} \qquad (3)$$

in which the quantity Λ^2 is the fundamental diffusion length appropriate to the shape and size of the plasma container. Different investigators employ different methods of obtaining recombination coefficients from fits of the observed electron decay to Eq. (3).

2.1 Measuring the electron density decay

There are essentially two experimental methods to measure the electron density decay during the afterglow: Langmuir probes and microwave techniques.

Some of the early experimenters were, perhaps unduly, concerned about the reliability of Langmuir probes and potential perturbations of the plasma by "sticking a wire into it" and drawing electrons out of the plasma. They preferred to use a "probing microwave" to detect the change of the resonant frequency of a microwave cavity due to free electrons. Microwave technology had just been developed and many investigators came from research programs on RADAR development. Somewhat simplified, the macroscopic polarization of the plasma by a microwave with frequency above the plasma frequency causes the dielectric constant of the plasma to become less than unity. Hence, the microwave wavelength exceeds its vacuum value, and, to bring the cavity back into resonance, the microwave frequency has to be increased. The change in frequency can be found by sweeping the probing microwave frequency over a small range and noting the time at which the reflected power goes to zero. The method is highly accurate and relies on well-known textbook physics. However, it returns a "microwave-averaged" electron density, rather than local values and this requires additional but fairly straightforward computations. A second, but minor, problem is that the frequency shift is proportional to the electron density only if the electromagnetic field in the cavity is only slightly perturbed by the plasma. This limits the maximum electron density (at 3GHz) to a few 10^9 cm^{-3}.

Later work, especially in flowing afterglow plasmas, usually employed Langmuir probes to measure n_e. It now appears that earlier concerns about Langmuir probes were unwarranted. They are reliable, provided one uses the probes in the electron collecting (rather than ion collecting) mode.

2.2 Afterglows in pure nitrogen

Early afterglow measurements suffered from incomplete knowledge of the "ion chemistry", which was seriously underdeveloped at that time. The microwave-afterglow measurements of Bryan et al. [3] were carried out in pure N_2. The experimenters observed the decay of the electron density and also recorded optical emissions. They concluded that recombination in the early N_2 afterglow occurred with coefficients from about 2.3×10^{-6} cm^3/s at a nitrogen pressure of 6.4 Torr and 6.1×10^{-6} cm^3/s at 29 Torr. The pressure dependence and the very large recombination coefficient suggest that their plasma contained N_4^+ rather than N_2^+ ions. They also found rather strong emissions of the second positive band system of N_2, which were later shown (Chao and Johnsen [4]) to arise from N_4^+ recombination.

The lesson to be learned from this early work is that afterglows in pure molecular gases are difficult to interpret. It is better to use a rare gas (the "carrier" or "buffer" gas) to slow down diffusion and to add only enough of the molecular gas to form the desired ion species. This strategy has been employed ever since. Of course, it helps a great deal to have the extensive data on ion-molecule chemistry that became available in the 1960s! Since one can never be sure that the plasma actually contains the desired species, and no impurities, one adds a mass spectrometer that sample ions from the plasma. Kasner and Biondi [5], then working at the Westinghouse lab in Pittsburgh, used a highly diluted mixture of N_2 in neon, added a radio-frequency mass spectrometer to their afterglow apparatus, and were able to establish conditions where N_2^+ ions were the dominant species. They observed a remarkably consistent set of N_2^+ recombination coefficients of close to 2.8×10^{-7} cm^3/s at T= 300 K over a significant range of neon and N_2 densities. For some unknown reason, the values are somewhat higher than those found in later work but they are essentially correct.

2.3 Measuring the temperature variation

Ionospheric regions can have temperatures in excess of several thousand degrees K. Several methods have been used to measure the variation of recombination coefficients with temperature. One can heat

(or cool) the entire afterglow apparatus to change both the electron and gas temperature. In practice, heating to temperatures above 600 K creates serious technical problems and achieves very little that cannot be done easier by the microwave heating method, to be discussed later.
Kasner, using a modified version of the apparatus as Kasner and Biondi, measured the recombination of N_2^+ in the gas temperature range from 205 to 480 K. The 300 K value agreed with that of Kasner and Biondi of 2.8×10^{-7} cm^3/s, but Kasner [6] did not observe any significant temperature variation. If one accepts the $T^{-0.39}$ variation of α that was found in later work, Kasner's coefficient at 480 K should have been about smaller by only ~30% than that at 205 K, perhaps too small a change for a reliable determination of the temperature variation.

This does not mean that changing the gas temperature never serves a useful purpose. For instance, cooling the apparatus to temperatures below 300 K is essential for studies involving weakly bound ions (e.g. clusters or weakly bound dimers).

2.4 Microwave heating of electrons

Much higher electron temperatures (up to 10000 K) can be reached by microwave heating of the electrons gas, a technique that was used extensively in stationary afterglows and that has provided data on many important ion species, including N_2^+. The basic principles of microwave heating of plasma electrons are well understood and are fairly easy to apply if the transfer of power from the electron gas to the neutral gas involves only elastic collision between electrons and atoms. Electrons, while colliding with gas atoms, absorb power from the microwave field, heat up and, once the steady-state energy has been reached, pass an equal power to the gas atoms. The mean electron energy can be calculated from the power balance. The heating mechanism is sometimes misunderstood: The energy contained in the oscillatory motion of the electrons in the microwave field constitutes only a small fraction of the mean electron energy! The collisional "friction" between electrons and the gas is an essential ingredient. It can be shown that the steady-state electron velocity distribution approaches a Maxwellian in the limit where the microwave frequency is much larger than the electron collision frequency.

The first microwave-heated experiments on recombination of rare-gas dimer ions were carried out in pure rare gases. In typical experiments, the plasma container (usually a quartz flask) occupied only part of the microwave cavity (or wave guide) such that the heating microwave field was fairly uniform in the plasma volume. In later work on molecular ions, molecular gases were added to form the desired ion species, and the instruments incorporated mass spectrometers. Now, the afterglow plasma filled the entire cavity, and uniformity of the microwave field became worse. The combined effects of adding molecular gases and non-uniformity of the heating field were not fully understood at the time. It was believed that the heat conductance in the electron gas would suffice to smooth out spatial temperature variations, and that the presence of small traces of molecular gases would not have serious consequences. The problematic nature of these assumptions became apparent in a study of NO^+ recombination (Dulaney et al, [7] and Johnsen [8]). I will skip the rather involved details. In essence, microwave heating does not produce a uniform electron temperature and the average electron temperature is reduced significantly by inelastic electron-molecule collisions. The electron velocity distributions are still close to Maxwellian locally, but are not so when averaged over the plasma volume. The effects can be moderate and sometimes negligible (for low concentrations of homonuclear diatomic molecules such as N_2), but addition of even small amounts of polar gases like NO, H_2O or NH_3 drastically reduces microwave heating. This happens because rotational excitation of polar molecules by electrons is an extremely efficient electron cooling mechanism. In extreme cases, ignoring this effect led to the erroneous inference [8] that the recombination coefficients of cluster-ions of the type $H_3O^+(H_2O)_n$ and $NH_4^+(NH_3)_n$ were independent of electron temperature.

Fortunately, the N_2^+ measurements by Mehr and Biondi [9] in microwave-heated afterglows were not much affected by the complications just discussed and the results have withstood the test of time. In the range from T_e= 300 to 5000 K, the measured recombination coefficients were found to fit the power law

$$\alpha(N_2^+) = (1.8 \pm 0.4) \times 10^{-7} (T_e / 300)^{-0.39}\ cm^3/s, \tag{4}$$

This often quoted benchmark result should still be considered reliable, even though the electron energy distribution probably was not exactly Maxwellian.

2.5 Shock heating to raise the neutral and vibrational temperatures

Higher gas temperatures T_{gas} can be reached by employing shock heating of the afterglow plasma. As in the stationary afterglow, a plasma is first generated by a pulse of RF power. Then the plasma is rapidly compressed and heated by a high-pressure driver gas that expands into the shock tube, after a diaphragm bursts that initially separates the two section of the shock tube. At the Mach numbers used in the experiments (up to about 6), translational gas temperatures from 500 K to about 10000 K can be reached. The degree of heating can be calculated from the shock-wave (Rankine-Hugoniot) equations. This experimental method was applied by Hobson and co-workers to several molecular ions, beginning with molecular rare-gas ions. It was found that for T_{gas}>700 K the recombination coefficient of Ne_2^+ [10] declined much faster with temperature (as ~$T_{gas}^{-3/2}$) than the T_e dependence observed in microwave-heated afterglows. This interesting result agreed rather well with a theoretical prediction by O'Malley that vibrational excitation should reduce the recombination coefficient.

However, when the shock-heating technique was applied to N_2^+ and O_2^+ ions the observed dependence on gas temperature [11] was found to be nearly the same as that on electron temperature found by microwave heating. Hence, the authors suggested that shock heating did not produce a vibrational N_2^+ population in equilibrium with the translational degrees of freedom. Of course, they could equally well have concluded that vibrational excitation has no significant effect on N_2^+ recombination. Later, an improved version of the theory (Cunningham et al [12]) showed that the vibrational effect should be significant only when the probability of auto-ionization subsequent to electron capture is large, and that this is true for Ne_2^+ and other rare-gas dimer ions, but not for N_2^+ and O_2^+.

In hindsight, it appears that in the Ne_2^+ shock-tube experiment vibrational equilibration of the ions was approached rapidly, presumably as a result Ne_2^++ Ne→Ne + Ne_2^+ ion-atom interchange, and that recombination of Ne_2^+ actually is reduced by vibrational excitation. However, no firm conclusion can be reached for N_2^+ and O_2^+ since the degree of thermal vibrational excitation in these experiments was not known. Probably, some vibrationally excited ions were produced during the discharge phase and persisted during the afterglow.

The shock tube technique provides a clean way to heat the gas to high temperatures, but is has the practical problem that after each experimental cycle the tube has to be partially disassembled to replace the diaphragm and then has to be pumped out again for the next cycle. Hence, data collecting is painfully slow and that may have contributed to the demise of this method. I will now turn to an interesting experiment in which the N_2^+ vibrational excitation was actually measured.

2.6 Laser induced fluorescence determination of the effect of N_2^+ vibrational excitation

The high density of recombining ions in an afterglow makes it possible to measure their concentrations by laser-induced fluorescence (LIF). Zipf [13], in a pioneering but unfortunately not definitive experiment, applied LIF to the first negative band system of N_2^+ and determined the rate at which N_2^+ ions in v= *0, 1, 2* vibrational states decay in an N_2-Ne afterglow at 300 K. The initial vibrational excitation occurs during the discharge, mainly by Penning ionization by neon metastables. Typically, the $v=1$ and $v=2$ ions had concentrations of ~1% or ~0.1 %, respectively, of that of $v=0$ ions. Zipf found that the decay rates of the $v=0, 1, 2$ ions depended only very weakly on vibrational state, combined his results with the T_e dependence found by Mehr and Biondi and expressed the variation of the recombination coefficient with vibrational and electron temperature by a power law:

$$\alpha(N_2^+) = (2.2) \times 10^{-7} (T_e / 300)^{-0.39}\ (T_{vib} / 300)^{0.04} cm^3/s \tag{5}$$

Obviously, such a formula is very useful for ionospheric models and it was accepted for some time until Johnsen [8] pointed out an unexpected complication in the interpretation of the experiment. The

problem is that vibrational excited N_2^+ ions are quenched rapidly by charge transfer with neutral N_2 and could not have persisted in the afterglow as long as Zipf had observed and if, as he had assumed, the neutral N_2 in the gas mixture was vibrationally cold. In reality, the neutral N_2 must have been excited during the discharge to a vibrational temperature of about 1500 K such that rapid back-and-forth ion-neutral charge transfer kept the N_2^+ ions in vibrational equilibrium with the hot N_2. The only conclusion that can be drawn is that an N_2^+ population with a vibrational temperature of ~1500 K recombines with nearly the same recombination coefficient as the majority $v=0$ ions. The measurements do not exclude the (however unlikely) possibility that the $v=1$ and $v=2$ minority ions recombine very slowly or not at all, but they do exclude the possibility that the majority $v=0$ ions recombine much more slowly than $v=1$ and $v=0$ ions. A more complete analysis of Zipf's and other experiments is given in a paper by Bates and Mitchell [14]. Even though Zipf's conclusions and formula (5) do not follow from his experiment, later work has confirmed that vibrational excitation has only a minor effect on N_2^+ recombination.
There were some further "twists and turns" in the history of N_2^+ *(v)* recombination. Model analyses of satellite ionospheric data seemed to imply that N_2^+ in v=0 recombined much more slowly than in $v=1$ and $v=2$. I will discuss them only briefly.

2.7 The "great afterglow in the sky".
During the 1970s extensive data on the ionospheric composition became available and computational models were developed that combined both laboratory and observational data. Obviously, ionospheric models are very valuable tools but their limitations were sometimes disregarded and apparent discrepancies were "resolved" by adjusting, for instance, recombination rates to fit the model. In one example, the modeling of Atmosphere Explorer data by Orsini et al [15] led to the inference that vibrationally excited N_2^+ should recombine much more rapidly (DR coefficients $>10^{-6}$ cm^3/s) than $v=0$ ions, which, as Biondi [16] quickly pointed out, contradicts both the basic physics of DR and available laboratory data. The ensuing controversy was eventually resolved in favor of the laboratory data when Breig et al [17] included in the model a previously ignored fast charge transfer reaction of metastable O^+ (^2D) with N_2. The details are not very important in this context. The positive aspect is that seeming discrepancies, if properly diagnosed, often lead to new insights.

3. Merged beam and storage-ring measurements

The foregoing discussions make it clear that the plasma environment in afterglow studies of recombination introduces some "interesting complications". One would like to be able to observe purely binary recombination between ions and electrons with well-defined relative velocity in an otherwise "empty universe". This "theorists' view" of recombination is closely approached in merged-beam and storage-ring methods, but the conceptual simplicity comes at a high cost and requires large-scale facilities.

The original single-pass merged-beam method developed at the University of Western Ontario [10] and storage-rings have in common that recombination of ions and electrons takes place between coaxial ion and electron beams of nearly the same velocity. In a single-pass merged beam the ion beam passes through the electron target beam once and is then discarded; in a storage ring the ions circulate in the ring and pass through the interaction region many times. Except in cases where the ions under study are "expensive" to make, it is not the recycling of ions in storage rings that makes them preferable but the fact that the longer storage time (up to 10 seconds) in a ring removes all instance infrared active vibrationally excited ions. The relative velocity between the two beams can be made very small and the velocity spread in the electron target gas is greatly reduced by accelerating the (initially "hot") electrons to a high velocity that closely matches that of the ions. In part, the narrowing of the electron velocity distribution in the direction of the beam (but not transverse to it) is a kinematic effect that follows from the classical equations of motion. In the single pass merged-beam apparatus, further improvements in the energy resolution were made by using trochoidal analyzers to merge electron and ion beams. Storage rings employ "electron coolers" in which the electrons are cooled by expansion in a magnetic guiding field. The coolers also cool the ion beam by a "friction" effect and reduce the diameter of the ion beam.

In the merged-beam techniques recombination events are detected by counting their products using an energy sensitive barrier detector. The technique has the advantage that product branching fractions and the kinetic energies released by recombination can be determined with high accuracy. Afterglow experiments can provide complementary information on branching fractions by detecting products via laser induced fluorescence of emission spectroscopy, but with few exceptions (e.g. Doppler broadening of emission lines) the kinetic energy of fragments cannot be determined.

3.1 Merged beam data on N_2^+ recombination

The first merged-beam measurements on N_2^+ (Mul and McGowan [18]) yielded recombination cross sections up to an energy of 1eV that declined with energy as E^{-1}, equivalent to a $T_e^{-0.5}$ dependence of the thermal rate coefficient, somewhat faster than the $T_e^{-0.39}$ dependence observed by Mehr and Biondi [9] in a microwave heated afterglow. The absolute values at 300 K agreed to about 30%. Mul and McGowan state that the ions may have been vibrationally excited. A later repletion of the merged-beam experiment (Noren et al [19] employed two different ions sources and led to the puzzling conclusion that N_2^+ *(v=0)* recombines much more slowly (by a factor of 5) than the afterglow value. As Bates and Mitchell [14] discuss in a later joint paper, this experimental result is incompatible with both theory and other experiments, but the precise cause of the discrepancy remains unclear.

3.2 Flowing afterglow data N_2^+ recombination.

As has been discussed before, exposing molecular gases to an intense discharge in stationary afterglows containing N_2 creates a reservoir of vibrationally hot molecules that keep the ions hot during the afterglow. The flowing afterglow technique gets around this problem by separating the discharge region from the recombination region. Since the N_2 downstream from the N_2 inlet is cold, charge transfer with N_2 now reduces the N_2^+ vibrational state to $v=0$ and one measures the recombination of this ion. Two such experiments have been performed. The 300 K rate coefficients were found to be 2.6×10^{-7} cm^3/s (Canosa et al [20]) and 2.0×10^{-7} cm^3/s (Geoghegan and Smith [21]) in reasonable agreement with the Mehr and Biondi value.

3.3 Storage-ring data on N_2^+ recombination

Storage rings are currently the method of choice for recombination studies and really deserve a more extended discussion of their outstanding capabilities than can be given in this brief historical review. The results obtained by this technique have been quite consistent over the years and the N_2^+ measurements by Peterson et al [22] are no exception. Converted to a thermal rate coefficient, the result (for a vibrational distribution of 46% in $v=0$, 27% in $v=1$) can be expressed as

$$\alpha(N_2^+) = (1.75\pm0.09)\times10^{-7}(T_e/300)^{-0.30}\ cm^3/s \qquad (6)$$

which is essentially the same as that given by Mehr and Biondi, except that the temperature exponent is slightly smaller. The storage ring measurements employed two different ion sources that produce N_2^+ ions with different vibrational populations. Both gave nearly the same recombination cross sections. The authors also determined the product fractions for three of the branches listed in Eq. (1). At a collision energy of zero and for $v=0$ ions, the fractions were found to be 37%, 11%, and 52% for $N(^4S) + N(^2D)$, $N(^4S) + N(^2P)$,and $N(^2D) + N(^2D)$, respectively.

3.4 Comparison with theoretical results

The extensive theoretical work by Guberman on N_2^+ recombination largely supports the observed of the recombination rate but his theoretical result [23] indicates a somewhat stronger dependence on vibrational state and the theoretical branching fraction [24] for the $N(^4S) + N(^2D)$ product channel is considerably larger than was observed by Peterson et al [22].

4. Conclusions

Peterson et al [22] sum up the long history of N_2^+ recombination measurements by saying *"We note that surprisingly good agreement is found between all reported experiments on* N_2^+*, using different techniques, and undoubtedly with a variation in vibrational composition. This can now be rationalized, since the DR rates from our two ion sources with markedly different vibrational populations were about the same".*

My quick tour through the history was intended to review the different experimental approaches using N_2^+ as an example. In this case, the story indeed has a "happy ending", at least as far as experiments are concerned, but one should not necessarily assume that the same is true for other ion species. Differences in vibrational distributions are not the only cause of discrepancies between measurements and I still believe that recombination of ions in plasmas can involve additional mechanisms that are not present in the rarefied environment of beam experiments. But the "bogus" era of DR studies certainly has clearly come to an end!

5. Acknowledgement

The author gratefully acknowledges travel support from the US National Science Foundation and NASA

References

[1] Bates, D R 1988 *Planetary Space Science* **36** 55
[2] Bates D R 1950 *Phys. Rev.* **78** 492
[3] Bryan R B, Holt R B, Oldenberg O 1957 *Phys. Rev.* **106** 83
[4] Cao Y S and Johnsen R 1991 *J. Chem. Phys.* **95** 7356
[5] Kasner W H and Biondi M A 1965, *Phys. Rev.* **137** A317
[6] Kasner W H 1967 *Phys. Rev.* **164** 194
[7] Dulaney J L, Biondi M A and Johnsen R 1987, *Phys. Rev. A*, **36** 1342
[8] Johnsen R 1987 *Int. J. Mass Spec Ion Proc.* **81**, 67
[9] Mehr F J and Biondi M A 1969 *Phys. Rev.* **181**, 264
[10] O'Malley T F, Cunningham A J and Hobson R M 1972 *J. Phys. B* **5**, 2126
[11] Cunningham A J and Hobson R M 1972 *J. Phys. B* **5**, 2328
[12] Cunningham A J, O'Malley and T F, Hobson R M. 1981 *J. Phys. B* **14**, 773
[13] Zipf E C 1980 *Geophys. Res. Lett.* 7, 645
[14] Bates D R and Mitchell J B A 1991 *Planet. Space Sci.* **39** 1297
[15] Orsini N , Torr D G, Brinton H C, Brace L H, Hanson W B, Hoffman and J H, and Nier A O, 1977, *Geophys. Res. Lett.* **4** 431
[16] Biondi M A 1978, *Geophys. Res. Lett.* **5** 661
[17] Breig E L, Hanson W H, Hoffman J H and Abdou W A, 1983, *J. Geophys. Res.* **A9** 7190
[18] Mul P M and McGowan J Wm 1979, *J. Phys. B* **9** 1591
[19] Noren C, Farouk B and Y. Mitchell 1989, *J. Chem. Faraday Soc* **85** 1697
[20] Canosa A, Gomet J C, Rowe B R and Queffelec J L, 1991, *J. Chem. Phys.* **94** 7159
[21] Geoghegan M, Adams N G, and Smith D, 1991, *J. Phys. B* **24**, 2589
[22] Peterson J R, Le Padellec A, Dunn G H, Larsson M, Larson A, Peverall R, Strömholm C, Rosen S, af Uglass R, and van der Zande W J 1998 *J. Chem. Phys.* **108** 1978
[23] Guberman S L 2003 in "Dissociative recombination of molecular ions with electrons" Kluwer/Plenum Academic Press, New York 2003) p. 187 (see corrected version: http://www.sci.org/kluwer.N2+.pdf)
[24] Guberman S L 1991 Geophys. Res. Lett. **18**, 1051

Journal of Physics: Conference Series **300** (2011) 012003 doi:10.1088/1742-6596/300/1/012003

Studies of electron-impact dissociative excitation and ionization using the crossed electron-ion beams technique

M. Fogle[1], E. M. Bahati[2], M. E. Bannister[2], S. H. M. Deng[2], C. R. Vane[2], R. D. Thomas[3], V. Zhaunerchyk[3]

[1] Department of Physics, Auburn University, Auburn, AL
[2] Physics Division, Oak Ridge National Laboratory, Oak Ridge, TN
[3] Department of Physics, Albanova, Stockholm University, Stockholm, Sweden

E-mail: fogle@physics.auburn.edu

Abstract. Absolute cross sections for heavy ion fragment production due to electron-impact dissociation of XH_2^+ (X = B, C, N, O, F), N_2D^+ and O_3^+ have been investigated with the crossed electron-ion beams technique in the energy range 3 – 100 eV. This energy regime covers the dissociative excitation and ionization processes. Two and three-body dissociation channels are observed in these systems and marked resonant-like structure is also observed in the dissociative excitation channels of some systems. It is unclear if direct excitation or resonant capture processes lead to these enhanced dissociative excitation cross sections.

1. Introduction

Plasma chemistry is driven by collisions of free electrons with atomic and molecular species. Low-energy electrons can particularly access dissociative channels that govern branching fractions, energy balance, neutral particle transport, and excited state dynamics. Such low-energy electron collisions take place in the low-temperature regions of fusion plasmas [1, 2] and planetary atmospheres [3]. The physics and chemistry of industrial interests, such as flames, lighting and gas lasers, are likewise areas where the knowledge of these collision processes are needed.

The production of heavy ion fragments from molecular cations due to electron scattering can proceed through various channels. The Direct Dissociative Excitation process proceeds via a vertical transition to either a dissociative excited state or a bound excited state that couples to a dissociative pathway. Both of these pathways result in a charged ion fragment and neutral fragments. The Resonant Dissociative Excitation process proceeds through the resonant capture of an electron to a neutral Rydberg state. This state then decays via autoionization and subsequent dissociation producing a charged ion fragment and neutral fragments. We will refer to both of these process jointly as Dissociative Excitation (DE).

The Dissociative Ionization (DI) process is similar to the dissociative excitation process with the exception that an electron is removed from the system resulting in at least two charged ion fragments. Given that two positively charged fragments are initially in close proximity at the time of dissociation, some amount of Coulomb force is experienced by the mutual repulsion of the charged fragments. This augments the kinetic energy release due to bond breaking in the dissociation process.

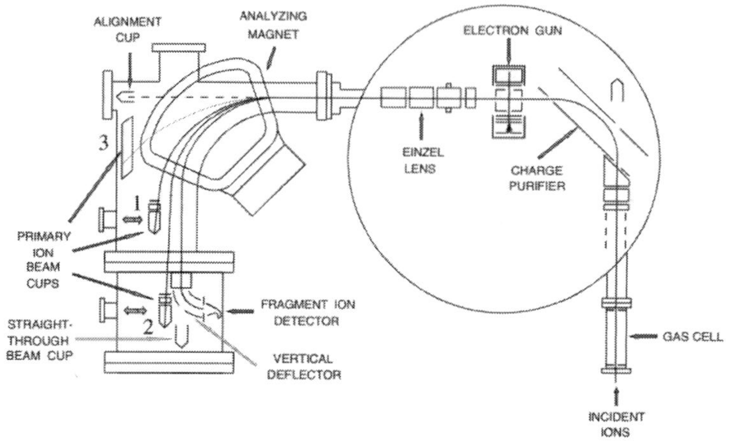

Figure 1. ORNL crossed electron ion beams apparatus. See text for details.

Two additional processes are not considered here; Resonant Ion-Pair Formation (RIP) and non-dissociative ionization. While the RIP process can lead to heavy ion fragments, the cross sections for this process are typically small [4] and aren't expected to contribute significantly to our investigation. Non-dissociative ionization of the molecular cation species likewise have small cross sections and do not contribute to singly charged ion fragment measurements.

2. Experiment

2.1. Ion Beams

The molecular ion beams were produced in the ORNL electron cyclotron resonance (ECR) ion source [5] using total working gas pressures of approximately 10^{-5} Torr and injected RF powers of a few Watts at 10 GHz. All of the XH_2^+ molecular ion beams, except $^{12}CH_2^+$, were produced by first establishing a plasma on deuterium gas fed into the source. Given that specific charge-to-mass ratios can be contaminated by various molecular species, the pure deuterium plasma allowed for a mass analysis to gauge possible contamination from ion chemistry with the source vacuum residual gas. Once the desired charge-to-mass was determined to be free of contamination, a second working gas was added to the source and the desired species produced via ion chemistry in the source plasma. $^{11}BD_2^+$ and $^{19}FD_2^+$ were produced by adding BF_3 to the already established deuterium plasma, $^{14}ND_2^+$ was produced by adding N_2 and $^{16}OD_2^+$ was produced by adding O_2. As discussed in reference [6], $^{12}CH_2^+$ was produced from methane gas alone. The N_2D^+ ion beam was also formed by first establishing a deuterium plasma to gauge m/q contamination and then N_2 was added to chemically produce N_2D^+. The O_3^+ ion beam was formed using only O_2 working gas. Deuterated species were used solely for the purpose of separating the ion fragments produced during dissociation from the parent ion beam.

Ions produced by the ECR source were extracted to form beams with energies of 7 – 10 keV and were subsequently momentum analyzed by a dipole magnet before being transported, via magnetic and electrostatic elements, to the crossed electron ion beams apparatus. Typical ion beam currents delivered to the crossed-beams apparatus range from 5-100 nA.

Given that the flight time from the ion source to the interaction region was on the order of a few microseconds, it is not expected that any vibrational excitations, with lifetimes ranging in the milliseconds, would relax before reaching the interaction region. The current experiment

Table 1. Absolute experimental uncertainties. These uncertainties are combined with the relative uncertainties at a 90% confidence level to determine the total uncertainty in each data point.

Source of Uncertainty	Uncertainty (%)
Product ion detection and pulse processing	±5
Transmission of product ion to detector	±4
Absolute value of form factor	±4
Ion current measurement	±2
Electron current measurement	±2
Ion velocity	±1
Electron velocity	±1
Quadrature Sum	±8.2

has no means to determine the initial rovibrational distribution of the parent molecular ions.

2.2. Crossed-beams Method

The ORNL crossed electron ion beams apparatus is shown schematically in Figure 1. Incident molecular ions passed through a 90° electrostatic analyzer to remove any ions that may have undergone charge exchange or dissociative interactions with the residual gas of the vacuum system during beam transport. The analyzed molecular ion beam then passed through the interaction region perpendicular to the magnetically confined electron beam. The center-of-mass interaction energy was tuned by adjusting the extracted electron beam energy.

The details of the custom electron gun have been discussed previously [7, 8]. Electrons are produced by an indirectly heated planar cathode and extracted over a uniform electrostatic field to define the electron beam energy. A 250 G magnetic field, coaxial to the electron gun direction, confines the electrons and together with the extraction apertures form a uniform rectangular beam (approximately 2 mm wide and 10 mm in height) over the 2 mm interaction region. Opposite the the interaction region, the electrons are collected by a Faraday cup constructed of stacked, tantalum, razor-edged blades oriented with the sharp edges towards the beam. The collector is biased by a +300 V battery to minimize electron loss that would lead to an inaccurate electron current measurement needed for absolute cross section determinations. Typical electron beam currents are ~10 μA at 10 eV and ~250 μA at 100 eV. The electrons are chopped at 1 kHz in order to separate fragment ion signals due to electron impact as compared to collisions with the residual gas.

Fragment ions produced by dissociation of the parent ion beam are separated from the parent ion beam by a double-focusing dipole magnetic. Depending on the mass ratio of the fragment ion to the parent ion beam molecule, the parent beam is collected in one of three different Faraday cups used to measure the incident ion beam current. The fragment ions of interest are directed towards a vertical 90° electrostatic deflector which steer the fragment ions into a discrete dynode electron multiplier detector. The detector (ETP model 14880) has an acceptance window of 25 × 10 mm with the widest dimension oriented in the dispersion plane of the analyzing magnet so as to collect all of the ion fragments regardless of kinetic energy release during dissociation. This detector is capable of counting particle impacts in excess of a million events per second with little deadtime. This capability was essential given that background signal rates for some of the ion fragment channels investigated were more than could be sustained by a conventional channel electron multiplier.

The absolute cross sections for heavy fragment ion production were determined from measured quantities by

$$\sigma(E) = \frac{R}{I_i I_e} \frac{q e^2 v_i v_e}{\sqrt{v_i^2 + v_e^2}} \frac{F}{\epsilon} , \quad (1)$$

where $\sigma(E)$ is the absolute cross section at the center-of-mass energy E, R is the fragment ion signal rate, qe is the charge of the incident ions, v_i and v_e are the incident ion and electron velocities, ϵ is the detection efficiency (90%), and F is the form factor.

The overlap of the ion and electron beams in the direction orthogonal to both beams, here labeled the z direction, was measured at each interaction energy with a rotatable slit probe scanned across both the ion and electron beams in the interaction region. Current profiles of the ion and electron beams, $I_i(z)$ and $I_e(z)$ respectively, were measured independently and a form factor, F, was determined by

$$F = \frac{\int I_e(z)\, \mathrm{d}z \int I_i(z)\, \mathrm{d}z}{\int I_e(z) I_i(z)\, \mathrm{d}z} . \quad (2)$$

The predominant systematic uncertainties associated with the experiment are listed in Table 1 with the estimated error for each component at a high confidence level (equivalent to a 90% confidence level for statistical uncertainties). These errors are treated as random sign errors, summed in quadrature, and combined with the statistical uncertainty of each data point (at a 90% confidence level) to determine the total uncertainty. Detailed discussions of the experimental uncertainties have been published previously by [9]. All absolute cross section figures in this report indicated uncertainties at a 90% confidence level.

3. Results and Discussion

3.1. e^- + XH_2^+ (X = B, C, N, O, F)

Figures 2 and 3 show the combined absolute cross sections for production of XH^+ and X^+ heavy ion fragments, respectively. Table 2 lists the various XH_2^+ DE and DI dissociation channels and corresponding thresholds. These thresholds were determined by using NIST [10] ionization energies and heats of formation. For all systems, except FD_2^+, the infinite dissociation limit of neutral fragments was verified by comparison to the ground state dissociation energies measured for dissociative recombination at storage rings [11, 12, 13, 14, 15]. The CH_2^+ data were measured with the ORNL crossed-beams apparatus previously and are reproduced here from Vane et al. [6].

The absolute cross sections plotted in Figure 2, for the dissociation channel leading to XH^+, show a clear similarity for the underlying DE and DI processes with the DI cross sections all converging around 100 eV. For the BD_2^+ and CH_2^+ species, a large resonant-like structure is seen in the DE region. ND_2^+ and OD_2^+ also exhibit small resonant-like peaks. It is unclear if this is a direct excitation to a dissociative state or if there is a contribution from a resonant capture process that proceeded via autoionization and subsequent dissociation. Lecointre et al. [16] have recently measured this same structure in the DE of CD_2^+, however, their results indicate a peak cross section about a factor of four smaller. Their results also do not exhibit a two-body X^+ DE process. This is likely due to unknown excited ro-vibrational states of the parent molecular ion in each experiment as source conditions are likely different in each case. The differences could also be representative of an isotope effect if indirect processes are governing the dynamics.

Figure 3 shows the absolute cross sections for X^+ fragment production along with the lowest energy DE thresholds that appear to contribute to the cross sections. If the thresholds for DE are used as a guide to determine if a two- or three-body process is contributing to an observed cross section, it appears that the BD_2^+ and CH_2^+ cations exhibit a two-body DE process while

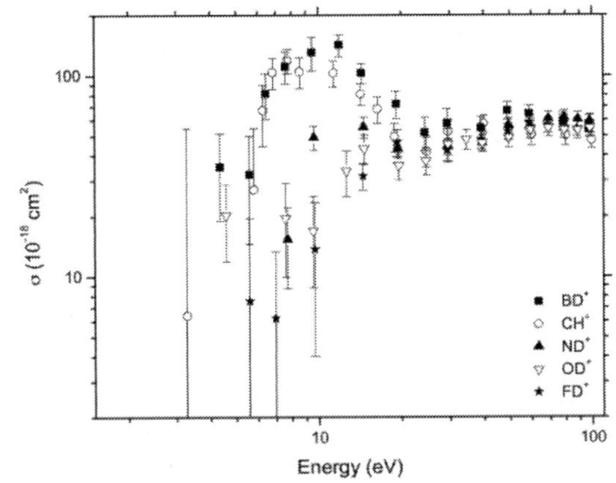

Figure 2. Absolute cross sections for DE and DI of XH_2^+ (X = B, C, N, O, F) resulting in XH^+ ion fragments. Thresholds are not shown in this figure for clarity. Refer to Table 2.

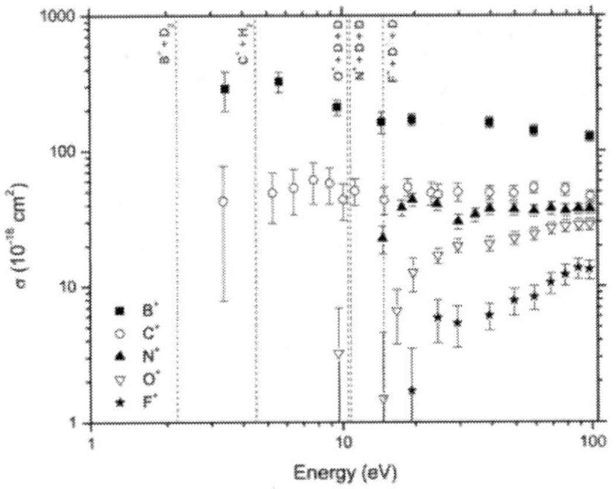

Figure 3. Absolute cross sections for DE and DI of XH_2^+ (X = B, C, N, O, F) resulting in X^+ ion fragments.

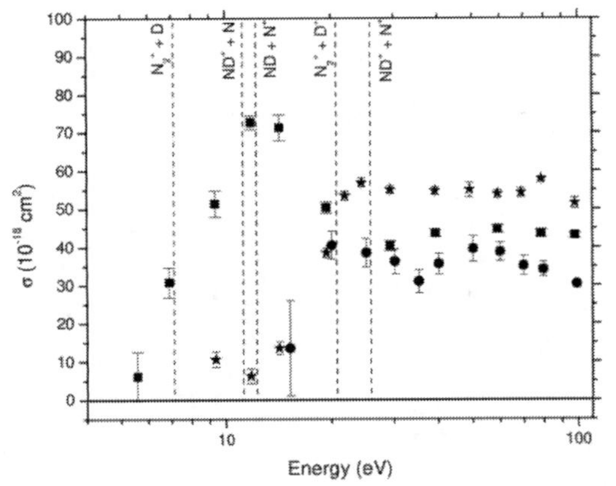

Figure 4. Absolute cross sections for DE and DI of N_2D^+ resulting in N^+ (filled stars), ND^+ (filled circles) and N_2^+ (filled squares) ion fragments.

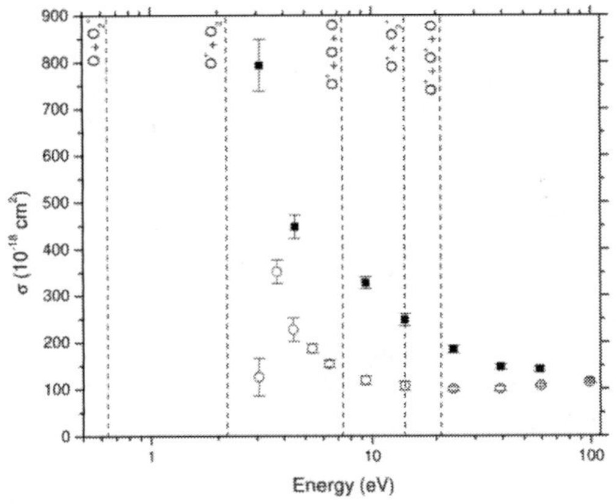

Figure 5. Absolute cross sections for DE and DI of O_3^+ resulting in O^+ (open circles) and O_2^+ (filled squares) ion fragments.

the remaining XH_2^+ species all exhibit a three-body break up when X^+ fragments are produced. This is mirrored by the observed two- and three-body dissociation dynamics seen in storage ring DR measurements for these species [11, 12, 13]. We also note that the rising slope of the cross sections near threshold vary depending on the threshold value, i.e., lower energy thresholds tend to exhibit a sharper increase while higher energy thresholds tend to show a more gradual increase in cross section as a function of collision energy. The lower energy thresholds that

Table 2. Dissociation channels producing heavy fragment ions and their corresponding thresholds in eV. See text for description of how thresholds were determined.

Dissociation Channel			X =	B	C	N	O	F
$e^- + XH_2^+$	$\rightarrow$	$XH^+ + H + e^-$		4.8	4.8	6.1	5.5	7.5
	$\rightarrow$	$X^+ + H + H + e^-$		6.7	9.0	10.8	10.5	14.7
	$\rightarrow$	$X^+ + H_2 + e^-$		2.2	4.5	6.3	6.0	10.2
	$\rightarrow$	$XH^+ + H^+ + 2e^-$		18.4	18.4	19.7	19.1	21.1
	$\rightarrow$	$X^+ + H^+ + H + 2e^-$		20.3	22.6	24.4	24.1	28.3
	$\rightarrow$	$X^+ + H_2^+ + 2e^-$		17.6	19.9			
$e^- + N_2D^+$	$\rightarrow$	$ND^+ + N + e^-$		11.2				
	$\rightarrow$	$ND + N^+ + e^-$		12.3				
	$\rightarrow$	$N_2^+ + D + e^-$		7.1				
	$\rightarrow$	$ND^+ + N^+ + 2e^-$		25.8				
	$\rightarrow$	$N_2^+ + D^+ + 2e^-$		20.7				
	$\rightarrow$	$N^+ + N + D^+ + 2e^-$		29.0				
$e^- + O_3^+$	$\rightarrow$	$O_2^+ + O + e^-$		0.6				
	$\rightarrow$	$O_2 + O^+ + e^-$		2.2				
	$\rightarrow$	$O^+ + O + O + e^-$		7.4				
	$\rightarrow$	$O_2^+ + O^+ + 2e^-$		14.3				
	$\rightarrow$	$O^+ + O^+ + O + 2e^-$		21.0				

initially contribute to DE also seem to produce larger relative cross sections.

3.2. e^- + N_2D^+

Figure 4 shows the absolute cross sections for heavy fragment ion production from N_2D^+ along with the thresholds for the various DE and DI dissociation channels. The corresponding threshold values are given in Table 2. The thresholds are determined in the same manner as discussed in the previous section. As can be seen in Figure 4, the DE region is dominated by N_2^+ production. This channel also shows a slight resonant-type feature as seen for the BD_2^+ and CH_2^+ species discussed in the previous section. The DE channels leading to ND^+ and N^+ can also bee seen in Figure 4. These channels are a result of N–N bond breaking that was previously a surprising result of DR measurements of N_2D^+ [14]. These two channels have similar DE thresholds and the relative cross sections appear to remain similar until the DI threshold is reached. These two channels share a common DI channel, i.e., $ND^+ + N^+$ and, assuming a two-body breakup model, should exhibit identical DI cross sections. However, The N^+ product channel shows a larger cross section. This suggests that a three-body breakup dynamic is contributing, namely the $N^+ + N + D^+$ channel at 29 eV. This three-body channel would increase the apparent N^+ cross section given that multiple dissociation channels could contribute. This increase is observed in Figure 4, as compared to the ND^+ DI cross section. A similar three-body effect has also been observed by Lecointre et al. [17] for C_2D^+.

3.3. e^- + O_3^+

Figure 5 shows the absolute cross sections for heavy fragment ion production from O_3^+ producing O_2^+ and O^+ fragments along with the various DE and DI thresholds. The corresponding thresholds are given in Table 2 and have been determined in the same manner as described previously. As can be seen in figure 5, the DE region is dominated by a large cross section for the production of O_2^+. The crossed-beam apparatus could not measure sufficiently low in electron energy, due to reduce electron currents at low energy, to establish the threshold behavior for this channel. This relatively large cross section could also indicate a resonant-like feature as observed for BD_2^+, CH_2^+ and N_2D^+.

The O^+ fragment channel threshold, however, could be reached and a sharp onset in cross section is observed. The presence of a contribution to the O^+ channel below the $O^+ + O + O$ threshold indicates that the two-body process contributes dominantly to the dissociation. The DI cross sections are observed to converge at around 100 eV. It is worth noting that the DR measurements for O_3^+ show a propensity for three-body breakup [15] while the DE process seems to follow a two-body dissociation dynamic.

4. Conclusion

The absolute cross section for the production of heavy fragment ions from the triatomic species XH_2^+, N_2D^+ and O_3^+ due to electron impact have been investigated in the energy range of 3 – 100 eV using the crossed electron-ion beams apparatus at Oak Ridge National Laboratory. Two- and three-body dissociation channels are observed as well and resonant-like contribution to DE. It is unclear if these resonant-like features are due to direct excitation or to resonant capture processes. More theoretical effort is needed to help clarify these effects.

References

[1] Tawara H 1995 *Atomic and Molecular Processes in Fusion Plasmas* ed Janev R K (Plenum, New York) pp 461–496
[2] Janev R K and Reiter D 2002 *Phys. Plasmas* **9** 4071–4081
[3] Dalgarno A 1976 *Atomic Processes and Applications* ed Burke P G and Moiseiwitsch B L (North-Holland, Amsterdam) pp 109–132
[4] Zong W, Dunn G, Djuric N, Larsson M, Greene C, AL-Khalili A, Neau A, Derkatch A, Vikor L, shi W, Padellec A L, Rosén S, Danared H and af Ugglas M 1999 *Phys. Rev. A* **83** 951
[5] Meyer F W 2001 *Trapping Highly Charged Ions: Fundamentals and Applications* ed Gillaspy J (Nova Science Publishers, Huntington, NY) p 117
[6] Vane C R, Bahati E M, Bannister M E and Thomas R D 2007 *Phys. Rev. A* **75** 052715
[7] Taylor P O, Dolder K T, Kauppila W E and Dunn G H 1974 *Rev. Sci. Instrum.* **45** 538
[8] Gregory D C, Meyer F W, Müller A and Defrance P 1986 *Phys. Rev. A* **34** 3657
[9] Bannister M E 1996 *Phys. Rev. A* **54** 1435
[10] 2010 NIST Chemistry WebBook, NIST Standard Reference Database Number 69 URL `http://webbook.nist.gov/chemistry/`
[11] Zhaunerchyk V, Vigren E, Geppert W D, Hamberg M, Danielsson M, Kamińska M, Larsson M, Thomas R D, Bahati E and Vane C R 2008 *Phys. Rev. A* **78** 024701
[12] Thomas R D, Hellberg F, Neau A, Rosén S, Larsson M, Vane C R, Bannister M E, Datz S, Petrignani A and van der Zande W J 2005 *Phys. Rev. A* **71** 032711
[13] Jensen M J, Bilodeau R C, Heber O, Pederson H B, Safvan C P, Urbain X, Zajfman D and Anderson L H 1999 *Phys. Rev. A* **60** 2970
[14] Geppert W D, Thomas R, Semaniak J, Ehlerding A, Miller T J, Osterdahl F, af Ugglas M, Djuric N, Paal A and Larsson M 2004 *Astrophys. J.* **609** 459
[15] Zhaunerchyk V, Geppert W D, Larsson M, Thomas R, Bannister M E, Fogle M R, Vane C R and Osterdahl F 2007 *Phys. Rev. Lett.* **98** 223201
[16] Lecointre J, Belic D, Jureta J, Janev R and Defrance P 2009 *Eur. Phys. J. D* **55** 557
[17] Lecointre J, Cherkani-Hassan H, Cherkani-Hassani S, Belic D S, Jureta J and Defrance P 2010 *Eur. Phys. J. D* **online ahead of print**

Journal of Physics: Conference Series **300** (2011) 012004 doi:10.1088/1742-6596/300/1/012004

Branching ratios in dissociative recombination of formyl and isoformyl cations

Christian Nordhorn[1], **Dennis Bing**[1], **Henrik Buhr**[1,2], **Manfred Grieser**[1], **Oded Heber**[2], **Claude Krantz**[1], **Mario B Mendes**[1], **Roland Repnow**[1], **Dirk Schwalm**[2,1], **Andrey Shornikov**[1], **Julia Stützel**[1] **and Andreas Wolf**[1]

[1] Max-Planck-Institut für Kernphysik, D-69029 Heidelberg, Germany
[2] Dept. of Particle Physics, Weizmann Institute of Science, Rehovot 76100, Israel

Oldřich Novotný and Michael Lestinsky

Columbia Astrophysics Laboratory, Columbia University, New York, NY 10027, USA

Åsa Larson and Michael Stenrup

Department of Physics, Stockholm University, SE-106 91 Stockholm, Sweden

E-mail: aasal@fysik.su.se

Ann E. Orel

Department of Applied Science, University of California, Davis, CA 95616, USA

Abstract. The energy dependence of the branching ratios in dissociative recombination of DCO^+ with a known small admixture of DOC^+ has been measured for collision energies from 0 to 25 eV using an energy- and position-sensitive surface barrier detector which is designed for the analysis of multi-fragment events occurring in a molecular fragmentation study. The measurements are compared with theoretical calculations on the direct mechanism of dissociative recombination of HCO^+ including the contribution from HOC^+ at the experimental abundance fraction. At low collision energies, dissociative recombination of HCO^+ is dominated by dissociation into H + CO. For collision energies above 2 eV there is a transition into dissociation to HC + O, which can be explained by electron capture into resonant states. Signatures of DOC^+ dissociative recombination are found experimentally and confirmed by the calculations for HOC^+. Three-body breakup becomes important for collision energies above 6 eV.

1. Introduction

Low energy dissociative recombination (DR) of HCO^+ is known to be a reaction crucial for the chemistry of interstellar space. The reaction has been studied extensively both experimentally and theoretically [1]. The measurements agree on a relatively large DR rate of about 2×10^{-7} cm^3s^{-1} at 300 K [1]. On the theory side there has been a debate on the mechanism for the low energy DR of HCO^+. Recent calculations indicate that at low energies the reaction is driven by the electron capture into Rydberg states situated below the ion. The Renner-Teller effect couples the electronic and nuclear motions. The calculated rate using the multichannel quantum defect theory is about a factor of 2–3 lower than measured rates [2]. At low energies measurements of the branching ratios showed that in this reaction the H-C bond will break [3, 4], in agreement with the prediction [5] that this is the only energetically accessible channel. Analysis of the CO band emission in flowing afterglows shows that with a yield of 0.23 ± 0.12 [5] the DR of HCO^+ also forms the long-lived excited $a^3\Pi$ state of CO.

Another ion formed in plasmas containing HCO^+ is the isoformyl cation HOC^+, whose ground state lies approximately 1.7 eV higher than that of the formyl ion. About the DR of HOC^+ much less is known, although it has been observed in both dense and diffuse molecular clouds [6, 7, 8] and it also was found in the plasmas used when studying HCO^+ [5]. To the best of our knowledge neither an experimental nor a theoretical determination of the DR rate coefficient of HOC^+ exists to date.

The present paper gives experimental results on the energy dependence of DR branching ratios for the deuterated species DCO^+ and DOC^+ in the energy range from 0 to 25 eV. The measurements were performed with fast merged ion and electron beams in the heavy-ion storage ring TSR using the energy-sensitive multi-strip detector EMU [9]. The EMU detector records the impact positions [10] of multi-fragment events with a simultaneous identification of the recombination products by their masses, which yields a determination of the fragmentation channel on an event-by-event basis.

In this fast-beam fragmentation experiment, the products undergo an angular deflection from the beam direction depending on the kinetic energy they gain in the rest frame of the reaction. At the large kinetic energy release (KER) of the DR reaction toward ground-state fragments in the channel H + CO from HCO^+ [5] (and similarly from HOC^+) the H fragments would reach the detector position with too large transverse offsets from the beam axis, outside the available detector size; hence, the deuterated species were chosen, yielding considerably smaller transverse velocities and, correspondingly, transverse positions falling within the detector size. The final product channels of the DR of DCO^+ and DOC^+ are listed in Table 1 together with their exothermicities at vanishing electron impact energy from the ground state of the cation to the ground states of the products.

The results for DCO^+ and DOC^+ are compared to theoretical results for HCO^+ and HOC^+. In a previous study [11] the electronic resonant states of HCO and HOC were computed by combining electron scattering calculations with structure calculations. We have here performed two-dimensional wave packet propagation on the resonant states to study the direct mechanism of DR. In the model autoionization is included, but no couplings between the neutral states. Also DR of the HOC^+ ion is theoretically investigated and by assuming a 9% contamination of HOC^+ ions in the TSR experiment, the measured branching ratios are compared with theoretical predictions.

2. Experiment

The DR of the isomeric systems DCO^+ and DOC^+ is probed in a molecular fragmentation study making use of the merged-beam facility provided by the electron target [12, 13] at the TSR at the Max-Planck Institute for Nuclear Physics in Heidelberg. The ion beam is produced by the same method as in a previous experiment at the TSR [14], which analyzed the isomeric composition of the stored ion beam as discussed below. A tandem Van-de-Graaff accelerator is applied, starting with D_3CO^- from a Cs sputter ion source operated with deuterated methanol. These ions are stripped to DCO^+ or DOC^+ at an energy of ~1.6 MeV in the terminal of the tandem accelerator and further accelerated to reach an energy of 3.1 MeV at their injection into the TSR. In the storage ring, the beam velocity of the electron target is matched to the ion beam velocity for phase-space cooling during 7 s after the injection. In the time interval from 7 to 17 s after injection, the electron beam energy is set to detune its velocity from that of the ions. For measuring intervals of 50 ms, switching with a duty cycle of 50% between detuned and matched beam velocities, variable detuning energies E_d in the ion beam rest-frame are applied. By using different values of E_d during the measurement intervals of subsequent injections, the energy dependence of the reaction is investigated.

Upon electron capture the molecule breaks into neutral fragments which travel forward to the detector at the ion beam velocity, undeflected by the magnetic field of the downstream dipole magnet of the storage ring; behind this deflector they impinge on the EMU detector [9] in a dedicated beamline, creating pulses proportional to their kinetic energy. The 128 vertical and horizontal stripes, respectively, on the front and back sides of the $10 \times 10\,\mathrm{cm}^2$ detector are read out individually. Thus the recombination products can be identified by their masses on an event-by-event basis by coincidence imaging and the corresponding fragmentation channel can be assigned. Furthermore, each fragment impact position is known so that one can obtain the distribution of (mass-weighted) fragment separations, which in turn represents the distribution of the KER under the assumption of an isotropic breakup [9, 10].

The relative contributions of each final DR channel (Table 1) resulting from the detector data need small corrections to be applied since the slight deviations of the detection efficiencies from unity do not have the same values for the various fragment channels. They are caused by dead areas separating the strips, geometrical limitations and finite mass resolution. Thus, events of special geometry where the fragments CO and D (mass 30 amu) arrive on a single stripe cannot be resolved from those of a single CO (mass 28 amu) on the same stripe; this slightly reduces the efficiency for identifying the channel CO + D. These efficiency corrections are determined by Monte-Carlo simulations taking all details of the experimental geometry into account [9]. In addition, a background from dissociative electron capture of DCO^+ and DOC^+ ions in the residual gas was observed, composed of ~91% CO + D, ~5.5% of D + C + O, ~2.5% of DC + O, and ~1% of DO + C. It could be analyzed by fragment imaging with EMU and amounts to a few percent of the rate at $E_d = 0$; at higher energies, it causes a significant uncertainty of the DR rate to be assigned to CO + D, while it only little affects the DR rates of the other channels, which are also clearly found to be caused by DR through their variation with the electron energy. The final analysis regarding the efficiency correction and the subtraction of this background is ongoing and could slightly modify the branching ratios at higher energies, but not their overall energy dependences.

In a previous experiment at TSR, the isomeric composition of a stored ion beam produced under essentially the same conditions as in the present case was investigated by Coulomb explosion imaging [14]. After a relaxation time of ~4 s following the injection into the TSR, the deuterated isoformyl cation DOC^+ was found at stable fractional abundance of 9% up to 12 s

Table 1. List of channels available in the DR of DCO^+ and DOC^+ and their respective exothermicities (in eV). The corresponding energy gains for HCO^+ and HOC^+ [5] are reduced by 0.06 eV and 0.08 eV for DCO^+ and DOC^+, respectively, to account for the shift in the zero point energies in these parent species. For channels which would imply a reconfiguration and which therefore are highly unlikely to occur, the exothermicities are marked by an asterisk.

	Channel	Exothermicity	
		DCO^+	DOC^+
(α)	D + CO	7.25	8.97
(β)	DC + O	−0.17	1.55*
(γ)	DO + C	0.57*	2.29
(δ)	D + C + O	−3.90	−2.18

of storage time. The previous study also showed that both isomers internally cooled down from a strong initial vibrational excitation during the above relaxation time; however, the bending vibration was not completely relaxed in DCO^+ with a remaining excitation energy of $\sim$0.2 eV, while a lower remaining excitation was found in DOC^+.

At low collision energy, only the final channels (α) and (γ) of Table 1 are energetically open for DR of DCO^+; of these, channel (γ) would need the removal of the central C atom while creating a new DO bond, which can be considered unlikely. For DOC^+ in low-energy DR these two channels are open as well, together with channel (β) which, however, reciprocally can be considered unlikely because of the strong re-arrangement required for its formation. Energetic thresholds at higher E_d are expected at only 0.17 eV for DCO^+ towards channel (β) and starting at $\sim$2–4 eV for the three-body fragmentation channel.

3. Theory

3.1. Calculation of resonant states

The details of the calculations of the energies and autoionization widths of the electronically resonant states of HCO and HOC have been outlined in a previous study [11] and hence only a short summary is given here. The potential energy surfaces of the electronically bound states are computed using the Multi-Reference Configuration Interaction (MRCI) method with natural orbitals of the ground state of the ion. In the calculations the $(1\sigma)(2\sigma)$ core orbitals are kept doubly occupied and the reference configurations are obtained by excitations of 11 electrons (10 for the ion) among eight orbitals. The MRCI wave function is constructed by allowing all single excitations out of the reference configurations. By using the Complex-Kohn variational method [15] the energy positions and autoionization widths of the resonant states are determined. The same MRCI wave function for the target ion is used in both the structure and electron scattering calculations.

The Rydberg states converging to the ground state of the ion as well as the ionization continuum have the same configuration as the ion plus an outer electron in a diffuse orbital. The resonant states are Rydberg states converging to an excited ionic core. These resonant states all have vacancies in the (5σ) or (1π) molecular orbitals. By following the states with these configurations a quasidiabatization of the resonant states relative to the Rydberg manifold can be performed. In the present model no electronic couplings between the diabatic states are included.

For the HCO system no electronic states that cross the ion close to its minimum is obtained. At collision energies larger than 2 eV several resonant states can be reached. As a function of the H-C coordinate the resonant states of HCO show clear indications of avoided crossing among each other. The resonant states are repulsive as a function of the C-O coordinate.

The resonant states of HOC are lower in energy relative to the ion compared with HCO. Also the HOC resonant states show indications of avoided crossings and are repulsive as a function of the O-C coordinate.

As a function of the bending angle the resonant states of both HCO and HOC are relatively flat in the Franck-Condon region. In the present study the two radial coordinates are included while the bending angle is kept at the linear geometry. The resonant states are constructed on a product grid where the C-O distance (O-C for HOC) is ranging from 1.5 to 10.0 a_0 and H-C (H-O for HOC) from 1.0 to 10.0 a_0.

3.2. Wave packet dynamics

The nuclear dynamics is studied using wave packet propagation with the Multi Configuration Time Dependent Hartree (MCTDH) method [16]. The direct mechanism of DR is investigated by numerically solving the time-dependent Schrödinger equation for the resonant states where autoionization is included as a local complex term added to the potential.

$$i\frac{\partial}{\partial t}\Psi = \left(T + V - i\frac{\Gamma}{2}\right)\Psi. \tag{1}$$

The initial condition for the wave packets is given by

$$\Psi(t=0) = \sqrt{\frac{\Gamma}{2\pi}}\chi_{\rm ion}, \tag{2}$$

where $\chi_{\rm ion}$ is the ground vibrational wavefunction of the ion.

The wave packets are propagated for 140 fs until they have reached the asymptotic regions of the resonant states. At large internuclear coordinates (8.0 a_0) complex absorbing potentials are used to determine the dissociating flux [17]. By applying projection operators of the one-dimensional eigenstates of the asymptotic Hamiltonian, the state-specific transition probabilities can be computed. Summarizing over the vibrational bound states, a separation between the two- and three-body breakup channels can then be performed.

For the HCO system, two resonant states of each of the symmetries $^2\Sigma$, $^2\Pi$ and $^2\Delta$ are included. For HOC, there are two states of $^2\Sigma$ and $^2\Delta$ symmetries and three states of $^2\Pi$ considered.

4. Results and Discussion

4.1. Measured branching ratios

The observed ratios of occurrence for the different final channels represent the detector data with small corrections as discussed in Sec. 2. They are shown in Fig. 1 as a function of the collision energy E_d and correspond to a known mixture of DCO^+ and DOC^+ in the parent ion beam. At near-zero collision energy the fragmentation channel CO + D (α) is dominant by more than 96%. Considering the low isomeric fraction of DOC^+ present, this shows that this channel dominates the low-energy DR of DCO^+, as expected (see Sec. 2). The small contribution of channel (γ) is likely due to DOC^+ ions that recombine into this channel already at low energy.

As the collision energy increases the branching into the originally dominating channel D + CO decreases, while both other two-body channels (β) and (γ) take a larger relative fraction of events. It is likely that the breakup into DO + C (γ) can be completely assigned to the DR of DOC^+, while DC + O (β) should be caused by DCO^+, considering the strong reconfiguration

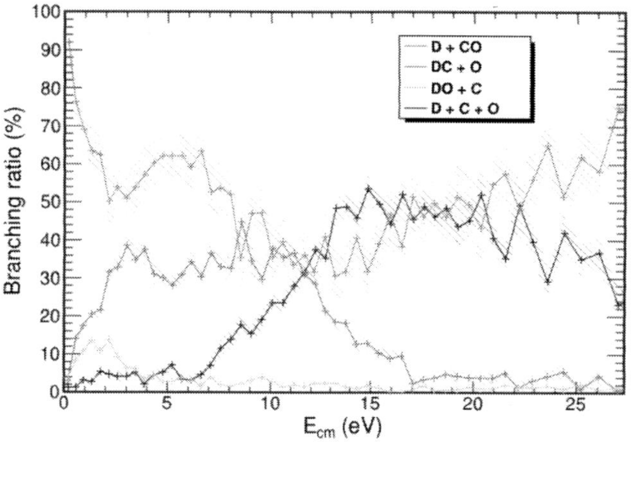

Figure 1. Energy dependence of the relative contributions from the indicated product channels to the DR of a DCO^+ ion beam having an admixture of 9% of DOC^+ ions as derived experimentally in ref. [14]. The shaded areas indicate the uncertainty from statistical errors and that estimated for the effiency correction. A preliminary subtraction of background rates from dissociative electron capture in the residual gas was performed (see Sec. 2).

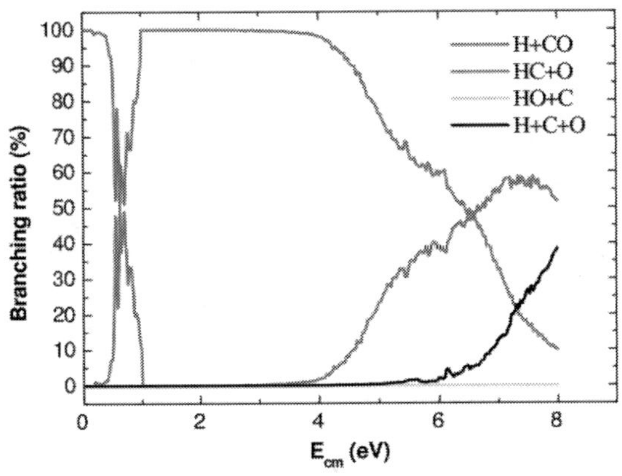

Figure 2. Calculated branching ratios for DR of HCO^+.

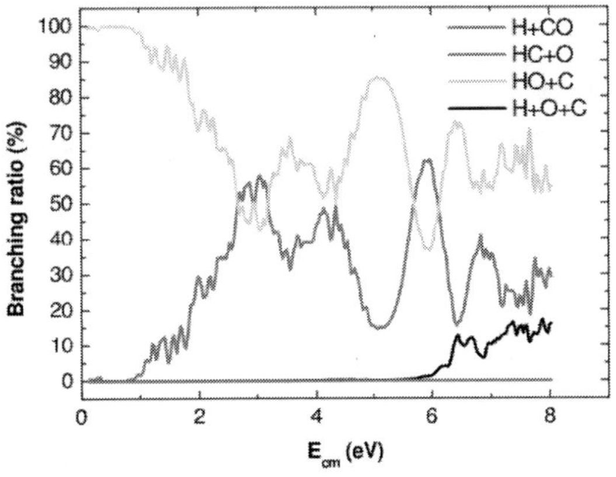

Figure 3. Calculated branching ratios for DR of HOC^+.

necessary to form either of these channels from the respective other isomer. Although the three-body channel (δ) is accessible for energies above $\sim$2–4 eV, it is only observed from $\sim$7 eV onward. It first seems to increase on the expense of the dominant channel (α), while above $\sim$12 eV it rather seems to be fed from channel (β). Above $\sim$18 eV only CO + D and D + C + O are significantly observed.

4.2. Calculations on HCO

The electronic resonant states of HCO do not contribute to the DR cross section at low energies (see Fig. 2 of Ref. [11]). An energy of about 1 eV is needed to capture into the resonant states. Above this energy the resonant states will produce a high-energy peak in the cross section. The largest contribution to this peak comes from the lowest $^2\Pi$ resonant states. The lower resonant states are dominated by a complete breakup into HC + O. For some of the higher resonant states ($1^2\Sigma^+$ and $2^2\Delta$) there are contributions of dissociation into H + CO. In Fig. 2 the calculated branching ratios for DR of HCO^+ are displayed. For low energy collisions we use the DR cross section calculated by Douguet *et al.* [2] and assume a 100 % breakup into H + CO. As can be seen in the figure there is a complete transition from H + CO dissociation into HC + O dissociation around 1 eV when the energy is high enough to capture into the electronically resonant states. Above 6 eV the branching ratio for breaking both bonds is increasing.

In the present study the two radial coordinates are included while the angle is fixed at linear geometry. This is motivated by the fact that in the Franck-Condon region the energies of the resonant states are relatively independent with respect to changes in the angle. We have begun calculations on the dynamics where all three internal coordinates are included and preliminary results indicate that this will not significantly change the results presented here.

4.3. Calculations on HOC

The resonant states of HOC (see Fig. 4 of Ref. [11]) are lower in energy relative to the ion compared to HCO. However, the nuclear dynamics reveal that the resonant states of HOC do not significantly contribute to the low energy DR cross section. The resonant states of HOC will produce a peak in the DR cross section centered around 0.6 eV. The largest contribution to this peak comes from the lowest resonant state of $^2\Delta$ symmetry. As can be seen in Fig. 3 at low energies, the HOC branching ratios are dominated by HO + C dissociation. In the present model the capture into the Rydberg states are not included and by including this indirect mechanism

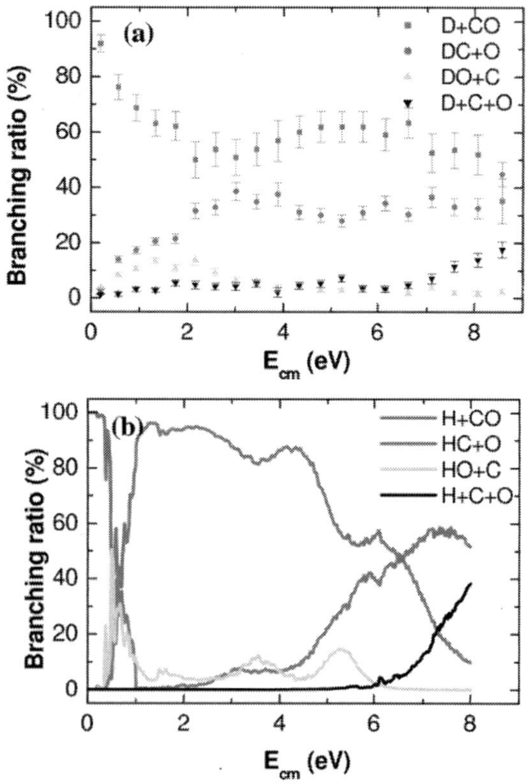

Figure 4. Measured branching ratios (a) are compared with the theoretical branching ratios (b) for a composite isomeric beam dominated by DCO^+ (HCO^+ in the theoretical model) with a 9% fraction of DOC^+ (HOC^+).

of DR the low energy branching ratios could indeed change. At larger energies the branching ratios oscillate around 50% dissociation into H + OC and 50% breakup into HO + C. For energies above 6 eV the breaking of both bonds can occur.

4.4. Comparison between experiment and theory

In Fig. 4 the measured branching ratios are compared with the computed branching ratios. In the theoretical model a 9% fraction for the HOC^+ component in the experiment is assumed. It should be noted that the experiments are carried out for the DCO^+ ion while in the theoretical modeling HCO^+ is studied. In the comparison, both the experimental and the calculated branching ratios show a transition into the dissociation channel equivalent to HC + O when the energy amounts to a few eV and electron capture into the resonant states of HCO can proceed. The calculated branching ratio for HC + O dissociation is larger than the measured one in the range up to about 7 eV on the expense of a strongly diminished observation of CO + H. Furthermore both experiment and theory show an opening of the three-body breakup for collision energies above 6 eV. There is some contribution of HO + C (DO + C) dissociation in both theory and experiment, which can be explained by the HOC^+ (DOC^+) contamination.

The remaining differences between the computed and measured branching ratios might be explained by several different factors. First, in the theoretical modeling only the direct dissociative recombination mechanism is considered. No electronic couplings between the neutral states are incorporated. By including these couplings, flux may be redistributed among the electronic states and this can change the branching ratios. Furthermore, in the theoretical study electron capture with ro-vibrationally relaxed ions is investigated. Experimentally the ions are known to be rotationally excited and for DCO^+ also vibrationally excited. Finally, for low energy collisions with HCO^+, we use the cross section computed by Doughet *et al.* [2] and assume a 100% breakup into H + CO. This branching ratio has not been theoretically calculated.

5. Summary

We present results on the energy dependence of the branching ratios in the DR of DCO^+ measured with a 9% contamination by DOC^+ ions known from a previous measurement. The experimental branching ratios are compared with computations on the direct mechanism of DR for the undeuterated corresponding ions HCO^+ and HOC^+. Applying a 9% admixture of HOC^+ ions in the calculations, a qualitative agreement between theory and experiment is found. In the present study the two radial coordinates are included in the study of dynamics; the angular dependence of the resonant states will be addressed in a future study.

Acknowledgments

We are grateful to V. Kokoouline for sharing unpublished data with us. Travel support for CN was provided from NSF and NASA. ÅL acknowledges support from the Swedish research council. AEO acknowledges support from the National Science Foundation under grant PHY-08-55092. ON was supported in part by the NSF Astronomy and Astrophysics Grant Program. ML was supported in part by the NASA Astronomy and Astrophysics Research and Analysis Program. HB and OH acknowledge partial support by the German Israeli Foundation for Scientific Research and Development (G.I.F.) under Grant I-900-231.7/2005 and by the European Project ITS LEIF (HRPI-CT-2005-026015). DS acknowledges support by the Weizmann Institute through the Joseph Meyerhoff program. Support by the Max-Planck Society is gratefully acknowledged.

References

[1] Larsson M and Orel A E 2008 *Dissociative recombination of molecular ions* (Cambridge university press, New York) p 244-7
[2] Douguet N, Kokoouline V and Greene C H 2009 *Phys. Rev.* A **80** 062712
[3] Adams N G, Herd C R, Geoghegan M, Smith D, Canosa A, Gomet J C, Rowe B R, Queffele J L, and Morlais M 1991 *J. Chem. Phys.* **94** 4852
[4] Geppert W D, Thomas R, Ehlerding A, Semaniak J, Österdahl F, af Ugglas M, Djurić N, Paál A and Larsson M 2004 *Faraday Discuss.* **127** 425
[5] Rosati R E, Skrzypkowski M P, Johnsen R and Golde M F 2007 *J. Chem. Phys.* **126** 154302
[6] Woods R C, Guderman C S, Dickman R L, Goldsmith P F, Huguenin G R, Irvine W M, Hjalmorson Å, Nyman L-Å and Olofsson H 1983 *Astrophys. J.* **270** 583
[7] Ziurys L M and Apponi A J 1995 *Astrophys. J.* **455** L73
[8] Liszt H, Lucas R and Black J H 2004 *Astron. Astrophys.* **428** 117
[9] Buhr H *et al* 2010 *Phys. Rev.* A **81** 062702
[10] Amitay Z, Zajfman D, Forck P, Hechtfischer U, Seidel P, Grieser M, Habs D, Repnow R, Schwalm D and Wolf A 1996 *Phys. Rev.* A **54** 4032
[11] Larson Å and Orel A E 2009 *Phys. Rev.* A **80** 062504
[12] Sprenger F, Lestinsky M, Orlov D A, Schwalm D and Wolf A 2004 *Nucl. Instrum. Meth.* A **532** 298
[13] Orlov D A, Krantz C, Wolf A, Jaroshevich A S, Kosolobov S N, Scheibler H E and Terekhov A S 2009 *J. Appl. Phys.* **106** 054907
[14] Wester R *et al* 2002 *J. Chem. Phys.* **116** 7000
[15] Rescigno T N, McCurdy C W, Orel A E, and Lengsfield III B H1995 *The Complex Kohn Variational Method* in *Computational methods for electron-molecule scattering* eds. Huo W H and Gianturco F A (Plenum New York)
[16] Beck M H, Jäckle A, Worth G H, and Meyer H-D 2000 *Phys. Rep.* **324** 1
[17] Jäckle A and Meyer H -D 1996 *J. Chem. Phys.* **105** 6778

Dissociative recombination and the decay of a molecular ultracold plasma

C. J. Rennick, N. Saquet, J. P. Morrison, J. Ortega-Arroyo, P. Godin, L. Fu, M. Schulz-Weiling and E. R. Grant

Department of Chemistry, University of British Columbia, Vancouver, BC V6T 1Z1, Canada

Abstract.
Double-resonant photoexcitation of nitric oxide in a molecular beam creates a dense ensemble of $51f(2)$ Rydberg states, which evolves to form a plasma of free electrons trapped in the potential well of an NO^+ spacecharge. The plasma travels at the velocity of the molecular beam, and, on passing through a grounded grid, yields an electron time-of-flight signal that gauges the plasma size and quantity of trapped electrons. This plasma expands at a rate that fits with an electron temperature as low as 5 K. Dissociative recombination of NO^+ ions with electrons provides the primary dissipation mechanism for the plasma. We have identified three dissociation pathways, and quantified their relative contributions to the measured rate: Two-body dissociative recombination competes with direct three-body recombination to neutral dissociation products, and with a process in which three-body recombination and electron-impact ionization form an equilibrium population of high-Rydberg states that decays by predissociation. Using available collision-theory rate constants for three-body recombination and ionization, together with quantum mechanical estimates of predissociation rates, we predict that the relaxation of the plasma to a high-Rydberg equilibrium outpaces direct three-body dissociative recombination, and, among second-order processes, the rate of two-body electron-cation dissociative recombination substantially exceeds the rate at which the high-Rydberg equilibrium dissociatively relaxes. The rate constant for dissociative recombination extracted from these data conforms with predictions drawn from theory for isolated electron-ion collisions. Methods based on the dissipation of molecular ultracold plasmas may provide a means for estimating rates of dissociative recombination for a variety of complex molecules.

1. Introduction

Ultracold neutral plasmas offer laboratory access to the important limit of a charged particle gas in which the average potential energy of electrostatic interaction approaches the thermal collision energy. Under such conditions, many-body forces govern dynamical properties, and spatial correlations can act significantly to determine the state of a system and its evolution in time (1; 2).

Most experiments prepare ultracold plasmas from atoms in a magneto-optical trap (MOT). Under typical conditions, radiation from a nanosecond pulsed dye laser excites a substantial fraction of the atoms contained in an initial volume of 1 mm^3 at a density of 10^{10} cm^{-3}.

For an excitation energy tuned just above the ionization threshold, the plasma forms directly (3). The photon energy in excess of the ionization limit sets the initial electron temperature in this plasma, while the mili-Kelvin trap temperature of the neutral precursor determines that of the ions. Below-threshold excitation to high-Rydberg states forms a Rydberg gas that evolves to a plasma (4; 5; 6; 7). Here, the dynamics of Rydberg-Rydberg and electron-Rydberg collisional ionization processes determine the initial electron temperature (8).

Following plasma formation in either case, super-elastic collisions of electrons with Rydberg atoms, initially present or formed by three-body recombination, rapidly raise the electron temperature to 30 K or more. The spatial relaxation of ions moving to positions of lower electrostatic potential energy elevates the ion temperature to the order of 1 K (9).

We have recently introduced a new method for the preparation and study of ultracold plasmas (10; 11). Our approach begins with a target atom or molecule entrained in a seeded supersonic expansion. Thus far, our experiments have focused on nitric oxide. For NO, seeded at 10 percent in He, expansion cools entrained molecules to a moving frame temperature of about 0.7 K, estimated from a measurement of rotational state populations.

Double-resonant excitation produces a Rydberg gas of $\approx 5 \times 10^{12}$ molecules cm^{-3} in a single selected state. This population evolves to produce prompt free electrons and a durable, cold, quasineutral plasma of electrons and intact NO^+ ions. Thereafter, a field of amplitude as low as 3 V/cm – far smaller than that required to field-ionize the precursor state – extracts a small signal of electrons, but pulses of amplitude as high as 200 V/cm fail to destroy the plasma. These observations signal two properties that confirm the evolution to a cold plasma. A low-voltage response establishes the presence of a surface charge of electrons that are trapped by the plasma space charge, but extractable by a weak field. The durability to high fields indicates a Debye screening length much smaller than the diameter of the plasma, shielding the charged particles at its core.

Downstream, transmission through a moveable grid samples the core electrons, profiling their spatial distribution as a function of time. We find that the volume increases at a rate that accelerates with time in accord with the behaviour expected for an electron-driven ambipolar expansion. Analysis of this expansion in the simple limit of the Vlasov equations returns electron temperatures as low as 5 K. Thus, the plasma created in a molecular beam seems both colder and denser than that formed in a MOT, putting this system further into the interesting regime in which the potential and kinetic energies of intermolecular interactions balance, and electrostatic correlations gain in importance.

Significantly, molecular cations carry the positive charge in this NO^+ - e^- plasma, which means it can decay by two-body, dissociative recombination. Dissociative recombination acts generally to regulate energy balance and composition in conventional plasmas (12). Neutralization reactions of NO^+ in particular have an important effect on NO_x chemistry in upper-atmosphere and combustion contexts (13), and have been the subject of significant theoretical interest (14; 15). Experimental studies have used both storage-ring and flowing-afterglow techniques to determine accurate thermal rate coefficients. It is interesting now to ask: To what extent can this information predict for the role of dissociative recombination in determining the properties of a nitric oxide molecular ultracold plasma?

Here, we describe techniques by which to monitor dissociative recombination in molecular ultracold plasmas, and characterize conditions under which DR represents the principal channel for decay.

2. Experimental Method

Experiments here, in which we study the rate of plasma dissipation, profile the electron density of the plasma volume as it passes through the (x, y) vertical plane of an imaging grid that caps a field-free region. Translation of this grid along the axis (z) of the molecular beam determines the distance traveled by the plasma from its point of preparation to the detection of its electron density. The electron yield detected as a function of flight time gauges the rate at which the plasma decays by recombination of its trapped electrons with NO^+ ions.

Figure 1 diagrams the apparatus. A molecular beam, formed by skimming the output of

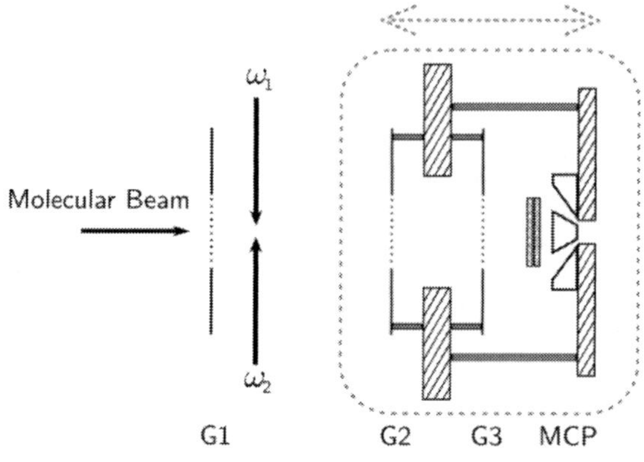

Figure 1. Experimental apparatus. A skimmed molecular beam passes through grid, G_1, where it intersects counter-propagating laser beams indicated as ω_1 and ω_2. As the excitation volume transits the plane defined by G_2, a microchannel plate detector (MCP) situated behind G_3 collects the signal of extracted plasma electrons. The components surrounded by the light-grey border translate together on a moving carriage.

a pulsed nozzle, passes through the first of three perpendicular grids (G_1) to enter a field-free region, where it crosses the collimated, pulsed output of counter-propagating dye lasers. The first dye laser pulse (ω_1, frequency doubled to 225 nm) excites an electronic transition from the ground rovibronic state of NO ($X^2\Pi_{1/2}$ $v = 0, J = 1/2$) to the ground rovibronic level of the A $^2\Sigma^+$ state. The second frequency-doubled dye laser pulse (ω_2) tuned over an interval from 330 to 340 nm, excites transitions from this intermediate A-state to Rydberg states of principal quantum number, n, from 35 to the ionization limit.

We operate these lasers in power regimes under which both transitions are saturated, driving 50 percent of the population in the minus parity component of the ground state to the A state, and 50 percent of the A-state population to a selected high-Rydberg state. Overlapping ω_1 and ω_2 pulses in time, stepwise saturated excitation with a molecular beam under our expansion conditions produces a population of high-Rydberg molecules with a peak density of 5×10^{12} cm^{-3}.

The A $^2\Sigma^+$ $v = 0, J = 1/2$ state decays with a lifetime of 210 ns (16). Thus, delaying the ω_2 pulse with respect to ω_1 yields a reduced number of A-state NO molecules for further excitation. We have no way at present to directly measure the fraction of NO* Rydberg states converted to free NO^+ ions and electrons, but on thermochemical grounds the observed electron temperature suggests that approximately of 50 percent of NO* - electron collisions drive population upward toward ionization. Simple rate-equation models, employing theoretical rate coefficients for collisional excitation and de-excitation of Rydberg states, support this estimate. We therefore assume that, with zero delay, saturated steps of double resonance ultimately yield a plasma with cations and electrons in approximately equal densities of 2×10^{12} cm^{-3}. Systematically delaying ω_2 with respect to ω_1 yields a set of well-defined reduced charged-particle densities.

Our excitation geometry produces a prolate ellipsoid excitation volume with an estimated aspect ratio of 4:1. The long axis of this ellipsoid, which is determined by the width of the molecular beam, lies transverse to the axis of propagation. The short axis conforms with the Gaussian intensity profile of a spatially filtered ω_1 laser beam, collimated to a full-width at half-maximum of 500 μm. The plasma produced in this volume travels with the velocity of the molecular beam over an adjustable distance to pass through the imaging grid, G_2.

As shown in Figure 1, the carriage supporting this grid, together with G_3 and the MCP detector, travels on linear bearings supported by four 0.5 inch diameter stainless steel rods. A bellows-isolated motorized actuator controls the position of this carriage to within 0.01 mm over a range of 10 cm.

We generally set G_2 to have the same potential as G_1 (nominally apparatus ground), creating a field-free region in which we vary the plasma time-of-flight. After G_2, the plasma encounters an electrostatic field determined by the potential applied to a third grid, G_3, spaced downstream by a fixed distance of 16 mm. For the present experiments, we apply 145 V to G_3. As the plasma traverses G_2, the field between G_2 and G_3 extracts electrons which, on the timescale of the measurement, appear instantaneously as signal at the multichannel plate detector. This signal represents the extracted electron density integrated in the transverse dimensions of the plasma, providing a time-dependent trace of this changing density as the plasma volume passes through G_2. We assume that the electrons extracted by G_2 supply a representative gauge of plasma width and relative charge density.

3. Results

Earlier studies in our laboratory have established that a gas of high-Rydberg NO molecules, entrained in a molecular beam with a longitudinal temperature of *ca.* 700 mK, evolves to form a quasineutral ultracold plasma of NO^+ ions and electrons (10; 11). Present measurements gauge the volume and lifetime of such a plasma, tracking its electron density as a function of time-of-flight over an adjustable interval of distance.

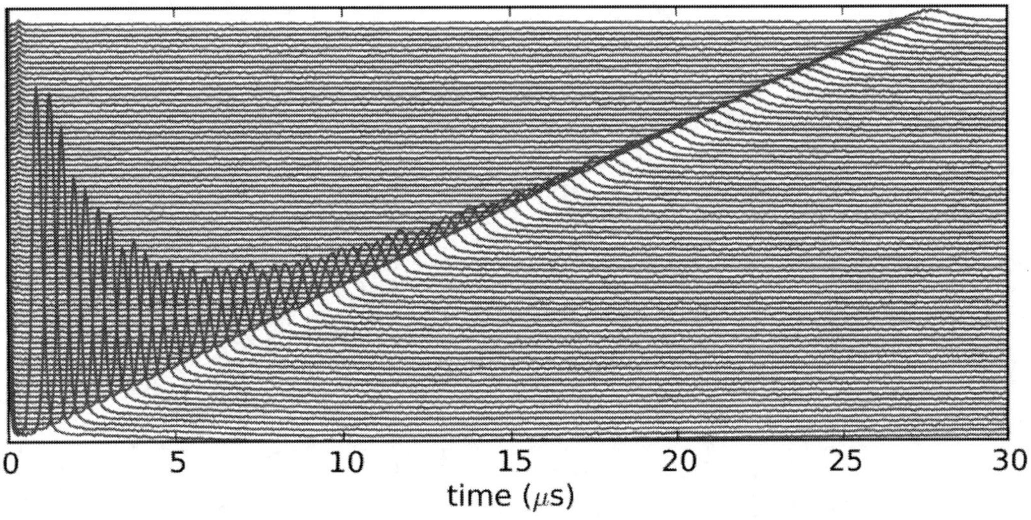

Figure 2. Waveforms captured by the MCP detector for different positions of the moving carriage. Traces have been offset vertically for clarity.

Figure 2 gives a sequence of electron signal traces recorded for an initial Rydberg state in the nf series converging to $N^+ = 2$ for which we have selected the principal quantum number $n = 50$ ($50f(2)$). Successive waveforms show electron signal amplitude as a function of time for increasing G_2 displacements. We see that the position of the late signal advances with the position of G_2, and that the arrival time of this feature corresponds in each case to the flight time of a neutral volume in the molecular beam from the excitation region to the perpendicular plane of G_2.

This signal of electrons extracted as the illuminated volume passes through G_2 broadens and decreases in amplitude with increasing flight time, which reflects the processes of expansion and decay of the plasma with time. Each trace fits well to a Gaussian function, and we use the parameters of such fits to extract the plasma arrival time, width and relative measure of the total number of electrons. Figure 3 plots the fitted width along the axis of propagation (the short axis of the ellipsoid excitation volume) measured as a function of flight time.

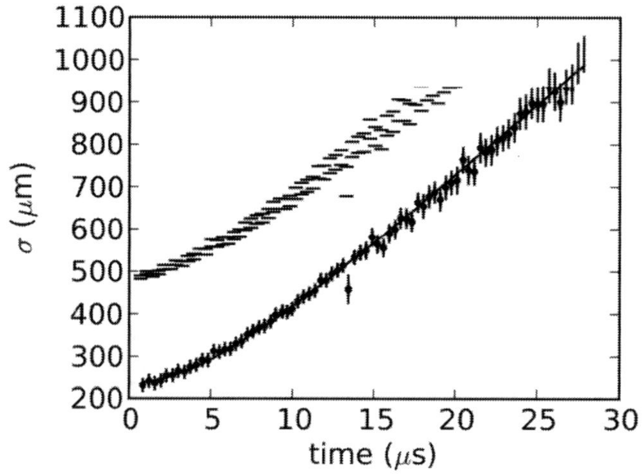

Figure 3. Fitted plasma width as a function of centre arrival time at G_2. Solid line is a fit of equation 3 to the experimental data, yielding a fitted temperature $T_e + T_I = 5$ K.

For any given time, we estimate the total number of electrons remaining in the plasma by integrating the area of the corresponding Gaussian fit. To express this in absolute terms, we use an extrapolation procedure that correlates the earliest-measured data points to the known initial plasma density to calibrate the detector response. For present experiments on the evolution of a Rydberg gas selected to populate the $50f(2)$ state, the signal extrapolated to $t = 0$ for the densest plasma we can create defines the starting point for decay from an estimated initial density $n_e(0) \approx 2 \times 10^{12}$ cm^{-3}. The introduction of a delay between the ω_1 and ω_2 excitation pulses, initiates decays for plasmas of systematically reduced charged-particle densities. Figure 4 shows a collection of plasma decay waveforms for initial densities varied in this way.

4. Discussion

An ultracold plasma composed of NO^+ ions and free electrons expands at a rate dictated by the electron temperature, and decays by dissociative electron-ion recombination. Electron signal, collected as the plasma volume transits the grid G_2 provides a measure of its width and relative electron density. By positioning this grid to sample the plasma for varying times of flight, we obtain data that gauge the rates both of plasma expansion and decay.

4.1. Expansion and Electron Temperature

The plasma volume increases as electrons drive radial hydrodynamic ion expansion with a local mean velocity, γr. For an ideal, spherical, Gaussian plasma, this velocity grows as:

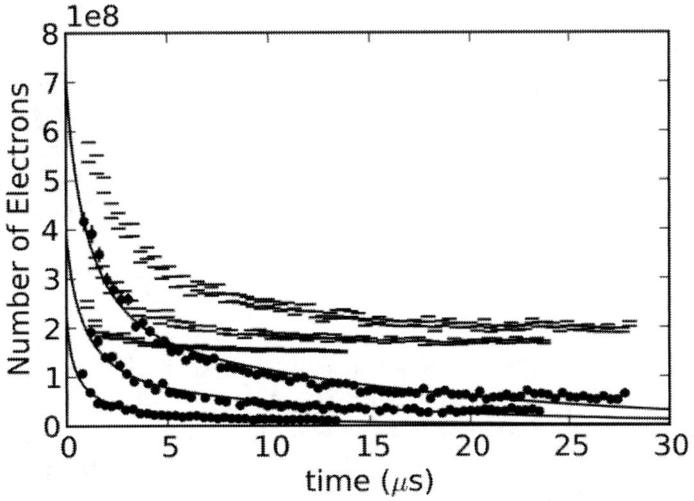

Figure 4. Integrated area of the plasma signal as a function of arrival time at imaging grid G2 for a range of initial densities. The lines are integrated rate equations fitted to the initial decay data. The calculations consider first-order electron evaporation, plasma expansion and second-order decay processes (c.f. Eqs. 8 and 9)

$$\frac{d\gamma}{dt} = \frac{kT_e}{m_i \sigma^2}, \tag{1}$$

where σ is the radius of the plasma and T_e is the electron temperature. The work of expansion, which converts electron thermal energy into cation hydrodynamic velocity, causes the electron temperature to fall. The initial plasma radius $\sigma(0)$, together with the electron (T_e) and ion (T_i) temperatures combine to determine a characteristic timescale for both expansion and cooling:

$$\tau^2 = \frac{m_i \sigma(0)^2}{k_B[T_e(0) + T_i(0)]}, \tag{2}$$

which then yield time-dependent radius and temperature as:

$$\sigma(t) = \sigma(0)[1 + t^2/\tau^2]^{\frac{1}{2}}, \tag{3}$$

and:

$$T_e(t) = \frac{T_e(0)}{[1 + t^2/\tau^2]}. \tag{4}$$

The present illumination geometry gives our plasma the form of a prolate ellipsoid, as opposed to the self-similar spherical shape described by Eqs.(2) - (4). However, Bergeson and coworkers have obtained fluorescence images of ellipsoidal plasmas showing differential rates of short- and long-axis expansion (17) described by expressions of the form of Eq. (3) for decoupled radial coordinates. Our experiment measures the rate of expansion along the short axis of the ellipsoid.

Assuming that a simple model for ambipolar expansion applies to describe the evolution of the electron density in this coordinate, we fit Gaussian radii, measured as a function of flight time, to Eq. (3), and, in this way, relate a measured rate of plasma expansion to the electron temperature. We note that the time-dependent width displayed in Figure 3 shows evidence of the acceleration characteristic of an electron-charge-driven ambipolar expansion. A Vlasov fit to this $\sigma(t)$ returns an initial electron temperature of 5 K.

At the electron density of our plasma, the Wigner-Seitz radius, a, is 500 nm, and such a T_e would indicate significant correlation ($\Gamma_e = 7$, where $\Gamma_e = q^2/4\pi a\epsilon_0 k_B T_e$). The Vlasov equations do not extend to describe the expansion of a charged-particle systems under such conditions. At high correlation, the positive field created by the cations contained in the volume of the plasma lowers the potential energy of free electrons. The depth of this well decreases with expansion, and this elevation of potential energy with expansion acts as a force that suppresses electron pressure. Thus, for the purposes of the following discussion, we can assume that the temperature estimated from rate of plasma expansion represents a lower limit.

4.2. Plasma Decay and Dissociative Recombination

The escape of electrons in the first few hundred nanoseconds of plasma formation creates a cationic space charge that acts as an attractive potential for the electrons that remain. Electrons trapped by this potential can decay only by electron-ion recombination. Subsequent plasma expansion, however, acts to reduce the depth of the electrostatic well, and this allows additional electrons to evaporatively escape.

A potential sufficient to trap the plasma electrons under our conditions of T_e requires an excess cation charge of less than one percent. The decaying plasma thus remains quasi-neutral, and we can represent the mass-action kinetics of charge recombination by differential equations written simply in terms of the time-dependent electron density, $n_e(t)$.

Elementary processes leading to the irreversible loss of charged particles include electron-electron-ion three-body recombination, followed by predissociation of the formed Rydberg state (collisional dissociative recombination):

$$\mathrm{NO}^+ + e^- + e^- \underset{k_{\mathrm{ion}}}{\overset{k_{\mathrm{TBR}}}{\rightleftharpoons}} \mathrm{NO}^* + e^-$$

$$\mathrm{NO}^* \xrightarrow{\mathrm{k_{PD}}} \mathrm{N} + \mathrm{O} \tag{5}$$

and two-body dissociative recombination:

$$\mathrm{NO}^+ + e^- \xrightarrow{\mathrm{k_{DR}}} \mathrm{N} + \mathrm{O} \tag{6}$$

Three-body recombination forms NO^* molecules in high-Rydberg states below the ionization threshold of nitric oxide. The rate at which these molecules predissociate determines the kinetics of electron loss by three-body recombination. In the limit where $k_{\mathrm{PD}} \gg k_{\mathrm{ion}} n_e$, irreversible decay via (5) occurs as a third-order process at the rate of three-body recombination:

$$-\frac{dn_e(t)}{dt} = k_{\mathrm{TBR}} n_e^3(t) + \frac{n_e(t)}{V(t)}\frac{dV(t)}{dt} + k_{\mathrm{ev}} n_e \tag{7}$$

where we include additional terms to account for the decrease in n_e owing to plasma expansion and electron evaporation.

In the opposite limit, for $k_{\mathrm{PD}} \ll k_{\mathrm{ion}} n_e$, decay proceeds via a steady-state population of NO^* with a second-order rate law:

$$-\frac{dn_e(t)}{dt} = \frac{k_{\mathrm{TBR}}}{k_{\mathrm{ion}}} k_{\mathrm{PD}} n_e^2(t) + \frac{n_e(t)}{V(t)}\frac{dV(t)}{dt} + k_{\mathrm{ev}} n_e. \tag{8}$$

Direct dissociative recombination occurs at a rate in accord with its elementary mechanism:

$$-\frac{dn_e(t)}{dt} = k_{DR}n_e^2(t) + \frac{n_e(t)}{V(t)}\frac{dV(t)}{dt} + k_{\text{ev}}n_e. \tag{9}$$

Considering these elementary mechanisms for irreversible electron loss by two- and three-body recombination, we can expect contributions to the time dependence for plasma decay that, allowing for expansion and electron evaporation, conform with either second- or third-order rate laws. These reactions proceed in parallel to determine the overall rate at which the electron density changes in time. Referring to late electron signal decays observed experimentally, we attempt now to judge the relative importance of each channel.

To predict $n_e(t)$ according to these rate laws, we represent the plasma volume by a cylinder of concentric shells with radial coordinate, r, reflecting the prolate ellipsoid of ions formed by the intersection of our excitation lasers with the molecular beam. We assign each shell an initial number of electrons according to a radial Gaussian density distribution with a peak density of 2×10^{12} cm^{-3}, and a radius equal to the value of $\sigma(0)$ we fit to the data, as shown in Figure 3. Numerical integration of expressions conforming to Eq. 7 or 9 for selected rate coefficients yields the number density within each shell as a function of time.

We vary the volume of each shell according to,

$$\frac{\partial V}{\partial t} = \frac{\partial V}{\partial r}\frac{\partial r}{\partial t}, \tag{10}$$

using the radial coordinate hydrodynamic mean expansion velocity to define the time dependence of r, viz: $\partial r/\partial t = \gamma r$, where we determine γ as a function of time from the fitted value of τ. Integrating the calculated number of electrons over the volume of the model gives the total number of plasma electrons as a function of time, which we compare with the integrated area of the plasma signal.

The rate of electron loss proceeds at third-order in charged particles only for the mechanism of three-body recombination in the limit for which $k_{PD} \gg k_{ion}n_e$. In a recombining atomic ultracold plasma, the balance between three-body recombination and electron-impact ionization creates a Saha equilibrium (18) between free charges and high-Rydberg states that populates a band of principal quantum numbers between a kinetic bottle neck at $n = \sqrt{R_y/4k_BT_e}$ and the thermochemical limit for bound states at $n = \sqrt{R_y/k_BT_e}$ (19). Reliable models enable the calculation of temperature-dependent electron-impact ionization rate coefficients in this range (8), on the basis of which our conditions yield an initial $k_{ion}n_e$ product of 4.6×10^9 s^{-1}.

Remacle and Vrakking (20) have modelled predissociation in optically populated NO Rydberg states with principal quantum numbers near 100. They find that fields as small as 40 mV cm^{-1} serve to mix fast-predissociating low-l zero-order states with non-penetrating, higher-l states of the same and different principal quantum numbers. This dilution of coupling strength to neutral decay channels produces biexponential decays with a short-component time constant on the order of 400 ns and a long-component decay time that exceeds 12 μs for $n = 105$. In contrast to optical excitation, three-body recombination directly addresses orbitals of high-angular momentum. For this reason, we can expect $k_{\text{PD}} \approx 1 \times 10^5$ s^{-1} for a collisionally populated Rydberg state with $n \approx 100$, falling as n^{-3} to 1×10^{-4} s^{-1} at $n \approx 180$, resulting in a relaxation rate to Saha equilibrium over this interval that far outpaces predissociation. Thus, we should not expect to observe a rate of decay that displays a third order dependence on the density of charged particles, and fits to the experimental data readily confirm this.

Solid lines through the experimental decays in Figure 4 represent fits to generalized second-order rate expressions of the form of Eqs. (8) and (9) for a set of systematically varied initial charged-particle densities. These fits yield a second-order electron-ion recombination rate constant of $k_{\text{eff}} = 7 \pm 5 \times 10^{-6}$ cm^3 s^{-1}, and first-order electron evaporation rate of $k_{\text{ev}} = 1 \pm 0.4 \times 10^5$ s^{-1}. This phenomenological second-order rate constant, k_{eff}, contains contributions from both dissociation mechanisms (5) and (6):

$$k_{\text{eff}} = \frac{k_{\text{TBR}}}{k_{\text{ion}}} k_{\text{PD}} + k_{\text{DR}} \,. \tag{11}$$

It remains now to decide which of these terms is more important, whether the rate process associated with electron loss is driven by three-body recombination accompanied by predissociation, or represents a conventional, two-body NO^+– electron recombination leading directly to neutral dissociation products.

To judge the influence of three-body recombination, we return to the population of high-Rydberg states in the principal quantum number range between $\sqrt{R_y/4k_BT_e}$ and $\sqrt{R_y/k_BT_e}$, which is sustained by the Saha equilibrium between collisional recombination and ionization. The corresponding equilibrium constant, which equates to the ratio of forward to reverse rate constants, is determined by statistical factors consisting simply of the density of bound Rydberg states and the partition function for free electrons:

$$\frac{k_{\text{TBR}}}{k_{\text{ion}}} = \frac{n_i^2 \Lambda^3}{\exp(-R_y/n_i^2 k_B T_e)}, \tag{12}$$

where $\Lambda = \sqrt{h^2/2\pi m_e k_B T_e}$ is the de Broglie wavelength for electrons with temperature T_e.

Over the relevant range of high-Rydberg principal quantum numbers, n_i, from approximately 90 to 180, this ratio varies slowly from 1×10^{-11} to 3×10^{-12} cm^3. Assuming a statistical distribution of population over these high Rydberg states and scaling the $n = 105$ predissociation rate by n_i^{-3}, we obtain an average value of the rate constant for collisional dissociative recombination, $(k_{\text{TBR}}/k_{\text{ion}})k_{\text{PD}}$, of 6×10^{-8} cm^3 s^{-1}. Thus, we can conclude that three-body recombination makes a minimal contribution to irreversible electron loss in this molecular ultracold plasma.

Experiments in afterglows and accelerators have measured rate constants for dissociative recombination of NO^+ at temperatures as low as 200 K (21; 22; 23). These observations conform well with *ab initio* R-matrix – multichannel quantum defect theory calculations of k_{DR} for electron temperatures from 1 to 5000 K (24). In the low-temperature range from 5 to 10 K, theory predicts a two-body rate constant for dissociative recombination from 2 to 3×10^{-6} cm^3 s^{-1}. Thus, even though the present conditions of electron correlation limit the certainty with which the observed expansion rate determines the plasma electron temperature, the corresponding variation in theoretically predicted rate constant does not exceed the present statistical uncertainty in our measurement. Perhaps with improved accuracy, determinations of the dissociative recombination rate can serve as a probe of plasma conditions.

Based on results obtained thus far, we can draw the following conclusions concerning the dissipation of a molecular ultracold plasma of nitric oxide. The decay of plasma density does not proceed as a rate process that is third-order in charged particles. Also, the population of high-Rydberg states maintained in equilibrium with ions by three-body recombination does not play a large role in determining the second-order decay rate. Instead, it appears that irreversible charged-particle loss occurs in simple two-body collisions by conventional electron-ion dissociative recombination. Even though this plasma seems demonstrably to be at or near conditions for Coulomb correlation, rate processes do not appear strongly perturbed by many-body non-ideal collisional effects. The rate constant for dissociative recombination extracted from these data conforms with predictions drawn from theory for isolated electron-ion collisions. Methods based on the dissipation of molecular ultracold plasmas therefore may eventually offer a practical means for measuring rates of dissociative recombination for a variety of complex molecules.

5. Acknowledgements
It is a pleasure to acknowledge helpful discussions with T. Pohl and J. M. Rost. This work was supported by the Natural Sciences and Engineering Research Council of Canada (NSERC), the Canada Foundation for Innovation (CFI) and the British Columbia Knowledge Development Fund (BCKDF).

References
[1] Ichimaru S 1982 *Rev. Mod. Phys.* **54** 1017–1059

[2] Dubin D H E and O'Neil T M 1999 *Rev. Mod. Phys.* **71** 87

[3] Killian T C, Kulin S, Bergeson S D, Orozco L A, Orzel C and Rolston S L 1999 *Phys. Rev. Lett.* **83** 4776–4779

[4] Robinson M P, Tolra B L, Noel M W, Gallagher T F and Pillet P 2000 *Phys. Rev. Lett.* **85** 4466–4469

[5] Li W, Noel M W, Robinson M P, Tanner P J, Gallagher T F, Comparat D, Laburthe Tolra B, Vanhaecke N, Vogt T, Zahzam N, Pillet P and Tate D A 2004 *Phys. Rev. A* **70** 042713

[6] Li W, Tanner P J and Gallagher T F 2005 *Phys. Rev. Lett.* **94** 173001

[7] Walz-Flannigan A, Guest J R, Choi J H and Raithel G 2004 *Phys. Rev. A* **69** 063405

[8] Pohl T, Vrinceanu D and Sadeghpour H R 2008 *Phys. Rev. Lett.* **100** 223201

[9] Killian T C, Lim M J, Kulin S, Dumke R, Bergeson S D and Rolston S L 2001 *Phys. Rev. Lett.* **86** 3759–3762

[10] Morrison J P, Rennick C J, Keller J S and Grant E R 2008 *Phys. Rev. Lett.* **101** 205005

[11] Morrison J P, Rennick C J and Grant E R 2009 *Phys. Rev. A* **79** 062706

[12] Bates D R 1950 *Phys. Rev.* **78** 492–493

[13] Thomas R 2008 *Mass spectrometry reviews* **27** 485–530 ISSN 1098-2787

[14] Schneider I F, Rabadán I, Carata L, Andersen L H, Suzor-Weiner A and Tennyson J 2000 *Journal of Physics B: Atomic, Molecular and Optical Physics* **33** 4849

[15] Motapon O, Fifirig M, Florescu A, Tamo F O W, Crumeyrolle O, Varin-Bréant G, Bultel A, Vervisch P, Tennyson J and Schneider I F 2006 *Plasma Sources Science and Technology* **15** 23

[16] Luque J and Crosley D R 2000 *The Journal of Chemical Physics* **112** 9411

[17] Cummings E A, Daily J E, Durfee D S and Bergeson S D 2005 *Phys. Rev. Lett.* **95** 235001

[18] Bates D R, Kingston A E and McWhirter R W P 1962 *Proc. Roy. Soc. A* **267** 297

[19] Stevefelt J, Boulmer J and Delpech J F 1975 *Phys. Rev. A* **12** 1246–1251

[20] Remacle F and Vrakking M J J 1998 *J. Phys. Chem. A* **102** 9507

[21] Mul P M and McGowan J W 1979 *Journal of Physics B: Atomic and Molecular Physics* **12** 1591

[22] Alge E, Adams N G and Smith D 1983 *Journal of Physics B: Atomic and Molecular Physics* **16** 1433

[23] Vejby-Christensen L, Kella D, Pedersen H B and Andersen L H 1998 *Physical Review A* **57** 3627–3634 ISSN 1094-1622

[24] Schneider I F, Rabadán I, Carata L, Andersen L H, Suzor-Weiner A and Tennyson J 2000 *Journal of Physics B: Atomic, Molecular and Optical Physics* **33** 4849

Eighth International Conference on Dissociative Recombination (DR2010) IOP Publishing
Journal of Physics: Conference Series **300** (2011) 012006 doi:10.1088/1742-6596/300/1/012006

Rotational Cooling of HD^+ by Superelastic Collisions with Electrons

D Schwalm[1,2], D Shafir[1], S Novotny[2], H Buhr[2], S Altevogt[2], A Faure[3], M Grieser[2], A G Harvey[4], O Heber[1], J Hoffmann[2], H Kreckel[2], L Lammich[2], O Motapon[5,6], I Nevo[1], H B Pedersen[2], H Rubinstein[1], I F Schneider[6], J Tennyson[4], F O Waffeu Tamo[5,6,7], A Wolf[2], D Zajfman[1]

[1] Department of Particle Physics, Weizmann Institute of Science, 76100 Rehovot, Israel
[2] Max Planck Institute for Nuclear Physics, 69117 Heidelberg, Germany
[3] Laboratoire d'Astrophysique de Grenoble, Universite Joseph Fourier, UMR 5571 CNRS, Grenoble, France
[4] Department of Physics and Astronomy, University Collage London, London WC1E 6BT, United Kingdom
[5] Physics Department, University of Douala, P. O. Box 24157, Douala, Cameroon
[6] Laboratoire Ondes et Milieux Complexes CNRS-FRE-3102, Université du Havre, 76058 Le Havre, France
[7] Laboratoire d'Etude du Rayonnement et de la Matiére en Astrophysique CNRS-UMR-8112, Observatoire de Paris, 92195 Meudon, France

E-mail: `schwalm@mpi-hd.mpg.de`

Abstract. Rotational cooling of HD^+ by superelastic collisions (SEC) with electrons was observed at the Heidelberg test storage ring by merging a beam of rotationally hot HD^+ ions with an electron beam at zero relative energy. Neutral fragments resulting from DR events were recorded at different electron densities using a high resolution imaging detector and a large-area, energy sensitive detector. The data allowed to deduce the time dependence of the population of three groups of rotational angular momentum states J built on the vibrational ground state of the ion together with the corresponding DR rate coefficients. The latter are found to be (statistical uncertainties only) $\langle\alpha\rangle_{0,1,2} = 3.8(1)$, $\langle\alpha\rangle_{3,4} = 4.0(2)$, and $\langle\alpha\rangle_{5,6,7} = 9.0(1.3)$ in units of 10^{-8} cm^3/s, in reasonable agreement with the average values derived within the MQDT approach. The time evolution of the population curves clearly reveals that rotational cooling by SEC takes place, which can be well described by using theoretical SEC rate coefficients obtained by combining the molecular R-matrix approach with the adiabatic nuclear rotation approximation. We verify the $\Delta J = -2$ coefficients, which are predicted to be dominant as opposed to the $\Delta J = -1$ coefficients and to amount to $(1-2)\cdot 10^{-6}$ cm^3/s, to within 30%.

1. Introduction

Collisions between electrons and molecular ions are of fundamental interest, both from a theoretical point of view as well as in connection with many applications. At low temperatures prevailing in a number of astrophysical environments, for example, inelastic collisions (IEC) between electrons and molecular ions are competing with dissociative recombination (DR), and a thorough knowledge of these processes is crucial to model the physics and chemistry taking place in these plasmas. While the DR process has been investigated both experimentally and theoretically in rather detail in the last two decades [1], only very few experimental studies have been performed so far dealing with IEC processes at very low electron energies. In particular, only one quantitative experimental investigation of the superelastic collisions (SEC) process, which results in a deexcitation of ro-vibrationally excited molecular ions, has been performed so far: Krohn et al. [2, 3] measured the $\Delta v = -1$ SEC rate coefficients between the lowest vibrational states of H_2^+. The SEC rate coefficients were found to be surprisingly large, amounting to approximately $1 \cdot 10^{-6}$ cm^3/s, about one order of magnitude larger than the corresponding DR rate coefficients. To explain these large vibrational SEC rate coefficients it turned out that detailed account has to be taken of the rotational structure and couplings of the H_2^++e$^-$ collision system [4].

As in H_2^+ and other symmetric molecular ions the cooling via radiative transitions is strongly hindered, and superelastic collisions seem to occur much more frequently than the destruction of the ions by the DR process, the SEC process may thus be used in storage ring experiments with merged electron-ion beams to vibrationally cool these ions. For infrared active molecular ions, on the other hand, the cooling of vibrational excitations by radiative transitions is quite fast and the molecules usually end up in their vibrational ground state within seconds. The situation is quite different, however, as far as the cooling of the rotational degree of freedom is concerned: Once the vibrational ground state is reached, the rotational cooling has to proceed along the ground state rotational band and several tens of seconds are needed to just reach equilibrium with the 300 K ambient radiation field. Assuming the SEC rate coefficients between rotational states to be of the same order of magnitude as those between vibrational states, it should be possible to use the SEC process to speed up the rotational cooling and to reach even subthermal rotational populations.

Although the occurrence of rotational cooling by SEC in storage ring experiments has been conjectured before (see e.g. [5]), no quantitative measurements of rotational SEC rate coefficients has been carried out so far. A dedicated experiment was therefore performed to study the rotational cooling of HD$^+$ by the SEC process [6]. This molecule is ideally suited for these investigations: Hot stored HD$^+$ ions were shown to vibrationally relax within < 0.5 s [7] thanks to their isotopic asymmetry, while radiative transitions along the rotational band built on the vibrational ground state are calculated to be considerably slower [8] such that the SEC process can be expected to dominate the rotational cooling process. Moreover, the small moment of inertia of HD$^+$ leads to a sizable spacing between the rotational states, which facilitates the determination of the rotational excitation of the ion by fragment imaging [9, 10]. And last but not least, the comparatively simple structure of HD$^+$ favors a theoretical description of the HD$^+$+e$^-$ collision system.

2. Experimental Procedure

The experiment uses the DR process to probe the time evolution of the population $P_J(t)$ of the ro-vibrational states $(v = 0, J)$ of HD$^+$ when merging a hot HD$^+$ beam circulating in a storage ring with cold electrons of the same mean velocity as the ions. As the relative ion-electron energies are smaller than the smallest difference between two ro-vibrational states, and the ions are vibrationally relaxed in less than 0.5 s, the only inelastic collisions which can take place thereafter are superelastic collisions between the rotational states, i.e.

$$\mathrm{HD}^+(0, J) + \mathrm{e}^-(E = 0) \rightarrow \mathrm{HD}^+(0, J') + \mathrm{e}^-(E' > 0) \tag{1}$$

with $J' < J$. The resulting time evolution of $P_J(t)$ is monitored by observing the kinetic energy release $E_k(J)$ carried by the two DR fragments applying fragment imaging [9]. Since the only open DR channel is

$$\mathrm{HD}^+(0, J) + \mathrm{e}^-(E = 0) \rightarrow \mathrm{H}(nl) + \mathrm{D}(ml') + E_k(J) \tag{2}$$

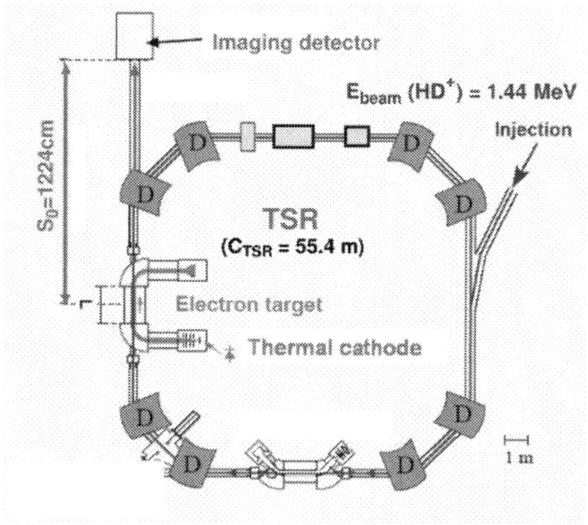

Figure 1. Schematic view of the experimental set-up. HD^+ ions were produced in a discharge ion source, accelerated to 1.44 MeV, injected into the storage ring for 30 μs, and stored for times $\geq$ 10 s. The ions are merged with a velocity matched electron beam supplied by the electron target station [12]. DR fragments are observed using a high resolution imaging system based on a microchannel plate, which is attached to a phosphor screen viewed by a CCD camera [10].

with $(n, m) = (1,2)$ or (2,1), $E_k(J)$ is given by $E_k(J) = E_0 + BJ(J+1)$, where $E_0 = 0.726$ eV denotes the energy difference between the ionic ground state and dissociation channel and $B = 2.72$ meV is the rotational constant of HD^+ [11]; thus $E_k(J)$ is a unique function of J.

A schematic view of the experimental set-up at the heavy ion storage ring TSR operated by the Max Planck Institute for Nuclear Physics in Heidelberg is shown in Fig. 1. The electrons were produced by a thermal cathode, accelerated and expanded to form a beam of 10 mm diameter; the resulting electron energy distribution in the comoving reference frame can be described by a flattened Maxwell distribution with $kT_\parallel \approx 45$ μeV parallel and $kT_\perp \approx 2.8$ meV transversal to the beam direction. The effective length of the electron target was $L = 1.22$ m, which leads to an overlap factor of $\eta = L/C = 0.022$ when compared to the circumference of the TSR of $C = 55.4$ m. The measurements were performed at three different electron densities $n_e = (0.28,\, 1.00,\, \text{and}\, 1.45) \cdot 10^7$ cm^{-3}, which we will refer to as "low", "medium", and "high", respectively. Within the first second after injection the HD^+ beam was phase space cooled by the electrons, which led to an ion beam of < 2 mm diameter and < 25 μrad divergence.

The two neutral fragments following the DR of HD^+ were recorded as a function of storage time t by the high resolution imaging system mounted 12.24 m downstream from the center of the electron target, which allowed to determine the projected (transverse) distance D between the two fragments with an accuracy up to $\pm 0.3\%$. A large-area energy sensitive Si detector could be temporarily placed about 1 m in front of the imaging detector to measure also the total DR rate as a function of t.

3. Analysis and Results

At matched electron-ion beam velocities the normalized projected distance distribution $f_J(D)$ resulting from the DR of a given rotational state J is only determined by the detection geometry and the kinetic energy release $E_k(J)$, the maximum projected distance being proportional to $(E_k(J))^{1/2}$ [9]. For an ensemble of rotationally excited HD^+ ions the normalized projected distance distributions $F(D, t)$ are thus given by

$$F(D,t) = (\alpha_{DR}(t))^{-1} \sum_J \alpha_J P_J(t) f_J(D), \tag{3}$$

where α_J denotes the DR rate coefficient of the rotational state J, $P_J(t)$ its population probability, and $\alpha_{DR}(t)$ the effective DR rate coefficient at time t.

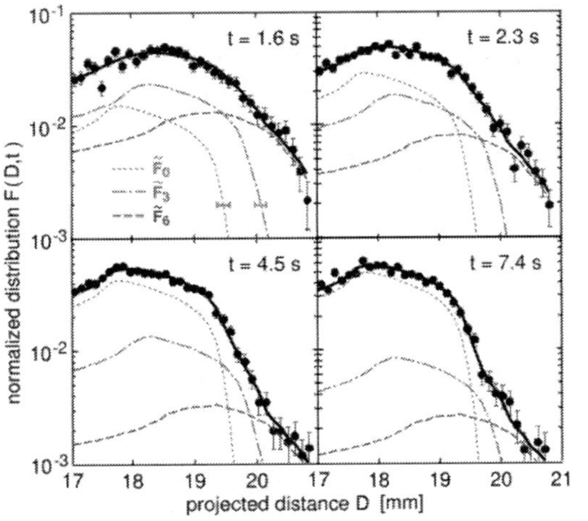

Figure 2. Projected distance distributions $F(D,t)$ recorded for distances $D \geq 17$ mm and several times t after injection. The electron density was $n_e = 1.0 \cdot 10^7$ cm^{-3}. The broken lines are the calculated $\tilde{F}_\mu(D,t)$ distributions reflecting the contribution of three groups of rotational states (μ=0 (J=0,1,2), μ=3 (J=3,4), μ=6 (J=5,6,7)), while the black solid line represents the sum of these distributions after adjusting their relative heights to the data. The horizontal error bars in the upper left panel reflect the calibration errors caused by uncertainties of the length of the merged beam region and the calibration of the imaging system.

Examples of normalized projected distance distributions (PDD) measured at "medium" electron density are shown in Fig. 2 for $D \geq 17$ mm, where these distributions are most sensitive to the rotational states populated. Because of counting statistics and the broad widths of the $f_J(D)$ distributions it was not possible to reliably deduce the individual amplitudes of $f_J(D)$ from the measured PDDs. Instead, we combined adjacent J states into three groups, denoting the groups ($J = 0, 1, 2$) by $\mu = 0$, ($J = 3, 4$) by $\mu = 3$, and ($J = 5, 6, 7$) by $\mu = 6$. Higher rotational states do not seem to contribute significantly to the DR rate at storage times larger than 2 s. Denoting the (unweighted) averages of the distribution functions $f_J(D)$ with $J \in \{\mu\}$ by $\tilde{f}_\mu(D)$, and the (unweighted) averages of the DR rate coefficients α_J with $J \in \{\mu\}$ by $\tilde{\alpha}_\mu$, we can approximate eq. 3 by

$$F(D,t) = \sum_\mu \tilde{F}_\mu(D,t) = \sum_\mu r_\mu(t)\tilde{f}_\mu(D), \tag{4}$$

$$\text{with } r_\mu(t) = \frac{\tilde{\alpha}_\mu}{\alpha_{DR}(t)}\hat{P}_\mu(t) \text{ and } \hat{P}_\mu(t) = \sum_{J\in\{\mu\}} P_J(t).$$

In a first analysis step we fitted Eq. 4 to the measured PDDs in the range 17 mm$\leq D \leq 21$ mm to determine the relative amplitudes $r_\mu(t)$. The measured distributions are well represented by these fits (see Fig. 2). The resulting amplitudes for the measurement performed at "medium" electron density are plotted in Fig. 3a. In a second step we used the relation

$$\alpha_{DR}(t) = (\sum_\mu r_\mu(t)/\tilde{\alpha}_\mu)^{-1}, \tag{5}$$

which follows from the definition of $r_\mu(t)$ and the normalization $\sum_\mu \hat{P}_\mu(t) = 1$, to determine the averaged DR rate coefficients $\tilde{\alpha}_\mu$ for the three groups of rotational angular momentum states from the total DR rate coefficient $\alpha_{DR}(t)$ measured at "medium" electron density (see Fig. 3b). Using the relative amplitudes $r_\mu(t)$ measured at this electron density we find

$$\tilde{\alpha}_0 = 3.8 \pm 0.1, \tilde{\alpha}_3 = 4.0 \pm 0.2, \text{ and } \tilde{\alpha}_6 = 9.0 \pm 1.3$$

in units of 10^{-8}cm^3/s. Only relative errors are given, the absolute scale is expected to be accurate to within ±20%.

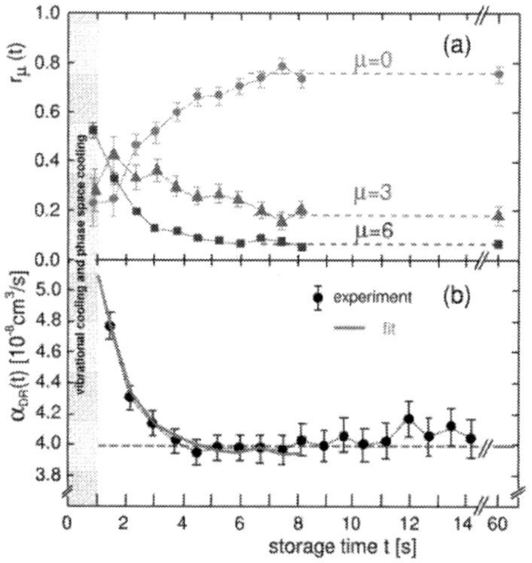

Figure 3. (a) Relative amplitudes $r_\mu(t)$ deduced from the measured projected distance distributions $F(D,t)$ and (b) the measured total DR rate coefficient $\alpha_{DR}(t)$ as a function of the storage time t. The data were taken at an electron density of $n_e = 1.0 \cdot 10^7$ cm^{-3}. The absolute scale of the DR rate coefficient was normalized to the steady state value of $4.0(8) \cdot 10^{-8}$ cm^3/s (dashed line), estimated for the present electron temperatures using the data collected in ref. [13]. The solid line in (b) connects the $\alpha_{DR}(t)$ calculated using Eq. 5 and adjusting the average DR rate coefficients $\tilde{\alpha}_\mu$.

In a final analysis step the $\tilde{\alpha}_\mu$ values were used to determine $\hat{P}_\mu(t)$ from the amplitudes $r_\mu(t)$. The resulting group populations are shown in Fig. 4 for the three electron densities employed. The data points in Fig. 4 clearly reveal that electron assisted rotational cooling of the HD^+ ions takes place, the cooling getting faster with increasing density n_e of the electron beam. At times shorter than about 2 s the separation between $\tilde{F}_0(D)$ and $\tilde{F}_3(D)$ was perturbed by the strong contribution of $\tilde{F}_6(D)$, which is partly covering up the high distance tails of $\tilde{F}_0(D)$ and $\tilde{F}_3(D)$ most sensitive to J (see Fig. 2).

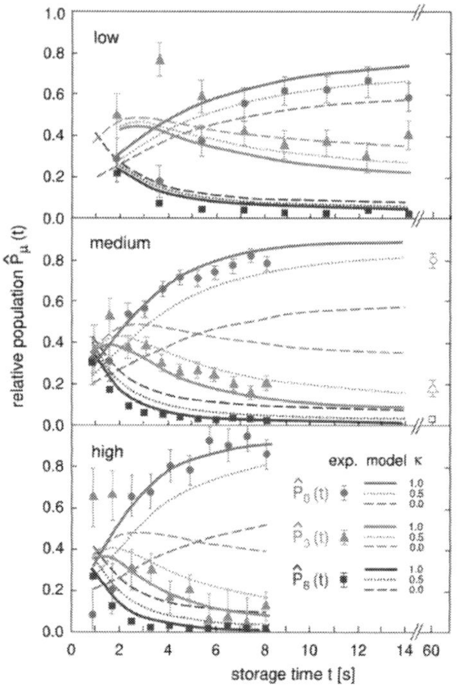

Figure 4. Relative group populations as a function of storage time t, which were extracted from the projected distance distributions measured at three different electron densities n_e of $0.28{\cdot}10^7$ cm^{-3} (low), $1.0{\cdot}10^7$ cm^{-3} (medium) and $1.45{\cdot}10^7$ cm^{-3} (high). The lines are the calculated group populations obtained by solving Eq. 6 with SEC rate coefficients given by $c_{J,J'} = \kappa \cdot c^{ANR}_{J,J'}$, where $c^{ANR}_{J,J'}$ are the theoretical SEC rate coefficient calculated within the R-matrix+ANR approach (see Fig. 5b), and κ is a constant factor. The solid (dashed) lines obtained for $\kappa = 1$ ($\kappa = 0$) are thus reflecting the cooling curves including (neglecting) the SEC process, while the dotted lines are the resulting curves assuming $\kappa = 0.5$.

To investigate the role of SEC in the rotational cooling process, we calculated the time evolution of the rotational populations $P_J(t)$ of the vibrational ground state by solving the coupled set of differential equations

$$\dot{P}_J(t) = \sum_{J'} [M^r_{JJ'} + \eta n_e M^c_{JJ'}] P_{J'}(t). \tag{6}$$

Here the matrix M^r is due to the exchange of photons with the ambient radiation field; the matrix elements $M^r_{JJ'}$ are the Einstein coefficients, which can be readily taken from literature [8] assuming the radiation field to be given by a Planck distribution with a temperature of 300 K. The matrix M^c describes the population change due to interactions with electrons; for $J > J'$ the matrix elements $M^c_{JJ'}$ are given by the SEC rate coefficients $c_{J,J'}$ leading from state J to state J', while for $J = J'$ the matrix elements are $-(\alpha_J + \sum_{J'<J} c_{J,J'})$. The initial rotational populations $P_J(t=0)$ are assumed to be given by a Boltzmann distribution for an initial temperature T_i.

We first calculated the time evolution of $P_J(t)$ assuming that only radiative transitions and cooling effects through J-dependent level depletion by DR are taking place. The DR rate coefficients α_J were set equal to the $\tilde{\alpha}_\mu$ for $J \in \{\mu\}$, and equal to $\tilde{\alpha}_6$ for $J \geq 8$, and the initial temperature was assumed to be $T_i = 1500$ K. The resulting group populations $\hat{P}_\mu(t)$ are shown in Fig. 4 by the dashed lines ($\kappa = 0$) and are clearly falling short in explaining the observed time evolution of $\hat{P}_\mu(t)$ and their dependence on the electron density n_e. In fact, the calculated $\hat{P}_\mu(t)$ are essentially independent of the electron density within the range of densities investigated, that means that radiative processes are dominating and contributions caused by state selective depletions via the DR process are small. The calculation also showed that the assumptions made for the initial level population are being washed out quite fast with increasing storage time; changes of T_i by ± 300 K affect the $\hat{P}_\mu(t)$ values at $t = 1$ s by $\pm 15\%$ and by less than 2% at $t = 7.5$ s. Thus neither rotational cooling by DR induced depletion nor radiative transitions alone can reproduce the observed group populations.

Superelastic collisions were included into the rate equations using inelastic electron collision cross sections calculated by combining molecular R-matrix wave functions with the adiabatic nuclei rotation (ANR) approximation. These calculations followed those performed on H_2^+ [14] for a fixed bond length of 2.0 a_0 except that they were carried out in $C_{\infty v}$ symmetry with a shifted center of mass, which results in a dipole moment of 0.85 D. As the ANR approach assumes that the rotational states are energetically degenerate, that is that the asymptotic kinetic electron energies E_k before and after an inelastic scattering event are the same, care has to be taken in using this approach when the electron energies are comparable with

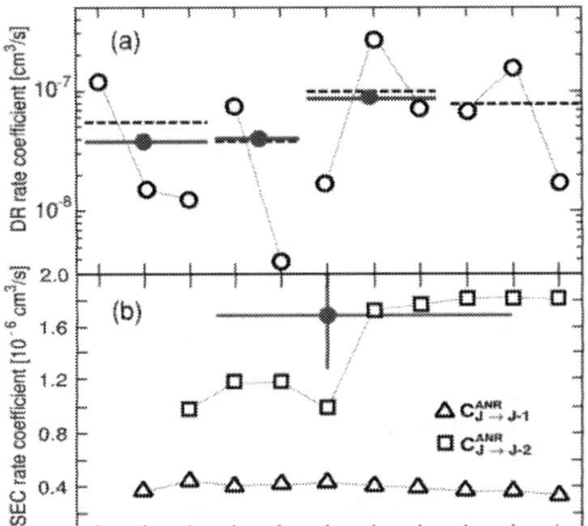

Figure 5. (a) Average experimental DR rate coefficients deduced for the three groups of J states built on the vibrational ground state of HD^+ (filled circles). They are compared to theoretical values (dashed lines) obtained by averaging the individual DR coefficients calculated within the MQDT approach (open circles). (b) SEC rate coefficients $c_{J,J'}$ calculated within the R-matrix+ANR approach and used to calculate the cooling curves shown by the solid lines in Fig. 4. The filled circle is the c_{-2} coefficient resulting from a fit of the experimental cooling curves assuming $c_{J,J-1} = 0$ and $c_{J,J-2} = c_{-2}$.

the rotational transition energies. It has been shown recently [15], however, that in view of the approximate E_k^{-1} dependence of the ANR cross sections for small kinetic energies, i.e. $\sigma_{i\to f}(E_{k,i}) \propto 1/E_{k,i}$, not only the expected threshold behavior for rotational excitation from E_{J_i} to E_{J_f} by inelastic electron scattering is secured by limiting the validity range of $\sigma_{i\to f}(E_{k,i})$ to kinetic energies $E_{k,i} > (E_{J_f} - E_{J_i})$, but that also the detailed balance relation $\sigma_{f\to i}(E_{k,f})(2J_f + 1)E_{k,f} = \sigma_{i\to f}(E_{k,i})(2J_i + 1)E_{k,i}$ is approximately fulfilled. Moreover, a quantitative assessment of the validity of the ANR approach has been performed in Ref. [15] for electron-impact rotational excitations of H_3^+ by comparing the ANR results with those obtained within the MQDT and rotational-frame-transformation approach.

The SEC cross sections calculated for HD^+ within the R-matrix+ANR approach show the expected $1/E_k$ dependence for all relevant electron energies, and – as already observed previously for ions with a moderate dipole moment [16] – the $\Delta J = -1$ cross sections are significantly smaller than those for $\Delta J = -2$. The SEC rate coefficients $c^{ANR}_{J,J+\Delta J}$ obtained by integrating the cross sections over the flattened electron distribution are shown in Fig. 5b for $\Delta J = -1, -2$ transitions; the SEC coefficients for higher transitions were found to be several orders of magnitude smaller. The resulting time evolution of $\hat{P}_\mu(t)$ obtained when taking the SEC coefficients for $\Delta J = -1, -2$ into account in the rate equation is shown by the thick lines ($\kappa = 1$) in Fig. 4. The agreement with the data is remarkable, in particular when taking into account that the only free parameter is the initial temperature T_i (assumed to be $T_i = 1500$ K), a parameter which is only moderately influencing the result within the first few seconds.

To estimate the sensitivity of the present experiment to the SEC rate coefficients several tests have been performed. The group populations expected when assuming $c_{J,J'} = \kappa \cdot c^{ANR}_{J,J'}$ with $\kappa = 0.5$ are depicted in Fig. 4 by the dotted lines; they clearly underestimate the observed cooling. The contribution of the $\Delta J = -1$ coefficients to the cooling is found to be small; the effect of setting, e.g., all $c_{J,J-1}$ coefficients equal to zero can be compensated by increasing κ by about 15%. Moreover, a fit of the cooling curves assuming $c_{J,J-1} = 0$ and $c_{J,J-2} = c_{-2}$ results in $c_{-2} = 1.7(4) \cdot 10^{-6}\text{cm}^3\text{s}^{-1}$, which agrees well with the average value of the calculated $c_{J,J-2}$ coefficients (see Fig. 5b), in particular when taking into account the expected increase caused by setting $c_{J,J-1} = 0$. The influence of heating processes on the cooling curves other than those induced by radiation were found to be negligibly small. Neither those caused by inelastic electron scattering in the bending section of the electron beam, where kinetic energies of up to 1.5 eV are available, nor those induced in the straight section by the tail of the Maxwellian energy distribution were found to change the group populations by more than a few percent. We conclude from these tests that our experiment verifies the dominant $\Delta J = -2$ SEC rate coefficients calculated within the R-matrix+ANR approach to within about 30%.

4. Summary and Conclusion

The average DR rate coefficients $\tilde{\alpha}_\mu$ deduced for the three rotational angular momentum groups μ are plotted in Fig. 5a together with the individual DR rate coefficients α_J calculated within the MQDT approach. These calculations were performed along the lines discussed in Refs. [17, 4], taking detailed account of the rotational structure and rotational couplings, and result in strongly J dependent DR rate coefficients. The values obtained by simply averaging the calculated α_J for $J \in \{\mu\}$ are consistent with the measured values, in particular when taking into account the $\pm 20\%$ uncertainty connected with the absolute scale of the measured values. Although the J dependent variations are strongly reduced by the averaging procedure, the observed increase of more than a factor of 2 between $\tilde{\alpha}_3$ and $\tilde{\alpha}_6$ constitutes another experimental verification of the strong J dependence of the DR cross sections for molecular hydrogen molecules [18, 10].

More importantly, the present experiment allows for the first time to obtain a quantitative estimate of rotational SEC rate coefficients. The observed cooling of the rotational degree of freedom induced by bathing rotationally hot HD^+ ions in cold electrons of $T \approx 10$ K was

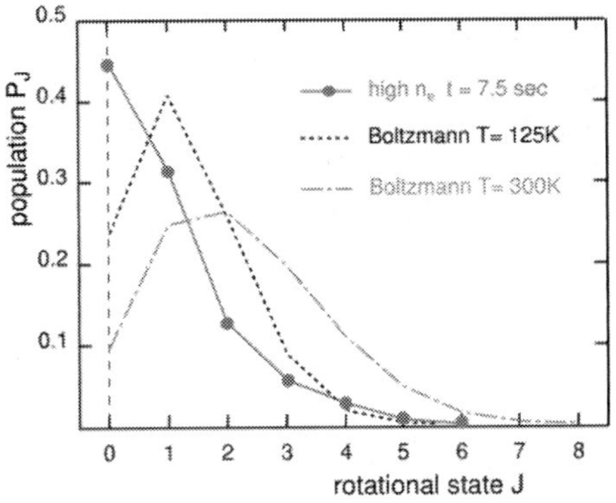

Figure 6. Population of the $(v = 0, J)$ states in HD^+ reached after 7.5 s of electron cooling at an electron density of $n_e = 1.45 \cdot 10^7$ cm^{-3} (filled circles). For comparison, Boltzmann distributions expected for rotational temperatures of 300 K (dashed dotted) and 125 K (dotted) are shown.

found to be well described by theoretical rate coefficients calculated by combining the R-matrix approach with the ANR approximation (see Fig. 5b); in particular, we could verify the size of the dominant $\Delta J = -2$ coefficients to within about 30%. These rate coefficients are of the order of $(1 - 2) \cdot 10^{-6}$ cm^{3}s^{-1} and are thus of the same size as the vibrational SEC rates observed for $\Delta v = -1$ transitions in H_2^+ [2, 3]. Superelastic collisions can thus be used to speed up the rotational cooling of molecular ions in storage ring experiments and to realize even subthermal distributions. This is demonstrated by Fig. 6, which displays the population of the individual $(v = 0, J)$ states of HD^+ reached after cooling for 7.5 s at the highest electron density investigated, which was calculated assuming the theoretical SEC rate coefficients shown in Fig. 5b. Although the resulting population cannot be represented by a Boltzmann distribution, a comparison of the centroids shows that the SEC controlled distribution corresponds to a rotational temperature of $\approx$ 125 K, which is well below the radiative equilibrium of 300 K; without electron cooling, moreover, the latter would only be reached after $\approx$ 30 s of storage.

Acknowledgments
This work was supported in part by the German-Israeli Foundation for Scientific Research and Development (GIF) under Grant No. I-900-231.7/2005. D. Sch. acknowledges support by the Weizmann Institute of Science through the Joseph Meyerhoff program.

[1] Larsson M and Orel A E 2008 *Dissociative Recombination of Molecular Ions* (Cambridge: Univ.Press)
[2] Krohn S *et al.* 2000 *Phys. Rev. A* **62** 032713
[3] Krohn S 2001 *Inelastic Collisions and Recombination between Electrons and Molecular Ions* (Ph.D. thesis, Univ. of Heidelberg)
[4] Motapon O *et al.* 2008 *Phys. Rev. A* **77** 052711
[5] Pedersen H B *et al.* 2005 *Phys. Rev. A* **72** 012712
[6] Shafir D *et al* 2009 *Phys. Rev. Lett.* **102** 223202
[7] Amitay Z *et al.* 1998 *Science* **281** 75
[8] Amitay Z, Zajfman D and Forck P 1994 *Phys. Rev. A* **50** 2304
[9] Amitay Z *et al.* 1996 *Phys. Rev. A* **54** 4032
[10] Novotny S *et al.* 2008 *Phys. Rev. Lett.* **100** 193201
[11] Zajfman D *et al.* 1995 *Phys. Rev. Lett.* **75** 814
[12] Sprenger F *et al.* 2004 *Nucl. Inst. Meth. A* **532** 298
[13] Al-Khalili A *et al.* 2003 *Phys. Rev. A* **68** 042702
[14] Tennyson J 1996 *At.Data Nucl.Data Tables* **64** 253
[15] Faure A *et al.* 2006 *J. Phys. B* **39** 4261
[16] Faure A and Tennyson J 2001 *Mon. Not. R. Astr. Soc.* **325** 443
[17] Schneider I F 1997 *J. Phys. B: At. Mol. Opt. Phys.* **30** 2687
[18] V. Zhaunerchyk V *et al.* 2007 *Phys. Rev. Lett.* **99** 013201

Eighth International Conference on Dissociative Recombination (DR2010) IOP Publishing
Journal of Physics: Conference Series **300** (2011) 012007 doi:10.1088/1742-6596/300/1/012007

A new technique to study plasma chemistry kinetics

Nicholas S Shuman, Thomas M Miller, Raymond Bemish and Albert A Viggiano[1]

Space Vehicles Directorate, Air Force Research Laboratory, Hanscom Air Force Base, MA 01731-3010, USA

e-mail: afrl.rvb.pa@hanscom.af.mil

Abstract. Accurate kinetics of plasma processes are necessary for modeling of the chemistry occurring in the upper atmosphere, re-entry, combustion, and discharges. While a great deal of data exists in the literature for most types of plasma chemistry processes, there remain gaps for reactions less amenable to traditional measurements. In particular ion-ion mutual neutralization reactions have received relatively little study, and essentially no detailed product branching fractions are known. Similarly, while hundreds of electron attachment rate constants to stable species have been reported, only a handful of measurements of electron attachment rate constants to unstable radical species exist in the literature. We report several measurements involving these classes of reactions using a novel flowing afterglow technique: Variable Electron and Neutral Density Attachment Mass Spectrometry (VENDAMS). The technique takes advantage of these processes occurring as secondary and tertiary chemistry in high density plasmas, and uses excess electrons as chemical ionization agents to monitor neutral product concentrations. Systems starting with a variety of neutrals have been studied over a temperature range of 300 to 550 K. We report several novel measurements including: the first temperature dependences of electron attachment rate constants to radical species; the first complete neutral product distributions of ion-ion mutual neutralization reactions along with the associated temperature dependences; evidence of a novel plasma chemistry process ($A^+ + B^- + e^- \rightarrow$ neutrals + e^-), which we term electron catalyzed mutual neutralization.

1. Introduction

Hundreds of reaction rate constants for plasma chemistry processes have been derived from flow tube measurements.[1] Typically rate constants are determined by maintaining pseudo-first order conditions while varying either reactant concentration or reaction time.[2] As long as a single reaction dominates the chemistry in the flowtube, extracting a reaction rate constant from this type of measurement requires only trivial analysis. However, these conditions are difficult to create for certain types of reactions. In particular, little reliable data exists for ion-ion mutual neutralization rate constants due to the challenge of accurately measuring initial ion concentrations in the flow tube, and, with the exception of the results reported here, no complete neutral product distributions exist. Additionally, while hundreds of thermal electron attachment rate constants to stable species have been reported,[3] only two measurements of electron attachment rate constants to unstable radical species exist in the literature.[4]

We have recently begun using a flowing afterglow Langmuir probe (FALP) apparatus in a novel manner (a technique dubbed variable electron and neutral density mass spectrometry (VENDAMS))[5] that trades the simplicity of the analysis described above for the ability to study other types of chemistry, including ion-ion mutual neutralization and electron attachment to radical species, which occur when an electron attaching neutral species is introduced to a weakly ionized plasma. The method has also yielded evidence that electrons act as third bodies to enhance the rate of ion-ion mutual neutralization; this plasma chemistry loss process has not been previously reported in the literature.

To date VENDAMS measurements have been made using a range of neutral precurosrs. These include several sulfur fluorides (SF_6, SF_5Cl, $SF_5C_6H_5$, SF_4, SF_3CN, $SF_3C_6F_5$) as well as $POCl_3$, $PSCl_3$, $(COCl)_2$, and

[1] To whom any correspondences should be addressed

$Fe(CO)_5$. The processes for which VENDAMS yields well-defined kinetic data are highly dependent on the specifics of the system being studied, but fall into the three broad categories outlined above. The remainder of this paper will detail the VENDAMS experimental technique, the method of analysis used to extract kinetic information from the data, and an overview of the results on systems studied to date ordered into each of those three categories of plasma chemistry processes.

2. Experimental

FALP apparatuses have been in use for several decades,[6] and the details of the instruments in general and the particular apparatus in use by the plasma chemistry group at Air Force Research Laboratory have been well-documented.[7] Briefly, a flow of helium gas is introduced to a stainless-steel lined, glass, 1m long, 7 cm diameter flow tube through an upstream mass-flow controller and removed downstream by a Roots pump. Shortly after being introduced, a portion of the helium passes through a microwave discharge cavity producing a weakly-ionized plasma consisting primarily of He^+, He_2^+, He^*, and e^-. Ar gas is flowed downstream of the microwave cavity at approximately 5% of the He flow, effectively converting He_2^+ and He^* to Ar^+. By the time the flow has reached a glass inlet port 20 cm downstream of the Ar inlet (roughly 2 ms at a typical ion velocity of 10^4 cm s^{-1}), an Ar^+/e^- plasma has been established with Ar^+ accounting for roughly 95% of the positive charge. The electron density is measured using the Langmuir probe over a dynamic range of 10^8 to $5x10^{10}$ cm^{-3}. The full range of initial electron densities ($N_e(0)$) at the downstream glass inlet is achievable by moving the microwave cavity further away or closer to the inlet and, to a lesser degree, by vary the microwave power. Charge densities are a small fraction of the neutral He density, which is typically $3.2x10^{16}$ cm^{-3} (1 torr at 300 K). At these pressures, ions undergo numerous collisions with the buffer gas and are, in most cases, assumed to be thermalized to the temperature of the flowtube. The flow is sampled 46 cm downstream of the neutral gas inlet by a 1mm on-axis aperture leading to a differentially pumped chamber and a quadrupole mass spectrometer with which to monitor the relative concentrations of ionic species. The Langmuir probe is used to measure the ambipolar diffusion rate of the plasma by monitoring the electron density as a function of distance along the flow tube for a known ion flow velocity. The velocity is also measured using the Langmuir probe, by pulsing the microwave discharge and noting the arrival time of the pulse on the probe as a function of the probe's distance along the flow tube.

Either one or two neutral gases are added through the downstream glass inlet at known rates via mass-flow controllers. A single gas is typically added in concentrations on the order of 10^9 to 10^{10} cm^{-3}, typically in a 0.1% to 1% concentration mixture in He at a flow rate of 1 to 10 std. cm^3 min^{-1}. Product anion abundances after the known reaction time (generally 4.6 ms) are monitored using the quadrupole mass spectrometer while varying $N_e(0)$ from well below to well above the neutral reactant density. At very low $N_e(0)$, the primary electron attachment to the neutral species quickly depletes the electron density, inhibiting all subsequent chemistry. These data yield product branching fractions of the primary electron attachment by measuring the relative abundance of product ions (the total electron attachment rate constant may be measured in the usual manner of measuring the decrease in the electron density above that of ambipolar diffusion as a function of reaction time). In the case of dissociative electron attachment, each ionic product is accompanied by a neutral co-product, generally an unstable radical species. At higher $N_e(0)$, sufficient concentrations of electrons survive to react with these radicals, assuming the attachment is energetically accessible. The ions resulting from the "secondary" chemistry are also monitored using the mass spectrometer and allow rate and product information of attachment to the radical species to be determined. At still higher $N_e(0)$, the anions produced in the primary attachment neutralize with Ar^+ present in the flowtube at a significant rate. If the resulting neutral products are exothermic for attachment, excess electrons in the flowtube act as chemical ionizing agents allowing for a sensitive probe of the concentrations of those products.

A second electron attaching gas may be added as well (a "2-gas" experiment) serving two purposes: 1) it is convenient to produce a monatomic anion from a primary attachment because the mutual neutralization of that anion with Ar^+ will proceed at a negligible rate, and as a result the concentration of the anion acts as an internal standard against which to measure the mutual neutralization rate constant of polyatomic anions with Ar^+. The second gas may be chosen to yield the monatomic anion in cases where attachment to the first species does not. 2) adding a large concentration of a fast electron attaching gas quickly depletes the electron density in the flowtube, inhibiting all of the "secondary" chemistry described above with the exception of mutual neutralization and allowing for a much simplified analysis in order to accurately determine mutual neutralization rate constants. Electron attachment to CCl_4 yields only Cl^- at the collisional rate constant. When added as the second gas in low concentrations ($<10^{10}$ cm^{-3}), CCl_4 serves purpose 1) above, and when added in high concentrations ($>10^{10}$ cm^{-3}) serves both purposes 1) and 2) above.

3. Analysis

Modeling the experimental anion abundances as a function of $N_e(0)$ in order to extract total and partial reaction rate constants requires accounting for the numerous reactions that can occur between all the species known (ions observed on the mass spectrometer) or inferred (corresponding product neutrals) to be present in the flowtube. The rate constant of each of these reactions is a parameter in the modeling. Product anion abundances may be calculated simply by iteratively solving the coupled differential equations describing the production and destruction of species as indicated by all the possible reactions. However, many of the reaction rate constants are largely or completely unknown. This creates concern that the data may be underdetermined.

The set of possible reactions is greatly reduced by maintaining experimental conditions in which only the fastest reactions affect the anion abundances outside of uncertainty. Specifically, neutral concentrations are held sufficiently low (generally less than 10^{10} cm^{-3}) that ion-molecule reactions, with typical upper limit rate constants of 10^{-9} cm^3 s^{-1}, are negligible relative to electron attachment and ion-ion mutual neutralization processes. The reduced set of relevant reactions may still number as high as 10-20 unknown rate constants depending on the system, creating a large parameter space. The full parameter space is explored by a Monte Carlo optimization in which 1) a set of reaction rate constants is randomly guessed; 2) the kinetics of the system are calculated from the known initial conditions throughout the known reaction time; 3) the resulting calculated anion abundances are compared to the experimental abundances via a weighted least-squares characterized by a goodness-of-fit (0 being a perfect fit); 4) the best-fit and uncertainty limits for each rate constant are determined by plotting the goodness-of-fit as a function of that rate constant. In some cases the goodness-of-fit may be a strong function of a particular rate constant, and that rate constant is then well-defined by the data. In other cases, the goodness-of-fit may be independent of a particular rate constant, and that rate constant is left completely undefined by the data.

4. Results and Discussion

Data derived from VENDAMS measurements fall across several disparate types of plasma processes. The remainder of this paper will discuss each of the types of processes studied in turn.

4.1 electron attachment to unstable radical species

Prior to the VENDAMS technique, only two reaction rate constants to unstable radical species had been reported in the literature. Each of those measurements (on CCl_3 and CH_2Br)[4] were derived by monitoring the electron depletion in a FALP as a function of reaction time above that explained by primary attachment to a neutral precursor, and each measurement was performed at room temperature. VENDAMS measurements to date on several systems have yielded electron attachment rate constants to several radical species: SF_5, SF_3, $PSCl_2$, and $POCl_2$. With the exception of $POCl_2$, each of these measurements extends across the temperature range of 300 K to 550 K, allowing for determination of the temperature dependence of the attachment. Reported rate constants appear in Table 1.

In this temperature range, most attachment rate constants to stable species show either no temperature dependence or a positive temperature dependence.[8] The exceptions are species which attach electrons with

Journal of Physics: Conference Series **300** (2011) 012007 doi:10.1088/1742-6596/300/1/012007

near unit efficiency, and are limited by the collision rate constant with an electron, which decreases as $T^{-1/2}$. For most species for which dissociative attachment is energetically accessible, this decrease along with the increase in the electron autodetachment rate with temperature is offset or swamped by an increased rate of stabilization through dissociation of the excited anionic species produced by electron capture.

Table 1. Measured electron attachment rate constants to the indicated radical species at the indicated temperatures

	k ($cm^3\ s^{-1}$)			
Species	**300 K**	**400 K**	**500 K**	**550 K**
SF_5	$2.5 \pm 1\ 10^{-8}$	$1.1 \pm x3\ 10^{-8}$	$9 \pm x3\ 10^{-9}$	-
SF_3	$4 \pm 3\ 10^{-9}$	$8 \pm x3\ 10^{-9}$	$8 \pm x3\ 10^{-9}$	-
$PSCl_2$	$1.4 \pm 0.6\ 10^{-7}$	$1.1 \pm 0.4\ 10^{-7}$	-	$9.1 \pm 5\ 10^{-8}$
$POCl_2$	$1 \pm x2\ 10^{-8}$	-	-	-

Both electron attachment to SF_5 and to $PSCl_2$ show possible steep negative temperature dependences (in both cases, a temperature dependence of zero is just within the experimental uncertainty). It is unclear if these dependences, if real, suggest a physical distinction between attachment to radical species and stable species, or if these happen to be two exceptional cases. Both SF_5^- and $PSCl_2^-$ are bound (although non-dissociative attachment is not observed for either system), and no differences in the potential surfaces between these species and those of other measured systems that could lead to different dynamics are immediately obvious.

4.2 ion-ion mutual neutralization

The dynamics of ion-ion mutual neutralization (MN) reactions involving only atomic species are fairly well-understood.[9-10] The likelihood of neutralization is a function of the location of curve crossings between the ionic potential and excited electronic states of the neutral atoms, with the large excess energy of the neutralization ending up in the only available degree of freedom: the kinetic energy of the separating neutrals. If either of the species is polyatomic, the situation is greatly complicated. Each curve crossing involves a quasi-continuum due to internal modes of the polyatomic species, and the exothermicity of the reaction has numerous degrees of freedom to be distributed amongst. VENDAMS measurements provide precise total MN rate constants and the first reported complete MN neutral product distributions against which theory of MN dynamics may be evaluated.

4.2.1 total rate constants

MN rate constants may be reasonably predicted from parameterized formulas based on the Landau-Zenner model of curve crossings.[11] Comparison of measured MN rate constants to those from theory has shown that the predicted dependences on electron binding energy (EBE) of the anion and on system temperature appear to generally hold.[12] However, the comparison suffers from two concerns: 1) rate constants for MN involving at least one structured species vary by less than an order of magnitude, requiring rather high precision in measurements in order to test theory and 2) many of the reported rate constants rely on ion concentrations measured using a Langmuir probe, a technique that has been shown to be unreliable.[13]

Using the "2-gas" procedure described above, VENDAMS yields MN rate constants to uncertainties as low as 10%. Importantly, the determination does not rely on determining the absolute concentrations of ions present in the flowtube, but rather on the relative concentrations of two anionic species, which may be measured much more precisely. In the condition where primary electron attachment to two neutral attaching gases depletes the electron density, all higher order chemistry is inhibited and only the anions produced by the initial attachment are observed in the flowtube. The ratio of the anions as a function of $N_e(0)$ varies almost solely as a function of the relative MN rate constants (in all cases here, the positive charge in the plasma is dominated by Ar^+, and any MN must occur with that species). If one of the anions undergoes MN with Ar^+ with a negligible rate constant, as Cl^- is known to do, the change in the relative abundance of the other anion yields the absolute MN rate constant.

MN rate constants have been measured for SF_6^-, SF_5^-, SF_4^-, $PSCl_2^-$, $PSCl^-$, $POCl_3^-$, $POCl_2^-$, c-$C_6F_5^-$, c-C_7F_{14}, $Fe(CO)_4^-$, Cl_2^-, FCl^-, and CN^- across the temperature range from 300 K to 550K. Rate constants appear in Table 2. In agreement with prior results, MN rate constants appear to follow the theoretical $T^{-0.5}$ dependence.[12] The dependence on EBE appears somewhat weaker than previously predicted, although larger EBE are weakly correlated with lower MN rate constants. MN involving a polyatomic species appear to have predictable rate constants in the range of 4-6 $x10^{-8}$ cm^3 s^{-1}. Measured MN rate constants for three systems involving a diatomic anion and a monatomic cation are nearly identical and are lower than those involving a polyatomic species. That these reactions involving simpler systems are slower is expected because they involve fewer curve crossings, and the likelihood of a curve crossing favorable to MN is proportionally lower. For the same reason, however, we would expect the simpler systems to have less predictable rate constants than those involving polyatomic species, but this is not borne out by the current data.

Table 2. Measured mutual neutralization rate constants of the indicated anions with Ar^+ at the indicated temperatures

	k (x10^{-8} cm^3 s^{-1})			
Species	**300 K**	**400 K**	**500 K**	**550 K**
SF_6^-	5.0 ±0.5	3.8 ±0.5	3.4 ±0.5	-
SF_5^-	3.5 ±0.5	2.9 ±0.6	2.6 ±0.6	-
SF_4^-	4.0 ±0.8	3.3 ±0.7	2.9 ±0.7	
$POCl_3^-$	5.6 ±0.6	4.5 ±0.6	4.1 ±0.7	-
$POCl_2^-$	4.6 ±0.7	3.8 ±0.7	3.0 ±0.6	-
$PSCl_2^-$	4.5 ±0.7	3.3 ±0.5	-	2.6 ±0.5
$PSCl^-$	5.4 ±1.1	4.3 ±0.8	-	3.9 ±0.8
c-$C_6F_5^-$	5.5 ±1.3	-	-	-
$(COCl)_2^-$	5.3 ±1	4.3 ±0.9	-	-
$Fe(CO)_4^-$	5.0 ±0.7	-	-	-
Cl_2^-	3.0 ±0.6	2.4 ±0.5	-	1.8 ±0.5
FCl^-	3.0 ±0.7	2.4 ±0.7	1.5 ±0.5	-
CN^-	3.0 ±1.0	-	-	-

4.2.2 product branching fractions

While overall reaction rate constants of MN appear reasonably predictable, the dynamics of the MN process are not. MN reactions are typically very exothermic, as typical ionization energies (8-20 eV) are much larger than typical electron affinities (0-5 eV). It is unclear what fraction of that available energy is funneled to internal modes of the products, what fraction remains in the kinetic energy of the separating products, and what aspects of the potential surface affect those fractions. The neutral product distributions of MN are clues to the dynamics; a larger amount of fragmentation indicates more energy placed in the internal modes of the neutralizing species. Prior to the VENDAMS technique, no complete product distributions were reported in the literature.

The MN reactions $SF_n^- + Ar^+ \rightarrow SF_m + (n-m)F + Ar$ (n=6,5,4) each have several energetically accessible product channels,[14] each of which may be characterized by the number of S-F bonds broken and the resulting SF_m neutral. Because the electron attachment rate constants of SF_m (m=2-6) are known (m=2,3,5 via VENDAMS, m=4,6 from the literature) the fraction of each SF_m neutral product may be quantified by measuring the amount of electron attachment product to each species. VENDAMS measurements using neutral precursors SF_6, SF_5Cl, and SF_4 each yield neutral product distributions of MN of the dominate anion produced by electron attachment (SF_6^-, SF_5^-, and SF_4^- respectively) with the dominate cation present in the flowtube, Ar^+.

Measured product distributions appear in Figure 1. The distributions of both SF_6^- MN and SF_5^- MN were measured over the temperature range of 300 K to 500 K, and no dependence was found. The fraction of fragmented product varies significantly between the three reactions. Differences in the product distributions are apparent, with a greater fraction of product from SF_5^- + Ar^+ undissociated than from SF_6^- + Ar^+, with SF_4^- + Ar^+ likely falling in between. The ordering is correlated to the total exothermicities of the three reactions (14.8 eV SF_6^- + Ar^+; 11.8 eV SF_5^- + Ar^+; 14.4 eV SF_4^- + Ar^+). While it may be initially unsurprising that the degree of fragmented product scales with available energy, in all three cases the available energy is well in excess of that needed to break several S-F bonds, and the energy in excess of that needed to break a single S-F bond is nearly identical in all three reactions (10.2, 10.1, 10.3 eV respectively). This suggests that factors beyond the total available energy dictate the likelihood of transferring a large amount of energy to the internal modes of the neutral products.

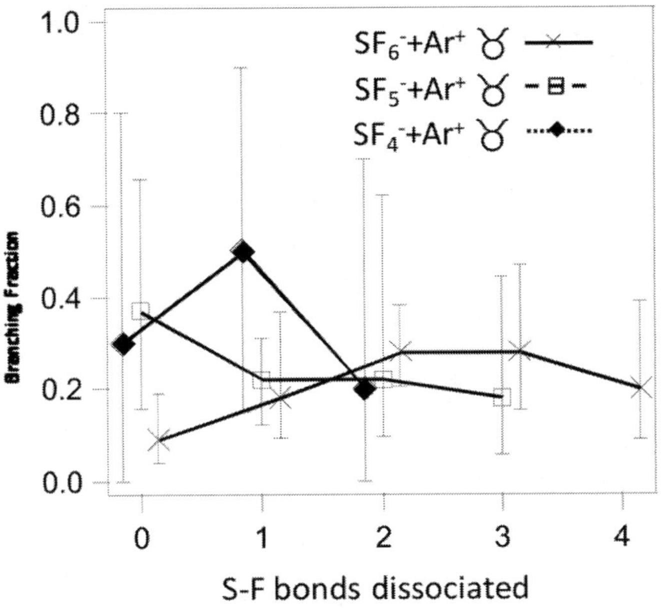

Figure 1. Product branching fractions of MN of the indicated anions with Ar^+. Product channels are defined by the number of S-F bonds broken in the reactant anion to yield the product neutral; i.e. channel 1 indicates SF_n + Ar^+ → SF_{n-1} + F + Ar.

4.3 electron catalyzed mutual neutralization

The effects of neutral third bodies enhancing the rate of plasma loss processes (radiative recombination, dissociative recombination, and mutual neutralization) have been well-documented and are important in understanding the chemistry of plasma at elevated pressures.[15-17] Electron third bodies have also been found to enhance the rate of radiative recombination,[18-19] and likely have a small, but difficult to measure, effect on much faster dissociative recombination processes. However, electron third bodies have not been previously suggested to increase the rate of mutual neutralization reactions. Despite the above observations, an enhancement would not be expected in MN because the previously recognized three-body processes all proceed by a collisional mechanism wherein the third body removes a significant amount of kinetic energy from the system, trapping the charged species in the attractive well, which can then only be exited by neutralization. Due to the extreme mass difference, an electron is a poor collisional energy sink for a system composed of two ions.

As stated above, the "2-gas" VENDAMS experiment determines an MN rate constant to within an uncertainty of 10-25% (Figure 2A). Surprisingly, modeling "1-gas" VENDAMS data using the measured MN rate constants consistently underestimates the loss-rate of the anion, with the magnitude of the underestimation scaling linearly with $N_e(0)$ (Figure 2B). This suggests that a plasma loss process unaccounted for in the modeling contributes significantly to the observed chemistry at higher $N_e(0)$. The negative charge can only be removed by neutralizing with Ar^+ (no other positive species is present in the flowtube in sufficient concentrations to have the measured effect, and any polyatomic positive species will show a decreasing effect with increasing $N_e(0)$ due to competing dissociative recombination). The only third bodies whose concentration scales appropriately with $N_e(0)$ are Ar^+ and e^-, and only assuming an e^- third body is consistent with both the "1-gas" and "2-gas" data. We suggest that the unaccounted for loss process is

$$M^- + Ar^+ + e^- \rightarrow neutrals + e^-$$

which we term electron catalyzed mutual neutralization (ECMN). The process appears general; measured rate constants for $M^- = SF_6^-$, $POCl_3^-$, and $POCl_2^-$ are on the order of 10^{-18} cm^6 s^{-1} and are non-zero outside of uncertainty as well as showing steep negative temperature dependences typical of three-body processes. Measured ECMN rate constants for several other species are similar in magnitude, but are not non-zero outside of uncertainty.

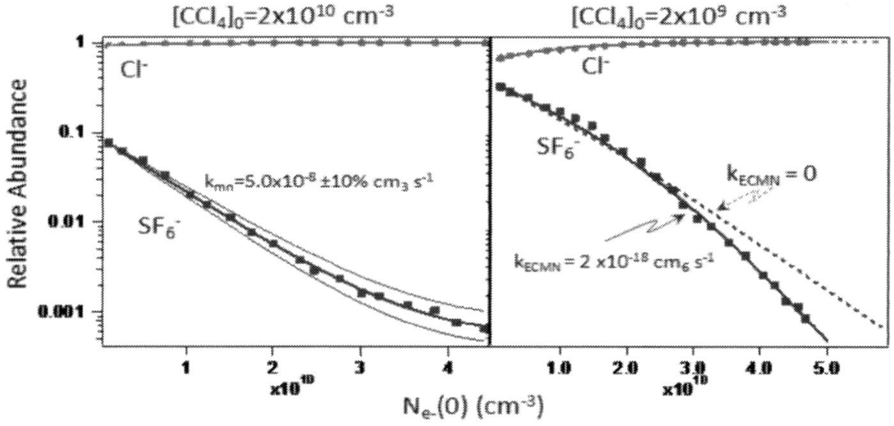

Figure 2. Measured (points) and best-fit calculated (lines) anion abundances from addition of 2×10^{-9} cm^{-3} SF_6along with the indicated concentration of CCl_4 as a function of $N_e(0)$. The high $[CCl_4]_0$ condition (A) determines k_{MN}. The low $[CCl_4]_0$ condition (B) is only fit by including a non-zero k_{ECMN}

A mechanism by which ECMN can occur is not obvious. Because ECMN is additive above MN, the mechanism must occur at inter-ion distances that would not otherwise lead to neutralization. While an e^- can not effectively remove kinetic energy of the system through collisions, if interaction with the e^- transfers kinetic energy of the system to the internal modes of the anion (the only readily available energy sink), the rate enhancing effect would be similar. Electrons loosely bound to a dipole have been shown to excite vibrations in the resulting neutral species when leaving the complex.[20] The leaving electron causes a fluctuation in the potential seen by the dipole, resulting in a time-dependent change in the dipole moment (a transition dipole) resulting in some probability of exciting a vibration, much as though a photon was interacting with the system. A similar effect could occur here as a third body electron passes between the anion and cation, shielding the ionic attraction and causing a fluctuation in the potential as the ionic attraction is decreased and then reestablished. For electrons moving at thermal velocities, the frequency of that fluctuation is on the order of hundreds of cm^{-1}, similar to the energy of low frequency vibrations. Although this is entirely speculative, the measured ECMN rate constants do show a correlation with the calculated IR strength of low frequency modes of the anion, as would be expected from the proposed mechanism.

5. Conclusions

A new flowing afterglow technique to study the kinetics of several plasma chemistry processes that are difficult or impossible to measure by other means has been described. The method, variable electron and neutral density attachment mass spectrometry, exploits the higher order chemistry that occurs when an electron attaching neutral species is added to a weakly ionized plasma. Unlike traditional FA methods, which maintain conditions where this other chemistry is negligible simplifying analysis, VENDAMS allows for a wide range of chemistry to contribute and relies on kinetic modeling of the system in order to extract rate constants and product branching fractions from the data. Several types of novel measurements are reported. Rate constants and product branching of electron attachment to unstable radical species along with temperature dependences are reported. Rate constants for mutual neutralization reactions are reported with uncertainties as low as 10%; importantly the measurements do not rely on unreliable determinations of the ion concentrations in the flowtube. The first complete product branching fractions for mutual neutralization reactions are reported. Finally, VENDAMS measurements yielded evidence of a new plasma loss process, electron catalyzed mutual neutralization.

Acknowledgments

This work has been supported by the Air Force Office of Scientific Research. N S S received funding through the National Research Council's Research Associateship Program.

[1] Y. Ikezoe *et al.*, *Gas Phase Ion-Molecule Reaction Rate Constants Through 1986* (Maruzen Company, Ltd., Tokyo, 1987).
[2] E. E. Ferguson, F. C. Fehsenfeld, and A. L. Schmeltekopf, in *Adv. At. Mol Phys.*, edited by D. R. Bates (Academic, New York, 1969), pp. 1.
[3] L. G. Christophorou, D. L. McCorkle, and A. A. Christodoulides, in *Electron-molecule interactions and their applications*, edited by L. G. Christophorou (Academic, New York, 1984), pp. 477.
[4] N. G. Adams, D. Smith, and C. Herd, Int J. Mass Spectrom. Ion Proc. **84**, 243 (1988).
[5] N. S. Shuman *et al.*, J. Phys. Chem. A **ASAP** (2010).
[6] E. E. Ferguson, F. C. Fehsenfeld, and A. L. Schmeltekopf, Advances in Atomic and Molecular Physics **5**, 1 (1969).
[7] T. M. Miller, J. F. Friedman, and A. A. Viggiano, Int J. Mass Spectrom. **267**, 190 (2007).
[8] L. G. Christophorou, *Electron-molecule interactions and their applications* (Academic, New York, 1984).
[9] R. E. Olson, Comb. and Flame **30**, 243 (1977).
[10] J. T. Moseley, R. E. Olson, and J. R. Peterson, Case Studies in Atomic Physics **5**, 1 (1975).
[11] A. P. Hickman, J. Chem. Phys. **70**, 4872 (1979).
[12] T. M. Miller, J. Chem. Phys. **72**, 4659 (1980).
[13] R. Johnsen *et al.*, Phys. Rev. E **50**, 3994 (1994).
[14] T. M. Miller, S. T. Arnold, and A. A. Viggiano, Int. J. Mass Spectrom. **227**, 413 (2003).
[15] M. R. Flannery, Philosophical Transactions of the Royal Society A **304**, 447 (1982).
[16] D. R. Bates, and I. Mendaas, Chem. Phys. Lett. **88**, 528 (1982).
[17] D. R. Bates, J. Phys. B. **14**, 3525 (1981).
[18] E. Hinnov, and J. G. Hirschberg, Phys. Rev. **125**, 795 (1961).

[19] D. R. Bates, and A. E. Kingston, Nature **189**, 652 (1961).
[20] C. G. Bailey, C. E. H. Dessent, and M. A. Johnson, J. Chem. Phys. **104**, 6976 (1996).

Eighth International Conference on Dissociative Recombination (DR2010) IOP Publishing
Journal of Physics: Conference Series **300** (2011) 012008 doi:10.1088/1742-6596/300/1/012008

Fast-beam fragmentation experiments on dissociative recombination

Andreas Wolf[1], **Henrik Buhr**[2,1]

[1] Max-Planck-Institut für Kernphysik, Saupfercheckweg 1, 69117 Heidelberg, Germany
[2] Department of Particle Physics, Weizmann Institute of Science, Rehovot 76100, Israel

E-mail: A.Wolf@mpi-hd.mpg.de

Oldřich Novotný

Columbia Astrophysics Laboratory, Columbia University, New York, NY 10027, USA

Abstract. An overview of recent developments is given for storage ring experiments on dissociative recombination using fast merged ion and electron beams. Dynamical aspects of the process are outlined, demonstrating that fast-beam reaction studies can advance the understanding of basic mechanisms holding generally for the fragmentation of electronically excited molecular systems. Experiments yield the collision energy dependence of the process, including rich resonant structure, product branching ratios of the fragmentation, and fragment momenta which also reveal the internal excitation of the fragments. Using cold electron beams and imaging detectors capable of fragment mass recognition at high readout speed, recent detailed fast-beam fragmentation experiments address the dissociative recombination of multielectron diatomic and polyatomic ions. Some of their results are presented together with an outlook on experiments at upcoming storage ring facilities.

1. Introduction

Dissociative recombination (DR) between a molecular cation and a free electron is one of the most important binary reactions in media governed by ion chemistry [1]. Exothermic in nature, it proceeds without an energy barrier and therefore very efficiently down to the lowest temperatures. In their collision, the electron and the ion form a neutral compound state with high electronic excitation, which then relaxes by molecular dissociation through a wide range of coupling mechanisms between electronic and nuclear motion. The cross section of the process and the internal states of the products are sensitive to the various steps in the fragmentation process and depend on the electron impact energy as well as, frequently, also on the initial excitation of the cation. Much information about the underlying reaction mechanisms can be gained from experiments at well-defined collision energy, if possible also for controlled initial states of the ion, and from the analysis of the final product states. Experiments of this type also offer most specific tests of theoretical calculations used to predict the rates and product properties for this reaction.

Comprehensive understanding of DR is, on one hand, motivated by the need of data about this process for modeling gas-phase chemistry in dilute ionized media of astrophysics [2] and low-temperature plasma physics [3]. Both, charged and neutral molecular abundances in these media often depend strongly on the DR cross section and its products. On the other hand, DR constitutes a test case for our ability to quantum mechanically predict the relaxation of electronic energy by molecular fragmentation, an area where the central importance of rovibronic and non-adiabatic coupling mechanisms is increasingly recognized.

Storage ring experiments with merged beams of molecular ions and electrons [4, 5, 6] offer powerful methods for observing DR reaction rates, branching ratios and fragment momenta. For the diatomic [7, 8] and triatomic [9] hydrogen molecules, experiments revealed many aspects of the quantum dynamics and pointed particularly to the crucial role of molecular Rydberg manifolds [10, 11, 12, 13]. Aside of the energy dependence of the DR cross section, also the outgoing atomic states and specific dynamics, such as angular correlations between the incident electron and the fragmentation direction [8], were investigated. The present paper addresses the extension of detailed dynamical studies to heavier multielectron molecules, both diatomic and polyatomic, where additional phenomena arise in the DR process and where both experiment and theory face the challenge of higher complexity, together with large opportunities for improving our understanding of molecular mechanisms with a wide occurrence in nature.

2. Processes

2.1. Resonant electron capture

The DR of a molecular ion is triggered by its resonant capture of an electron. The basic mechanisms that can occur in a multi-electron system are illustrated in Fig. 1 for the diatomic case. This diagram shows the lowest few potential energy curves of the ion on a common scale with the energy levels of the recombined neutral system. Also indicated is the Franck-Condon (FC) zone that defines the possible initial states of the binary collision. The kinetic energy E of the incident electron is added to the energy of the parent molecular ion in its electronic and vibrational ground state (near label 1 in Fig. 1), while the internuclear distances are defined by the vibrational wave function of the colliding ion. Thus, the initially populated state lies within the indicated area at a height given by the electron collision energy. Pathways for DR are entered from this initial height in the FC zone and, via dissociative neutral potential curves, can terminate in various neutral product states. The difference in total internal energy between the initial and final states defines the kinetic energy gained by the fragments. For a

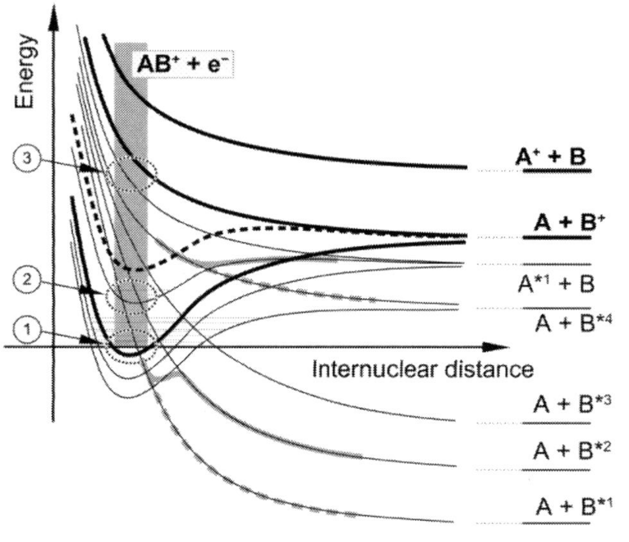

Figure 1. Typical potential curves in low and medium-energy electron collisions (up to ~30 eV) with a diatomic ion. Thick dark lines: ionic energy levels (dashed: excited bound ionic state); thinner lines: neutral (diabatic) levels and some neutral vibrational resonances. Grey rectangle: initial energy levels and internuclear distances; grey curves: fragmentation pathways with coherent splitting at diabatic curve crossings. Asterisks and indices: excited final atomic levels (schematic). Numeric labels: energy regions discussed in the text.

polyatomic ion, an equivalent map can be constructed with potential energy surfaces (PES) in the multidimensional nuclear coordinate space.

The dissociating potential curves crossing the initial FC zone are electronically doubly excited states of the recombined neutral system. Potential surfaces for such autoionizing states [14] can be obtained in the diabatic representation of the molecule [15], where specific configurations of electronic orbitals are used and configuration mixing is added in later steps of the model. (In this sense, also the initial electronic continuum states in the FC zone are diabatic states.) A diagram such as Fig. 1 can never individually show all neutral PES, as each potential surface of the ion is accompanied by a full series of neutral Rydberg states.

The configuration mixing between diabatic electronic states is crucial for DR: On one side, it can lead to the capture of the free electron – without any dynamic participation of the nuclei required in this step, thus being similar to dielectronic recombination of atomic ions [16]; the neutral states driving these capture processes are also called (electronic) resonances. On the other side, configuration mixing is important in the further evolution of the fragmentation, as any crossings between the diabatic potential curves are converted to avoided crossings in the adiabatic limit, and for moving nuclei dynamical transitions [17] of the Landau-Zener type can occur between the dissociating PES.

As sketched in Fig. 1, such transitions can lead to the splitting of a fragmentation pathway over two or more potential surfaces in the sense of a coherent superposition of the probability amplitudes. This quantum mechanical evolution – distributed over several coherently populated PES – has a crucial effect on the reaction as it is mirrored by the internal (electronic) states of the outgoing fragments and, in the polyatomic case, by their momentum correlations and by the vibrational excitations of molecular fragments; strictly, it can even be a source of quantum mechanical entanglement between the internal states of the outgoing fragments.

The schematic diagram in Fig. 1 translates into a typical DR cross section as a function of the collision energy E as outlined in Fig. 2. The large-scale structure with smooth energy variations in a low-energy region 1 and a high-energy region 3 ($\sim$8-20 eV) results from *direct* DR [18] into dissociative PES within the initial FC zone. Thus, in region 3 the numerous neutral Rydberg states (or Rydberg resonances) below each of the high-lying dissociative ionic surfaces open many direct pathways as determined by the FC overlap with the initial state. Similarly, in region 1 at small impact energy E, direct electron capture into a neutral (resonant) PES intersecting that of the cation near its equilibrium nuclear geometry results in a smooth energy dependence of the cross section, representing the product of the asymptotic threshold energy dependence E^{-1} and a further slowly varying function expressing, among other factors [19, 20], the FC overlap between the incident cation and any resonant states at $E = 0$. The intersection of the cationic PES by a neutral dissociating (diabatic) PES near the cationic potential minimum is a condition

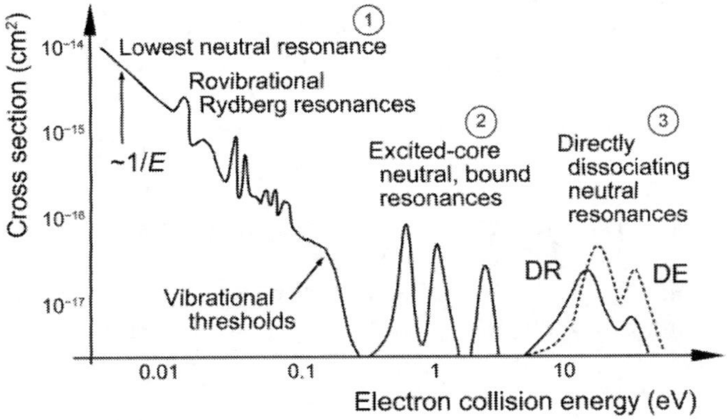

Figure 2. Typical schematic energy dependence of the DR cross section $\sigma(E)$ as a function of the collision energy E. Dissociative excitation (DE), the process leading to $A^+ + B$ or $A + B^+$, opens at the dissociation energy of the cation and has the cross section sketched by the dashed line.

for this direct DR process to occur at low impact energy and at low initial vibrational excitation; however, mechanisms going beyond the purely electronic coupling assumed in this *direct* elecron capture process can in many cases lead to strong low-energy DR even in the absence of an intersection of this type [21].

Apart from dissociating neutral states, also vibrationally bound neutral states can occur whose total energy matches that of the initial collision state; some of such states, attached to neutral bound (diabatic) PES, are indicated in Fig. 1. If suitable mechanisms exist to populate these states, resonant with respect to the total energy and not only its electronic part, sharp structures result in the collision cross section. For contributing to DR, the states of this type must be both autoionizing (in order to appear as a resonance in the electron collision) and predissociating (in order to lead to a fragmentation). Predissociation can occur by any dissociating neutral PES that crosses the neutral bound PES within the range of the vibrational wave function in this resonant state. It can be seen in Fig. 1 that this range can be considerably more extended than the FC zone of the colliding ion, so that the resonant state can act as a doorway to fragmentation paths inaccessible by the direct process. Moreover, the width of the resonant state is determined by the autoionization and predissociation rates, and complicated spectral structures can arise in the continuum if resonances become less separated from each other than by their width. The DR via such vibrationally bound resonances [18, 10] is denoted as the *indirect* process.

At low collision energy, *vibrationally excited Rydberg resonances* can be caused in particular by the Rydberg series of neutral PES below the potential surface of the incident cation (see Fig. 1). The mechanism populating such resonances in the electron collision is the vibrational excitation of the ion by the colliding electron, forming an autoionizing, vibrationally excited molecular Rydberg state whose subsequent predissociation leads to DR [18]. This indirect DR mechanism is generally important for diatomic as well as polyatomic ions, reflecting different types of coupling between nuclear and electronic motion. Its particular importance stems from the fact that it leads to strong DR at low impact energy even in the absence of a resonant potential crossing in the FC zone of the ion, as required for the direct process; thus, for many important species (H_3^+, H_3O^+, etc.) DR with low-energy electrons proceeds only through the indirect process. High-resolution measurements of the resonant energy dependence could be performed in merged-beams experiments [7, 8, 22], but the observed spectra still resist their definite assignment and detailed reproduction by theoretical calculations, even in such cases as HD^+ and H_3^+.

Measurements of the DR cross section as a function of the impact energy for more complex multi-electron ions, in particular diatomic ions with L-shell atoms such C, N, and O, have revealed further narrow structures in the intermediate energy region 2, typically at $\sim$0.3–2 eV (Fig. 2). The additional resonances are caused by the more complicated electronic structure of these ions, which often have vibrationally bound excited potential surfaces (see Fig. 1). Each of these states is accompanied by a series of electronically doubly excited neutral states, which lie in the initial electronic continuum. Energetically lower members of such series can lead to well separated capture resonances; since their first experimental observations [23, 24, 25] and theoretical [26, 27] studies they are referred to as *excited-core Rydberg resonances*. The various vibrational levels that lead to capture resonances can again, similar to the low energy case, open dissociation pathways outside the initial FC region, and interesting collision dynamics and energy dependences of the cross section can arise in case of fast, multi-path predissocitation. Thus, aside of the interest in DR data of this type for the ion chemistry in warm molecular plasma, the interplay of theoretical and experimental studies regarding such structures also provides a test field for our understanding of coupled electronic and nuclear motion in excited, fragmenting molecules and related quantum phenomena.

2.2. Fragmentation dynamics

Even at near-zero collision energy, DR often leads into various alternative final channels with different chemical compositions and different internal product states. The final-state branching ratios reveal the correlations between the resonant capture state and the product energy levels, which are not uniquely fixed, but depend on possible dynamical transitions between intersecting potential surfaces on the fragmentation pathway. The fragmentation route involves propagation of a quantum wave packet of nuclear motion on one or several potential surfaces, including its coherent splitting at crossings between diabatic potential surfaces (see Fig. 1). Localized multiple dynamical transitions on the way to fragmentation can introduce fast variations of the branching ratios as a function of the collision energy through multi-path interference, so far mainly observed in the related process of ion-pair production in electron-cation collisions [28, 29, 30, 31].

For polyatomic ions, illustration of the reaction by a map similar to Fig. 1 rapidly becomes more difficult as the number of nuclei increases. Experimental observations of branching ratios and, recently, their variations over a wider range of collision energies bring the opportunity of developing theoretical models for the electron capture process and for the relevant potential surfaces along the fragmentation path even in polyatomic cases, up to the propagation of multi-dimensional nuclear wave packets representing the fragmenting molecule. Related work is presented at this conference [32]. The neutral potential surfaces governing fragmentation in the DR of polyatomic ions were surveyed quite extensively over the multidimensional molecular coordinate space for only few cases, including in particular H_3^+ [33] and H_3O^+ [34]; however, even for this case many aspects of the underlying fragmentation dynamics are still only little understood. Ongoing theoretical and experimental studies aim at a much more complete picture of the process for this system and similar polyatomic species.

Merged beams experiments are capable of analyzing the DR process also by measuring the fragment momenta and the resulting kinetic energy release [35]. For polyatomic ions breaking up into three or more products, the momentum correlations between them give further information about the potential landscape leading to fragmentation [36]. Thus, in three-body breakup linear and triangular geometries of the fragment momenta can be distinguished and the full space of possible geometries is spanned by the Dalitz plot representing the kinetic energy sharing between the individual fragments. A number of fragment imaging studies for the three body breakup by DR have been performed. For H_3^+, singled out as the most fundamental polyatomic ion, symmetric linear momentum geometries of three H atoms were observed after DR by fragment imaging [37, 38]. For this case, the relation to the fragmentation pathway was studied in considerable detail. Interesting comparisons become possible to three-body fragmentation measurements following the electron capture by H_3^+ ions in an atomic gas target, which so far focused on laser-excited levels of neutral H_3 [33], but recently also addressed the electron capture into high-lying neutral Rydberg levels of H_3 [39]. The momentum geometries for three-body breakup following this type of electron capture are found [39] to strongly resemble those seen after DR at low electron impact energy [38]. This suggests that similar initial neutral states precede the fragmentation in both capture processes. A map of numerous excited potential surfaces was calculated for the neutral H_3 system and helps to identify, for the conditions of low-energy DR, a plausible fragmentation pathway [33] via several of them, transforming the molecule from its equilateral triangular shape before the collision to the linear geometry found after its breakup.

3. Experimental method

3.1. Merged beams in ion storage rings

Fast beam fragmentation experiments on DR are based on merging an electron and an ion beam in a collinear overlap region at similar average velocities (Fig. 3), which can be fine adjusted

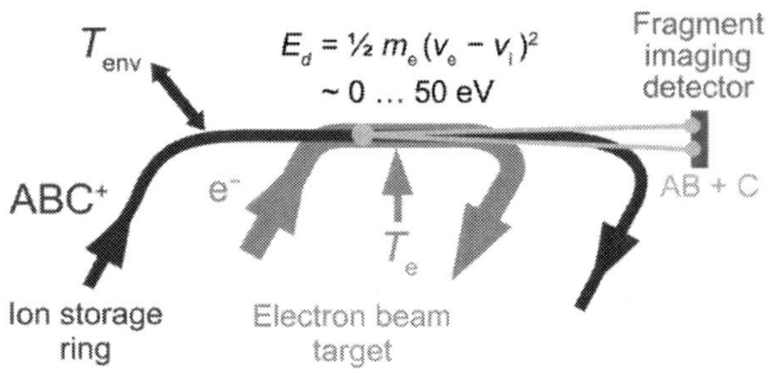

Figure 3. DR measurements in an ion storage ring. The collision energy E_d is controlled by the ion and electron beam velocities (v_e, v_i). The electron temperature T_e represents the spread of the electron velocities, while the internal states of the ions are coupled to the temperature T_{env} of the storage ring walls by blackbody absorption and emission.

by varying the acceleration voltage of the electron beam [4, 5, 6]. They are most efficiently performed inside an ion storage ring: among other benefits, the ion velocities can be made very uniform by phase-space cooling, various methods can be applied for controlling the initial state of the reacting molecules, and the ions can be recycled continuously even if they are produced by a pulsed source with a low duty cycle. In the overlap region, impact energies down to $\sim$1 meV (corresponding to temperatures as low as 10 K) can be realized, as determined mainly by the velocity spread of the electrons. The neutral fragments from the overlap region, typically $\sim$1 m long, proceed to a fragment detector installed typically $\sim$10 m downstream, behind magnetic or electric elements for separating any co-moving charged particles. The neutral fragments of each DR reaction have approximately the velocity of the incident ion, which also defines the motion of the center-of-mass (c.m.) during their propagation to the detector. The mutual distances increase during the drift to the detector and follow from the fragment velocities of the breakup reaction in the c.m. frame, reaching up to several cm at the DR detector. For $\sim 10^6$ stored ions in a continuous coasting beam and an electron density of 10^6 electrons per cm^3, DR reactions occur at rates up to $\sim 10^4\,\mathrm{s}^{-1}$ and, through their high energies in the laboratory frame, the neutral fragments create distinct signals in the various types of counting and imaging detectors used. The measured rates are normalized to the stored ion number and the electron density to derive the beam-beam rate coefficient of the reaction. It is obtained as a function of the collision energy E_d between the two particle beams (see Fig. 3) and represents the average $\langle v\sigma(E)\rangle$ of the DR rate coefficient over the distribution of collision velocities v. In contrast to thermal rate coefficients, the beam-beam rate coefficient is obtained at a nearly uniform collision energy, strongly peaked around E_d, except for low-energy DR when E_d is set to zero and the average represents the electron temperature.

Depending on the segmentation or position-sensitive readout capabilities of the detectors and their time resolution, fragment multiplicities, transverse fragment positions in the detector plane and arrival time differences can be obtained for individual DR events [35, 36]. Moreover, detectors which are also sensitive to the kinetic energies deposited by the impinging fragments are particularly useful in fast-beam measurements, as the laboratory velocities of all fragments are essentially given by the initial ion velocity and thus, the fragment kinetic energies directly reflect their masses. For DR events in which all fragments are detected and their individual masses can be assigned, the c.m. position in the detector plane can be identified and the possibility of analyzing the momentum geometries in polyatomic breakup is opened up.

The reaction data extracted from fast-beam DR measurements span a wide range. They reach from total DR cross sections as a function of the impact electron impact energy or other parameters, such as the internal states of the parent ion (and thus their storage time since the excited state populations often depend on it) to the analysis of DR branching ratios towards specific fragments, possibly to specific final states within them, and to detailed multi-coincidence fragment momentum measurements. Recent developments in fragment detection and imaging

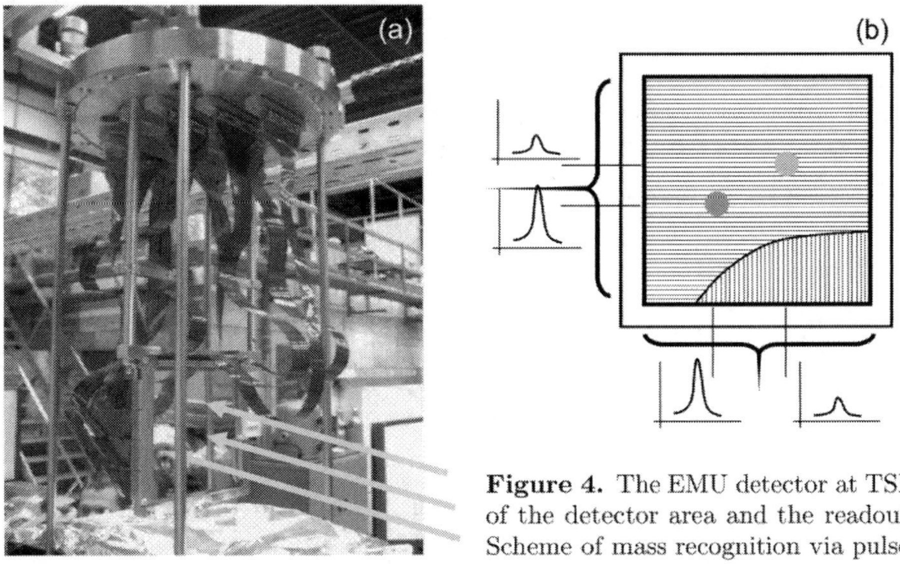

Figure 4. The EMU detector at TSR. (a) View of the detector area and the readout lines. (b) Scheme of mass recognition via pulse heights.

methods will be discussed in Sec. 3.2.

The recent results presented in this paper were obtained at the Test Storage Ring (TSR) facility operated by the Max-Planck Institute for Nuclear Physics in Heidelberg, Germany, which applies a photocathode electron target [40] to create the co-propagating electron beam [41]. Molecular ion beams with masses up to ~40 amu are produced with electrostatic or radio-frequency accelerators and stored in the TSR at typical kinetic energies of a few MeV. Within fragment imaging experiments, the c.m. positions of DR events are applied to monitor the ion motion in the overlap region with the electron beam [42]. After careful adjustment of the electron beam velocity and its direction during the phase-space cooling, the c.m. positions are in most cases found in a narrow region with a diameter of order 1 mm, indicating very low transverse divergence and size for the stored ion beam. It was shown that at TSR such conditions can be maintained up to high molecular masses of order 30 amu and the extension of efficient phase space cooling to even heavier molecular beams is under continuous study.

3.2. Fragment momentum imaging and branching ratios – recent advances

The first storage-ring fast-beam experiments on DR applied unsegmented surface-barrier detectors and analyzed the integral ionization signals from the coincident impact of neutral fragments from individual breakup events, occuring at spatially separated positions of the detector [4]. Signals representing the full kinetic energy of the parent ions in the incident beam signify a DR event. These detectors record each fragment with essentially unit probability; hence, they directly measure the total DR rate provided they fully cover the fragmentation cone around the beam axis. They are also able to reject the rather large background from dissociative excitation of the parent ions in collisions with the residual gas; the neutral fragments of this reaction represent less than the full kinetic energy of the parent ions and thus yield a smaller ionization signal in the detector, which can be used to discriminate the events they create. (A much smaller background from electron capture and dissociation in collisions with the residual gas occurs in addition; this cannot be discriminated directly, but often derived from the collision energy dependence of the measured rate.)

The widely used multichannel plates (MCP) can be applied to read out the fragment impact locations (and with particular upgrades [35] also the impact time differences) for coincident fragments of individual DR events, but are blind to their laboratory kinetic energies. While these detectors yield very good spatial resolution, their probability to record the individual fragment impacts is clearly below 1 (typically 0.7). In spite of the missing independent fragment mass recognition and the limited observation efficiency, the "monochromatic" impact signals from an MCP can under certain conditions [43] be used to reconstruct the masses belonging to individual fragment hit locations. In a fast-beam experiment with a well cooled ion beam [42] the c.m. motion, conserved in the DR reaction, is very uniform among all stored ions. Thus, the c.m. position of each correctly assigned DR event is expected to lie in a narrow region, defined by the ion beam motion, and by testing all possible mass assignments the one where the correct mass is associated with each hit can be recognized. A high (although not perfect) event-by-event reliability of this method could be achieved for many cases; demonstrations are given by three-body breakup measurements in the DR of D_2H^+ and H_2D^+ [44] as well as CH_2^+ [45]. Since MCP detectors are wide spread and practical to use, the limits of this method for analyzing multiparticle fragmentation with "monochromatic" imaging detectors warrant careful further exploration.

"Color" imaging in the sense of mass recognition for individual fragments is provided by detectors which are sensitive to fragment kinetic energies while also offering position sensitive readout or segmentation. The first detector of this type introduced for DR experiments is the Energy-sensitive MUltistrip detector system (EMU) [46] at the TSR (see Fig. 4), a multistrip surface-barrier detector with 128 stripes each in horizontal and vertical direction and a kinetic energy resolution shown to be suitable for imaging of separate DR channels on molecules of masses up to $\sim$30. By its high segmentation and its mass resolution, EMU can also be applied to determine branching ratios in polyatomic fragmentation. The data acquisition in EMU measurements proceeds with high count rates, so that new possibilities for the determination of branching ratios as a function of the electron impact energy are opened up, as demonstrated below and in Ref. [46, 47].

Non-imaging methods for the determination of DR branching ratios in fast-beam experiments were applied to already a rather large number of species. In the most widely used technique [36], an unsegmented surface-barrier detector is preceded by a metal plate with a dense pattern of sub-millimeter holes, yielding a grid mask of known geometrical transmission. Since the grid mask randomly blocks individual fragments of a DR event, the integral ionization signal no longer always appears at the full kinetic energy of the parent ion, but is reduced by an amount varying event by event, which corresponds to the mass of the blocked fragments. In a large sample of detected reactions, the relative probabilities of events with various missing fragment masses behind the grid can be analyzed [48] to extract the probablities at which the product mass combinations occur in DR. Often this reveals the branching ratios. The method was extended to rather large polyatomic ions, including systems such as protonated methanol (in its deuterated equivalent), $CD_3OD_2^+$ [49]. The background from dissociative excitation of the parent ions in residual gas collisions cannot be directly rejected event by event in the "grid method", as it cannot be distinguished from events where a neutral DR fragment was blocked by the grid; it is determined separately and subtracted in the analysis of the probabilities. The branching ratio measurements with EMU [46, 47, 50] can reject this dissociative excitation background event by event, reducing possible uncertainties.

Fast ion beams experiments on cluster fragmentation have demonstrated also another method [51] for analyzing the molecular composition of breakup products with an unsegmented surface barrier detector. Here, both the pulse integral and the rise time (the current amplitude measured at high temporal resolution) were analyzed for ionization signals from carbon cluster fragments. The pulses from fragments of the same total mass, having a fixed pulse integral, turned out to show different discrete values of the current amplitude; it was considerably smaller for a molecular impact at a single position than for the distributed impact of several fragments of the same total mass. It appears interesting to explore the applicability of this method for DR branching ratio measurements.

4. Recent experimental results

4.1. Excited-core neutral Rydberg resonances

Motivated by the strong interest in low-temperature media, such as in astrophysics, many studies of DR focused on the lowest collision energies achievable. However, substantial DR rates are also found at higher collision energies up to several eV (see Fig. 2). In this context, for multi-electron ions, excited-core neutral Rydberg resonances become important, as they can contribute to significant DR rates in a temperature range where molecular ions survive under plasma conditions [3, 53]. In fast-beam measurements, such contributions were identified for the DR of, e.g., OH^+ [24], CH^+ [25], N_2^+ [54], NO^+ [55] and, in a recent measurement at much higher energy resolution, CF^+ [52] (see Fig. 5). In CF^+, the lowest peaks from excited-core neutral Rydberg resonances occur near 1.6 eV as a prominent doublet in which the single peaks have an energy width below 0.2 eV; above, a highly structured resonance pattern continues up to ~8 eV. The beam-beam rate coefficient for DR drops at the vertical excitation energy to the $a\,^3\Pi$ bound excited state of the CF^+ parent ion, which probably indicates the series limit of neutral Rydberg levels below this state, as well at the adiabatic dissociation energy of CF^+, indicating the competition of ionic dissociation with DR. This demonstrates the high complexity and the large amount of information on molecular levels and their dissociation dynamics that are revealed by the DR cross section at higher impact energies. Similar high-resolution studies, complementing earlier experiments [54, 55], are desirable for other multi-electron molecules such as the atmospheric species N_2^+, NO^+, O_2^+ and others. Since higher temperatures are involved, also studies for ions in specific excited vibrational states are of interest. Theoretical studies of the CF^+ DR spectrum appear highly useful in the development of reliable methods for predicting DR via excited-core neutral Rydberg resonances.

4.2. Fragment imaging studies

4.2.1. Fragment and parent excitation in low-energy collision studies Since its implementation at the TSR, fragment imaging using the EMU detector, combined with mass recognition, has brought out important aspects regarding the DR of polyatomic ions as well as the conditions under which such ions are investigated in ion storage rings. This will be discussed on the example of fragment imaging for the DR of D_3O^+ with cold electrons, corresponding to a thermal

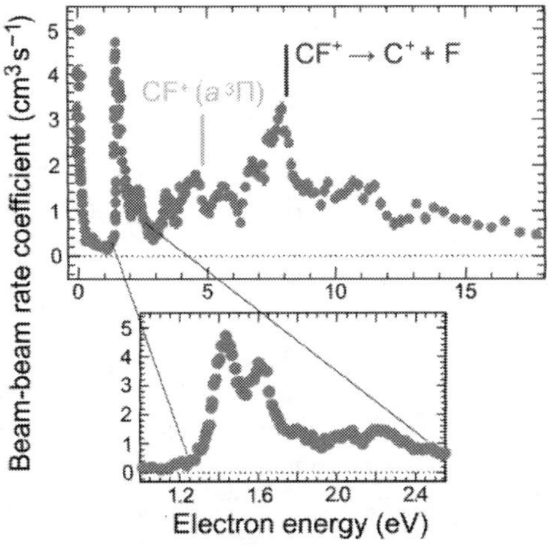

Figure 5. DR rate coefficient for CF^+ measured at TSR [52] with structure from excited-core Rydberg resonances. The vertical excitation energy from the CF^+ ground state to $CF^+(a\,^3\Pi)$ and the ionic dissociation threshold to C^+ + F are marked. A strong doublet peak, shown at a higher resolution in the lower plot, occurs near 1.6 eV. Fragment imaging reveals that for the lower component of the doublet the product channel $C(^3P)$ + F is by far the dominant one, while for the upper component the dominant product channel is $C(^1D)$ + F [52].

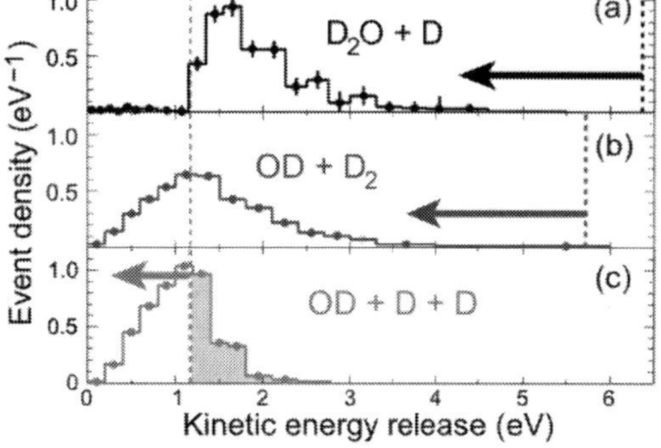

Figure 6. Distribution of the KER in three fragment channels (a)–(c) for low energy DR of D_3O^+ measured at TSR [50], using the EMU detector for channel-specific fragment imaging. Dashed lines: reaction energies from parent ground-state to product ground-states; arrows: product excitation; shaded area: parent excitation.

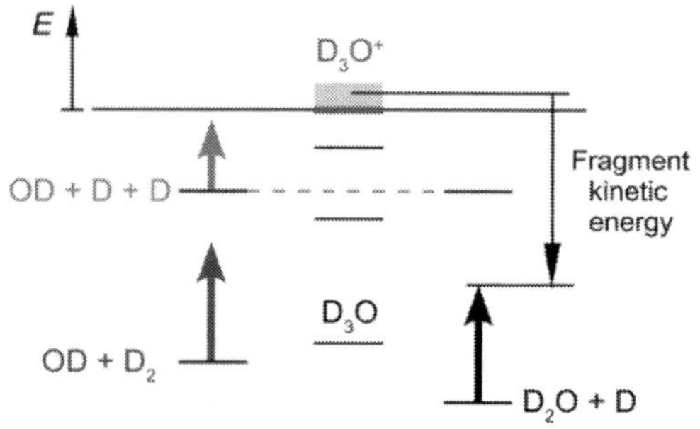

Figure 7. Energy level scheme (not to scale) for the product channels in low energy DR of D_3O^+, illustrating the observed KER in relation to product excitation (arrows) and parent excitation (shaded area). The reaction energy to OD + D + D also marks the onset of dissociation for the D_2O and D_2 products, respectively.

distribution of impact energies of ~10 K. Similar to previous ion storage ring experiments on this species [56, 57] as well as on other polyatomic ions, a discharge ion source is used to produce the parent ion beam, which is then kept circulating in the ion storage ring for several seconds before the start of the merged-beams measurement, giving excited ions the opportunity to relax by radiative emission. Since substantial lifetimes of many seconds or more can occur for rotational decay, the population of excited levels in this degree of freedom cannot generally be excluded after the initial storage period. Thus, it is desirable that DR measurements with such parent ions are accompanied by sensitive tests for their internal excitation, similar to those applied for H_3^+ and its isotopomers [38, 44].

In the low-energy DR of D_3O^+, three fragmentation channels occur with sets of branching ratios, slightly differing between the experiments, obtained already in earlier measurements [56, 57]. Thus, about 70% of the reactions lead into the three-body channel OD + D + D, while the remaining reactions produce the channels D_2O + D or OD + D_2 at, roughly, the same probabilities. While "monochromatic" fragment imaging with an MCP detector cannot distinguish between both two-body channels and, in addition, records a significant fraction of the three-body fragmentations as two-body breakup events, imaging with the EMU detector uniquely identifies all three product channels. Moreover, it is operated at a high readout speed and can determine the dependence of the product branching ratios on the electron impact energies over a wide range [47]. Furthermore, it also can unambiguously identify the background from ion collisions with residual gas molecules.

The fragment position distributions in the detector plane, as measured for the DR with cold electrons, can be analyzed [50] to determine the distributions of the kinetic energy release (KER) shared by the fragments, shown for the three breakup channels in Fig. 6. As illustrated in Fig. 7, considering the initial and final energies of the reaction, the observed fragment kinetic energies result from several contributions. They basically result from the reaction energies starting in the ground state of the parent ion and proceeding to the ground-state products in the three relevant product channels, as shown. However, any internal excitation energies of the molecular products are subtracted from these values, leading to a lower KER, while an internal excitation energy of the parent ion is added and increases the KER. Thus, the conclusions from the measured KER distributions are different for the three channels. For D_2O + D, Fig. 6(a), the difference between the maximum KER of 6.3 eV and the measured one reveals very high internal excitation of the D_2O molecular products; the reaction energy towards OD + D + D also appears in the KER distribution for this channel as it marks the internal energy above which D_2O molecules can dissociate into OD + D. For OD + D_2, Fig. 6(b), the interpretation is similar [50], but the internal energies of the two molecular products add up. Finally, the KER distribution for OD + D + D, Fig. 6(c), reflects the internal excitation of the OD fragment as it extends to lower energies. On the other hand, however, it also extends to significantly higher values than given by the reaction energy; this tail is interpreted as the distribution of internal excitation energies of the D_3O^+ parent ions, corresponding to an internal temperature as high as ~3000 K. The observed tails can be shown unambiguously not to result from background reactions of the D_3O^+ ions in the residual gas, contradicting the result of previous similar studies [58] which concluded that the stored D_3O^+ ions were internally cold.

Analysis of the transverse fragment position distributions for three-body breakup, as discussed above, has in recent years been used to probe the parent ion excitation in DR experiments for a number polyatomic species. Three-body breakup is considered since, in the smaller systems studied, all or most of the fragments are atomic and therefore (if any) have only few, well defined excited states. Thus, the transverse kinetic energy distribution extends up to the limit given by the reaction energy between the ground-state species and can be sensitively probed for any tails. For H_3^+, this fragment imaging method clearly reveals the high parent-ion temperatures around 3000 K obtained with standard discharge ion sources [38] and the effect of buffer-gas cooling obtained by using a cold ion source for H_3^+, down to the level of ~300 K [59, 22]. The asymmetric isotopologues of this ion are found cold (~300 K) even with the standard sources by radiative relaxation during a few seconds of storage [44]. Similar completion of radiative relaxation has been confirmed for other dihydrides, such as CH_2^+ [60, 45] and, very recently, D_2F^+ [61]. In contrast, high internal temperatures were observed in stored ion beams from discharge ion sources over up to 30 s of storage time for D_3O^+ [50], D_2HO^+ [62] and $DCND^+$ [63]. Hence, although all these species have a permanent dipole moment in their ground state and their vibrational modes are infrared active, a number of significantly populated, slowly decaying rotationally excited levels is revealed for them by the observations. Extended calculations for D_3O^+ and D_2HO^+ [64] suggest the co-existence of short and much longer lived rotational levels for both these systems; long radiative lifetimes (due to the large moment of inertia) are also derived [63] for the linear $DCND^+$ ion. By the method of Coulomb explosion imaging, incomplete internal relaxation was moreover observed for DCO^+ [65], in this case finding the bending vibration to remain excited at roughly 2000 K during ~12 s. This survey of observations demonstrates the need to consider parent excitation in DR measurements with great care for many species. Even in larger polyatomic ions, long-lived ro-vibrationally excited levels appear to become relevant, although they may co-exist with much faster decay of infrared active excitations.

4.2.2. Product channels traced along increasing collision energy Through the parallel readout of many detection channels, the new EMU detector also allows product channel branching ratios to be efficiently traced along the electron impact energy E. Thus, as the energy available for the molecular breakup increases, a much more complete picture of the DR process and of the excited potential curves of the molecular system is accessed. On one hand, additional final levels or chemical compositions of the reaction products are reached energetically. On the other hand, also new electron capture resonances become effective, and the nuclear velocities during the breakup change, together with the transition probabilities at crossings between diabatic potential surfaces and the phases in possible multi-path interferences; moreover, additional crossings leading to

Journal of Physics: Conference Series **300** (2011) 012008 doi:10.1088/1742-6596/300/1/012008

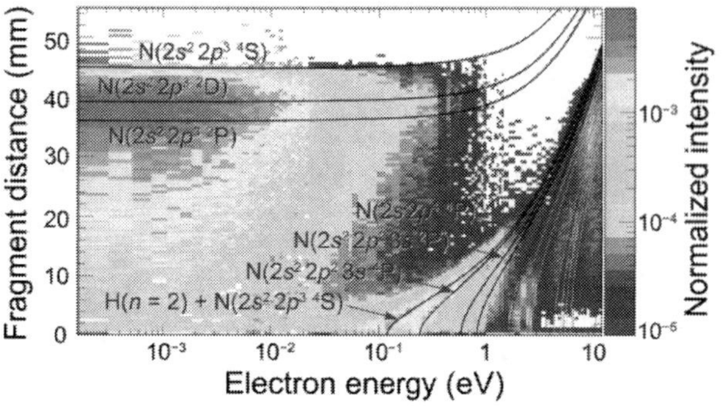

Figure 8. Scatter plot of fragment distance distributions and electron impact energy E_d for the DR of NH^+ measured with EMU at TSR (preliminary [66]). The signal intensity is normalized to the stored ion current. Curves show the peaks of calculated distance distributions for the indicated product states from ground-state NH^+.

alternative breakup pathways can be accessed (see Fig. 1). While earlier experimental work on the quasi-continuous energy dependence of fragment branching ratios are scarce (H_3^+ was studied [48]), measurements with the new fast detector system can be performed at much higher energy resolution and investigate the systematics for a much wider range of species, including D_2H^+ [46] and D_3O^+ [47]. The ion DCO^+ has become the first test system for experimental and theoretical studies on the resonant potential surfaces leading to DR at higher impact energy [32].

Also fragment *imaging measurements* under continuous variation of the impact energy E have become possible with the fast EMU detector. The data samples are sufficiently large and detailed that, for two-body breakup, two-dimensional representations of the fragment distance distributions can be obtained with the collision energy as a second variable. Preliminary results are given in Fig. 8 for the DR of NH^+ ions. Although the fragment distance in this measurement is projected onto the detector plane, events for which the dissociation is approximately transverse to the beam direction clearly indicate the size of the KER. In this diatomic case, the KER represents the various atomic excited levels of the outgoing N and H fragments as indicated. Energetic openings of new final states, their relative populations, and also the occurrence of new electron capture resonances are reflected by the structures in these two-dimensional data, which will reveal an extraordinarily complete picture of the DR process for this ion in the ongoing analysis.

5. Outlook: upcoming facilities

Fast beam fragmentation experiments on DR at ion storage rings have seen a strong development and recently further advances in the development of electron sources and fragment detection techniques. They shed new light on the rich diversity of phenomena that exist in the DR of polyatomic ions and of multi-electron molecular systems. While studies of DR cross sections and product branching ratios for cold electron collisions meanwhile cover a large number of molecular ions, assessing and quantifying the relevance of the reaction in ion chemical environments for the specific species, these data are far from being complete. For heavier polyatomic ions, the branching ratio results become less comprehensive [67] and the masses too high for efficient use of the storage ring technique. Moreover, the internal states of the products and the collision energy dependences of the process, tracing new electron capture routes and changes in the final channels of the reaction, were much less investigated, even for the lighter systems. Furthermore, the initial states of the parent ions are far from being controlled precisely in these studies; the excitation temperatures mostly correspond at least to room-temperature through the blackbody radiation field in the laboratory devices, and often to much higher values. This holds for light ions as well as for heavier polyatomics, where the uncertainties about the internal states in the experiment increase as the configuration space grows.

New storage ring facilities are becoming available for future studies at ion storage rings along these considerations. The large magnetic storage and cooler rings at high-energy heavy-ion and radioactive beam facilities, such as the GSI in Darmstadt, Germany, and the Heavy Ion Research Facility in Lanzhou (HIRFL), China, offer much higher magnetic deflection power than the ion storage rings so far used in DR experiments. Hence, with suitable molecular ion injectors, merged electron-ion beam experiments at similar beam velocities can be performed with significantly higher molecular masses. The molecular physics program [68] under planning at the storage ring CSRe at HIRFL [69] can be expected to extend the accessible mass range by a factor 6, corresponding to the gain in magnetic deflection power for this machine.

A different approach is taken by the project of the Crogenic Storage Ring (CSR) at the Max-Planck Institute for Nuclear Physics in Heidelberg, Germany [70, 71, 72]. Using compact electrostatic deflection and focusing devices, the storage ring is operated in a low-temperature surrounding with wall temperatures of ∼10 K and cryopumping at 2 K. The ion beam kinetic energy is reduced from typically 2 MeV at the TSR to 300 keV, where electrostatic acceleration from well accessible ion source platforms with a wide range of cold molecular ion sources is possible. The CSR includes a merged-beam electron target for phase-space cooling and electron collision studies, further developing the technology of the photocathode electron target of the TSR at envisaged electron beam energies down to 1 eV. Internally relaxed molecular ions from the cold sources can be kept cold during their storage, avoiding their heating by black-body radiation via infrared-active modes. With the low-energy electron beam, DR studies on stored and cryogenically cooled molecular ions are expected to become possible for molecular masses up to ∼160 amu with this new facility.

Acknowledgments

This overview is based on long-standing fruitful collaboration on fast-beam molecular fragmentation experiments at the TSR with O Heber, B Jordon-Thaden, C Krantz, H Kreckel, M B Mendes, C Nordhorn, S Novotny, D A Orlov, H B Pedersen, A Petrignani, M L Rappaport, D Schwalm, A Shornikov, J Stützel, X Urbain and D Zajfman, as well as recent common experimental work with W D Geppert, M Hamberg and D W Savin. The valuable contributions of M Grieser, R Repnow and the accelerator staff at the Max-Planck Institute for Nuclear Physics by running TSR and the ion sources are gratefully acknowledged. Great thanks are due to C H Green, P Jensen, C Jungen, Å Larson, A E Orel, I D Petsalakis, I F Schneider, G Theodorakopoulos and S Yurchenko for discussions and support with theory. HB acknowledges partial support by the German Israeli Foundation for Scientific Research and Development (G.I.F.) under Grant I-900-231.7/2005 and by the European Project ITS LEIF (HRPI-CT-2005-026015). ON was supported in part by the U.S. NSF Astronomy and Astrophysics. The continuous support of the activities by K Blaum and the Stored and Cooled Ion Division of the Max-Planck Institute for Nuclear Physics and by the Max-Planck Society is gratefully acknowledged.

References

[1] Larsson M and Orel A E 2008 *Dissociative Recombination of Molecular Ions* (Cambridge University Press, Cambridge)
[2] Herbst E 2001 *Chem. Soc. Reviews* **30** 168–176
[3] Goodings J, Bohme D and Ng C W 1979 *Combustion and Flame* **36** 27–43
[4] Larsson M 1997 *Annu. Rev. Phys. Chem.* **48** 151
[5] Phaneuf R A, Havener C C, Dunn G H and Müller A 1999 *Rep. Prog. Phys.* **62** 1143–80
[6] Wolf A, Buhr H, Grieser M, von Hahn R, Lestinsky M, Lindroth E, Orlov D A, Schippers S and Schneider I F 2006 *Hyperfine Interact.* **172** 111–24
[7] Al-Khalili A, Rosén S, Danared H, Derkatch A, Källberg A, Larsson M, Le Padellec A, Neau A, Semaniak J, Thomas R, af Ugglas M, Vikor L, Zong W, van der Zande W, Urbain X, Jensen M, C Bilodeau R C,

Heber O, Pedersen H B, Safvan C, Andersen L H, Lange M, Levin J, Gwinner G, Knoll L, Scheffel M, Schwalm D, Wester R, Zajfman D and Wolf 2003 *Phys. Rev. A* **68** 042702

[8] Novotny S, Rubinstein H, Buhr H, Novotný O, Hoffmann J, Mendes M B, Orlov D A, Krantz C, Berg M H, Froese M, Jaroshevich A S, Jordon-Thaden B, Lange M, Lestinsky M, Petrignani A, Shafir D, Zajfman D, Schwalm D and Wolf A 2008 *Phys. Rev. Lett.* **100** 193201

[9] Kreckel H, Novotný O, Crabtree K N, Buhr H, Petrignani A, Tom B A, Thomas R D, Berg M H, Bing D, Grieser M, Krantz C, Lestinsky M, Mendes M B, Nordhorn C, Repnow R, Stützel J, Wolf A and McCall B J 2010 *Phys. Rev. A* **82** 042715 and references therein

[10] Giusti-Suzor A, Bardsley J N and Derkits C 1983 *Phys. Rev. A* **28** 682–91

[11] Schneider I F, Dulieu O and Suzor-Weiner A 1991 *J. Phys. B: At. Mol. Opt. Phys.* **24** L289–97

[12] Kokoouline V, Greene C H and Esry B D 2001 *Nature* **412** 891–94

[13] Čurík R and Greene C H 2007 *Phys. Rev. Lett.* **98** 173201

[14] Guberman S L 1983 *J. Chem. Phys.* **78** 1404–13

[15] Smith F T 1969 *Phys. Rev.* **179** 111–123

[16] Müller A 2008 *Adv. At. Mol. Opt. Phys.* **55** 293–417

[17] Thorson W R, Delos J B and Boorstein S A 1971 *Phys. Rev. A* **4** 1052–66

[18] Bardsley J N 1968 *J. Phys. B: At. Mol. Opt. Phys.* **1** 365–80

[19] Giusti A 1980 *J. Phys. B: At. Mol. Opt. Phys.* **13** 3867–94

[20] Motapon O, Backodissa D, Waffeu Tamo F O, Tudorache D, Chakrabarti K, Mezei J Z, Lique F, Bultel A, Tchang-Brillet L, Dulieu O, Tennyson J, Wolf A, Urbain X and Schneider I F these proceedings

[21] Guberman S L 1994 *Phys. Rev. A* **49** R4277–80

[22] Petrignani A, Altevogt S, Berg M H, Bing D, Buhr H, Greene C H, Grieser M, Hoffmann J, Jordon-Thaden B, Kokoouline V, Krantz C, Kreckel H, Mendes M B, Novotny O, Novotny S, Orlov D A, Repnow R, Sorg T, Stützel J and Wolf A submitted (arXiv:1012.1441)

[23] Forck P, Broude C, Grieser M, Habs D, Kenntner J, Liebmann J, Repnow R, Schwalm D, Wolf A, Amitay Z and Zajfman D 1994 *Phys. Rev. Lett.* **72** 2002–5

[24] Amitay Z, Zajfman D, Forck P, Heupel T, Grieser M, Habs D, Repnow R, Schwalm D, Wolf A and Guberman S L 1996 *Phys. Rev. A* **53** R644–7

[25] Amitay Z, Zajfman D, Forck P, Hechtfischer U, Seidel B, Grieser M, Habs D, Repnow R, Schwalm D and Wolf A 1996 *Phys. Rev. A* **54** 4032–50

[26] Carata L, Orel A E, Raoult M, Schneider I F and Suzor-Weiner A 2000 *Phys. Rev. A* **62** 052711

[27] Guberman S L 2007 *J. Phys. Chem. A* **111** 11254–60

[28] Larson Å, Djurić N, Zong W, Greene C H, Orel A E, Al-Khalili A, Derkatch A M, Le Padellec A, Neau A, Rosén S, Shi W, Vikor L, Danared H, af Ugglas M, Larsson M and Dunn G H 2000 *Phys. Rev. A* **62** 042707

[29] Kalhori S, Thomas R, Al-Khalili A, Ehlerding A, Hellberg F, Neau A, Larsson M, Larson A, Huneycutt A J, McCall B J, Djurić N, Dunn G H, Semaniak J, Novotny O, Paál A, Österdahl F and Orel A E 2004 *Phys. Rev. A* **69** 022713

[30] Hoffmann J 2009 Ph.D. thesis Univ. of Heidelberg URL `http://www.ub.uni-heidelberg.de/archiv/9026`

[31] Larson Å, Roos J B and Orel A E 2009 *Journal of Physics: Conference Series* **192** 012019

[32] Nordhorn C, Bing D, Buhr H, Grieser M, Heber O, Krantz C, Mendes M B, Repnow R, Schwalm D, Shornikov A, Stützel J, Wolf A, Novotný O, Lestinsky M, Larson Å, Stenrup M and Orel A E these proceedings

[33] Galster U, Baumgartner F, Müller U, Helm H and Jungen M 2005 *Phys. Rev. A* **72** 062506

[34] Ketvirtis A E and Simons J 1999 *J. Chem. Phys.* **103** 6552–63

[35] Zajfman D, Schwalm D and Wolf A 2003 *Many-Particle Quantum Dynamics in Atomic and Molecular Fragmentation* ed Ullrich J and Shevelko V P (Berlin: Springer) chap 3, pp 59–70

[36] Larsson M and Thomas R 2001 *Phys. Chem. Chem. Phys.* **3** 4471–80

[37] Strasser D, Lammich L, Krohn S, Lange M, Kreckel H, Levin J, Schwalm D, Vager Z, Wester R, Wolf A and Zajfman D 2001 *Phys. Rev. Lett.* **86** 779–82

[38] Strasser D, Lammich L, Kreckel H, Krohn S, Lange M, Naaman A, Schwalm D, Wolf A and Zajfman D 2002 *Phys. Rev. A* **66** 032719

[39] Fechner P and Helm H 2010 *Phys. Rev. A* **82** 052523

[40] Orlov D A, Krantz C, Wolf A, Jaroshevich A S, Kosolobov S N, Scheibler H E and Terekhov A S 2009 *J. Appl. Phys.* **106** 054907

[41] Wolf A, Orlov D A, Krantz C, Lestinsky M, Shornikov A, Novotný O, Stützel J, Buhr H, Mendes M, Petrignani A, Grieser M, Schippers S, Müller A, Ferro F and Lindroth E 2009 *J. Phys.: Conf. Ser.* **194** 012024

[42] Wolf A, Krohn S, Kreckel H, Lammich L, Lange M, Strasser D, Grieser M, Schwalm D and Zajfman D 2004 *Nucl. Instrum. Methods Phys. Res. A* **532** 69–78

[43] Beckert M and Müller U 2000 *Eur. Phys. J. D* **12** 303–310
[44] Strasser D, Lammich L, Kreckel H, Lange M, Krohn S, Schwalm D, Wolf A and Zajfman D 2004 *Phys. Rev. A* **69** 064702
[45] Nevo I, Novotny S, Buhr H, Andrianarijaona V, Altevogt S, Heber O, Hoffmann J, Kreckel H, Lammich L, Lestinsky M, Pedersen H B, Schwalm D, Wolf A and Zajfman D 2007 *Phys. Rev. A* **76** 022713
[46] Buhr H, Mendes M B, Novotný O, Schwalm D, Berg M H, Bing D, Heber O, Krantz C, Orlov D A, Rappaport M L, Sorg T, Stützel J, Varju J, Wolf A and Zajfman D 2010 *Phys. Rev. A* **81** 062702
[47] Novotný O, Buhr H, Stützel J, Mendes M B, Berg M H, Bing D, Froese M, Grieser M, Heber O, Jordon-Thaden B, Krantz C, Lange M, Lestinsky M, Novotny S, Menk S, Orlov D A, Petrignani A, Rappaport M L, Shornikov A, Schwalm D, Zajfman D and Wolf A 2010 *J. Phys. Chem. A* **114** 4870–74
[48] Datz S, Sundström G, Biedermann C, Broström L, Danared H, Mannervik S, Mowat J R and Larsson M 1995 *Phys. Rev. Lett.* **74** 896–899
[49] Geppert W D, Hamberg M, Thomas R D, Österdahl F, Hellberg F, Zhaunerchyk V, Ehlerding A, Millar T J, Roberts H, Semaniak J, af Ugglas M, Källberg A, Simonsson A, Kaminska M and Larsson M 2006 *Faraday Discuss.* **133** 177
[50] Buhr H, Stützel J, Mendes M B, Novotný O, Schwalm D, Berg M H, Bing D, Grieser M, Heber O, Krantz C, Menk S, Novotny S, Orlov D A, Petrignani A, Rappaport M L, Repnow R, Zajfman D and Wolf A 2010 *Phys. Rev. Lett.* **105** 103202
[51] Chabot M, Negra S D, Lavergne L, Martinet G, Wohrer-Béroff K, Sellem R, Daniel R, Bris J L, Lalu G, Gardès D, Scarpaci J A, Désesquelle P and Lima V 2002 *Nucl. Instrum. Methods Phys. Res. B* **197** 155–64
[52] Novotný O, Motapon O, Berg M H, Bing D, Buhr H, Fadil H, Grieser M, Hoffmann J, Jaroshevich A S, Jordon-Thaden B, Krantz C, Lange M, Lestinsky M, Mendes M, Novotny S, Orlov D A, Petrignani A, Schneider I F, Orel A E and Wolf A 2009 *J. Phys.: Conf. Ser.* **192** 012021
[53] Prager J, Riedel U and Warnatz J 2007 *Proc. Combustion Inst.* **31** 1129–37
[54] Peterson J R, Padellec A L, Danared H, Dunn G H, Larsson M, Larson A, Peverall R, Strömholm C, Rosén S, af Ugglas M and van der Zande W J 1998 *J. Chem. Phys.* **108** 1978–88
[55] Padellec A L, Mitchell J B A, Al-Khalili A, Danared H, Källberg A, Larson Å, Rosén S, af Ugglas M, Vikor L and Larsson M 1999 *J. Chem. Phys.* **110** 890–901
[56] Neau A, Al Khalili A, Rosén S, Le Padellec A, Derkatch A M, Shi W, Vikor L, Larsson M, Semaniak J, Thomas R, Någård M B, Andersson K, Danared H and af Ugglas M 2000 *J. Chem. Phys.* **113** 1762–70
[57] Jensen M J, Bilodeau R C, Safvan C P, Seiersen K, Andersen L H, Pedersen H B, and Heber O 2000 *Astrophys. J.* **543** 764–74
[58] Zhaunerchyk V, Geppert W D, Rosen S, Vigren E, Hamberg M, Kaminska M, Kashperka I, af Ugglas M, Semaniak J, Larsson M and Thomas R D 2009 *J. Chem. Phys.* **130** 214302–5
[59] Petrignani A, Kreckel H, Berg M H, Altevogt S, Bing D, Buhr H, Froese M, Grieser M, Hoffmann J, Jordon-Thaden B, Krantz C, Mendes M, Novotný O, Novotny S, Orlov D A, Reinhardt S and Wolf A 2009 *J. Phys.: Conf. Ser.* **192** 012022
[60] Thomas R D, Hellberg F, Neau A, Rosén S, Larsson M, Vane C R, Bannister M E, Datz S, Petrignani A and van der Zande W J 2005 *Phys. Rev. A* **71** 032711
[61] Novotný O, Hamberg M, Geppert W D *et al.* in preparation
[62] Buhr H *et al.* in preparation
[63] Mendes M 2010 Ph.D. thesis Univ. of Heidelberg URL `http://www.ub.uni-heidelberg.de/archiv/10793`
[64] Jensen P and Yurchenko S N private communication
[65] Wester R, Hechtfischer U, Knoll L, Lange M, Levin J, Scheffel M, Schwalm D, , Wolf A, Baer A, Vager Z, Zajfman D, Mladenović M and Schmatz S 2002 *J. Chem. Phys.* **116** 7000–11
[66] Novotný O *et al.* in preparation
[67] Plessis S, Carrasco N and Pernot P 2010 *J. Chem. Phys.* **133** 134110
[68] Cai X *et al.* these proceedings
[69] Xia J W, Zhan W L, Wei B W, Yuan Y J, Song M T, Zhang W Z, Yang X D, Yuan P, Gao D Q, Zhao H W, Yang X T, Xiao G Q, Man K T, Dang J R, Cai X H, Wang Y F, Tang J Y, Qiao W M, Rao Y N, He Y, Mao L Z and Zhou Z Z 2002 *Nucl. Instrum. Methods Phys. Res. A* **488** 11–25
[70] Zajfman D, Wolf A, Schwalm D, Orlov D A, Grieser M, von Hahn R, Welsch C, Crespo Lopez-Urrutia J R, Schröoter C D, Urbain X and Ullrich J 2005 *J. Phys.: Conf. Ser.* **4** 296–9
[71] Lange M, Froese M, Menk S, Varju J, Bastert R, Blaum K, López-Urrutia J R C, Fellenberger F, Grieser M, von Hahn R, Heber O, Kühnel K U, Laux F, Orlov D A, Rappaport M L, Repnow R, Schröter C D, Schwalm D, Shornikov A, Sieber T, Toker Y, Ullrich J, Wolf A and Zajfman D 2010 *Rev. Sci. Instrum.* **81** 055105
[72] Krantz C, Berg F, Blaum K, Fellenberger F, Froese M, Grieser M, von Hahn R, Lange M, Laux F, Menk S, Repnow R, Shornikov A and Wolf A these proceedings

Journal of Physics: Conference Series **300** (2011) 012009 doi:10.1088/1742-6596/300/1/012009

The Molecular Ion Research Facility in Lanzhou (MIRFL)

X. Cai, F. Ruan, J. Yang, L. Mao, R. Lu, C. Shao, M. Song, D. Yu

Institute of Modern Physics, Chinese Academy of Sciences, Lanzhou 730000, China

Abstract:

Exploiting the advantage of the high magnetic rigidity (9.4Tm) of the HIRFL-CSRe, the recently launched Molecular Ion Research Facility in Lanzhou (MIRFL) project at the Institute of Modern Physics in Lanzhou can open a new window for dissociative recombination research by extending the mass range of molecular ions up to 150 amu.

1. Introduction

Dissociative recombination (DR) of polyatomic molecular ions plays an important role in the production of molecules and free radicals (molecules with unpaired electrons) in many types of plasma, natural or man-made. In interstellar molecular clouds, networks of bimolecular ion-molecule reactions are terminated by dissociative recombination. In addition to molecular clouds, dissociative recombination plays an important role in fusion edge plasmas, planetary ionospheres, plasma-enhanced combustion, plasma etching of microelectronic structures, and in mass spectrometry of biological molecules [1]. Due to the fundamental importance of the DR reaction in astrophysics, atmosphere physics, plasma physics, chemical physics, and biophysics, and many other areas of research, numerous works can be found in the literature [1-13].

The reason cooler-storage rings have been proven so powerful in the study of dissociative recombination derives from the fact that the ions are stored at a high energy (~MeV), while at the same time the electron-molecular ion collisions can be studied at meV energies owing to the merging of the electron and ion beams. The interaction region is about one meter in length, and neutral reaction products from this region come out with an energy of MeV. Magnetic separation is used to extract the neutral products from the stored ion beam, and the high beam energy makes it possible to separate neutral products based on their energy. Since the neutral products move with essentially the same velocity, the energy is a direct measure of their mass, and hence the neutral-particle detector works, in principle, as a mass spectrometer. However, the time resolution in the detector is not sufficient to differentiate between particles that arrive within a few nanoseconds, and instead a grid technique is used to obtain the branching ratios. A grid with known transmission is placed in front of the detector and blocks some particles from being detected. This leads to a redistribution of the pulse-height spectrum, which can be related to the branching ratios by means of a transmission matrix. In order to separate peaks in the pulse-height spectrum, high beam energy is required [14].

There are presently three cooler-storage rings in the world which are used for molecular ion physics; ASTRID in Aarhus, Denmark, CRYRING, Stockholm, Sweden, and the Test Storage Ring, Heidelberg, Germany. These three storage rings are characterized by a magnetic rigidity of 1.4-1.5 Tm and they all have electron coolers which serve the dual purpose of phase-space cooling the stored ion beam and acting as targets for electron-molecular ion collisions. Very active

Journal of Physics: Conference Series **300** (2011) 012009 doi:10.1088/1742-6596/300/1/012009

molecular ion physics programmes have been conducted at these cooler-storage rings during the last decade, and they have been the core facilities for a European Commission research training network during the last four years. One of the most important results reported from these studies is the DR reaction rate of the H_3^+ molecular ion [2, 15-19]; since these data are used to determine the cosmic-ray ionization rate, ζ, of H_2 in diffuse molecular clouds [20]:

$$\zeta = k_e\, n(e) n(H_3^+)/n(H_2),$$

where k_e is the DR rate for the H_3^+ molecular ion, $n(e)$ is the electron density, and $n(H_3^+)$ and $n(H_2)$ are the number density of H_3^+ and H_2, respectively.

However, the low magnetic rigidity of the three European rings is now becoming a bottleneck which cannot be overcome. For molecular ions exceeding a mass of 100 amu, the present generation of small, 1.4-1.5 Tm, storage rings run into considerable problems, and these problems are principle rather than technical. The only solution is to go to cooler-storage rings with much higher rigidity, and the only option on a world-wide scale is the HIRFL-CSRe in Lanzhou which has a magnet rigidity of 9.4Tm.

The HIRFL-CSR facility in Lanzhou is shown in Fig.1. This facility is mainly composed of two parts, the main ring CSRm and the experimental ring CSRe. The Molecular Ion Research Facility (MIRFL) at CSRe includes a molecular ion source, the 320 kV platform, an injection beam line, an ultra-cold electron target, a weak beam measurement device, and a neutral particle detection system. Taking into account the high magnetic rigidity of 9.4Tm of CSRe, and combining the new injection beam line, the new platform may offer unique opportunities to dissociative recombination (DR) research by extending the mass range of molecular ions up to 150 amu and, especially for ions with m>70 amu. Since the present HIRFL-CSR facility has only a single injector SFC, adding a new injector MIRFL may increase significantly the running efficiency of the HIRFL-CSR system. The designs of the injection beam line, the bumper for injection into CSRe, and the ultra-cold electron target have been finished. The technical design of the MIRFL will be presented in detail in this paper.

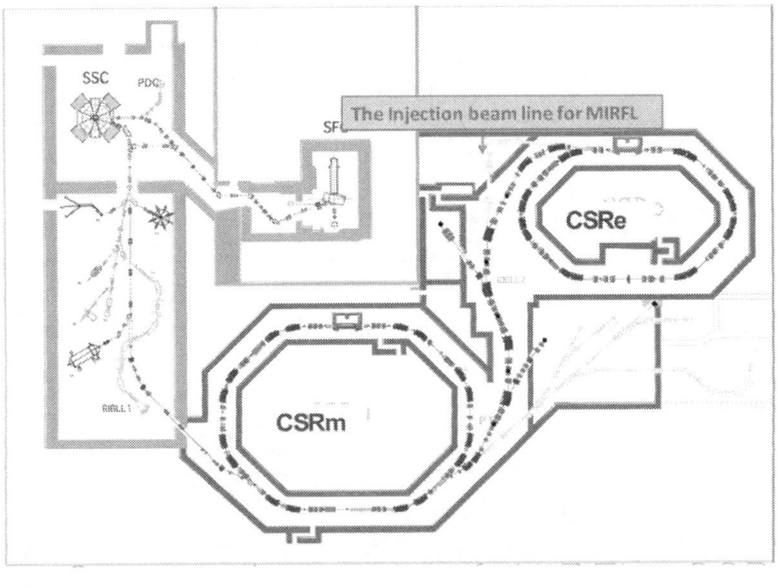

Fig. 1 The layout of HIRFL-CSR

2. General design of the MIRFL

The layout of the MIRFL is shown in Fig.2. It consists of a molecular ions injection beam line, an electron target, a neutral particle detector system, and the beam diagnostic system. The molecular ions of 40~150 amu at 320keV can be injected into CSRe by using a special designed bumper together with the existing injection coils in the ring, and then will finally be accelerated to 10^1~10^2MeV. The electron target with a photoncathode can produce an ultra-cold electron beam to serve as the electron target for DR study. The neutral particle created in the DR reaction between the molecular ions and the electron target will be detected by a neutral particle detector system.

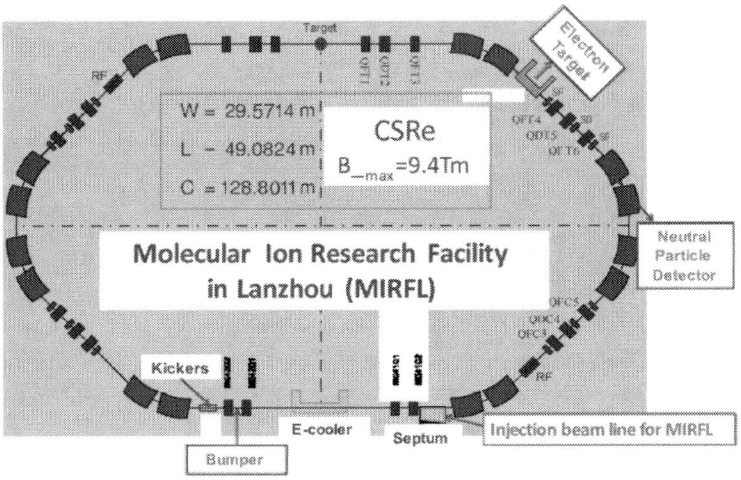

Fig. 2 The layout of MIRFL

2.1 The Molecular ions injection beam line

The molecular ion injection beam line includes the molecular ion source, the 320kV platform, the dipole and the quadrupole magnets, the vacuum system and the beam diagnostic devices. The ion-optical transmission of the injection beam line is shown in Fig. 3.

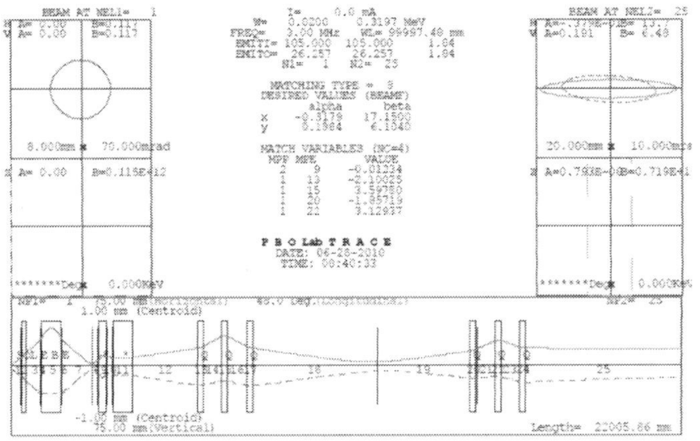

Fig. 3 The optical transmission of the injection beam line

2.2 The injection into CSRe

The existing injection system of HIRFL-CSRe consists of one magnetic septum, four dipole coils, and three quick pulse kickers (see Fig. 2). During the single-turn injection, the auxiliary coils in the four main dipoles are used to create a bump in the stored-ions orbit to allow injection. The injected beam is then deflected into the central orbit by the downstream kicker (see Fig. 2). In the newly designed MIRFL, with the pre-acceleration of the 320kV high voltage platform, the circulation time of the stored molecular ions in CSRe is expected to be about 100~250μs. However, in the currently available injection system, the flat-top time of the kickers is about 0.4~0.8μs, much shorter than the circulation time of the stored molecular ions. Thus, a new injection system including a slow bumper should be used.

The new MIRFL injection system includes a new bumper together with the existing injection coils in the CSRe. The bumper will be installed at the position indicated by BD42Pxy1 shown in Fig. 4, and is used to change the orbit of the stored molecular ion beam so that to match with the downstream displaced orbit.

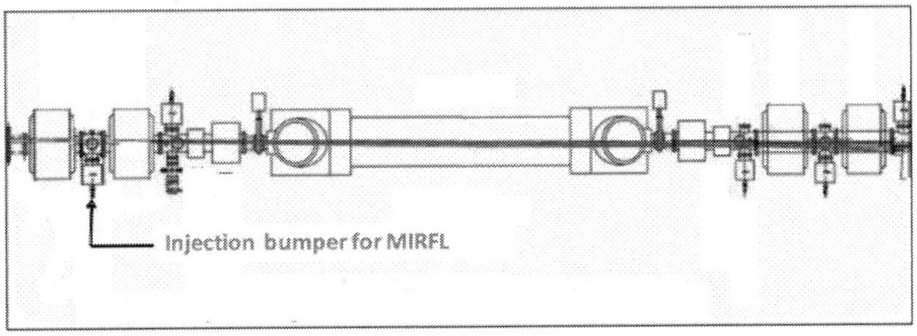

Fig. 4 The position of MIRFL's injection bumper

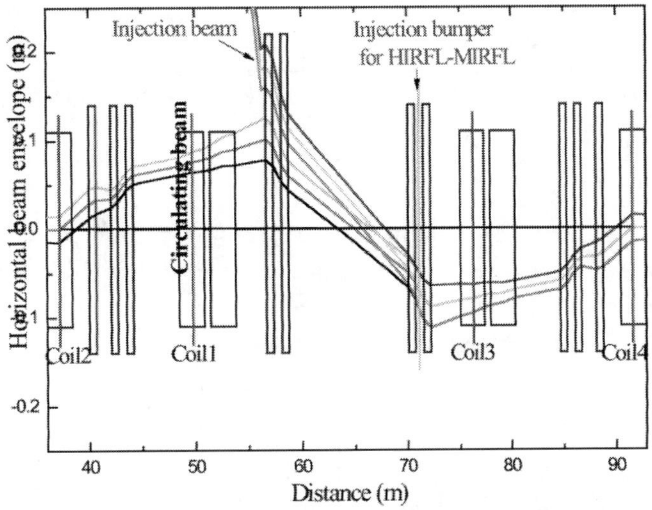

Fig. 5 Injected beam orbits

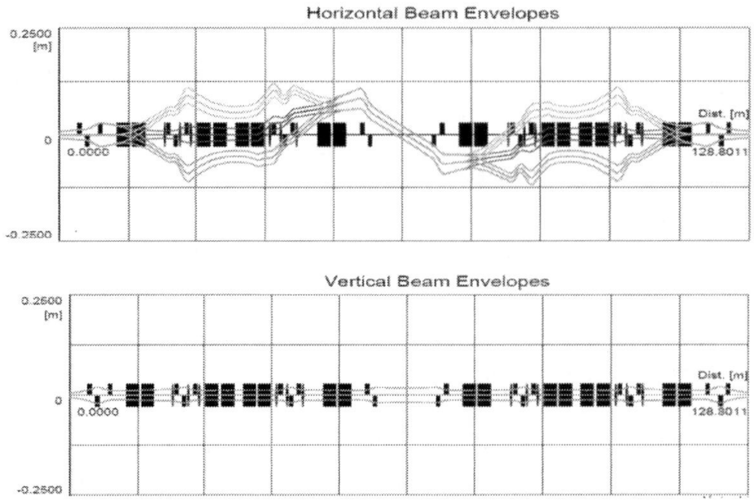

Fig. 6 The horizontal and vertical beam envelopes for stored molecule ions

Table 1 Parameters of injection components during injection

Septum	Coil1	Coil2	Coil3	Coil4	New bumper
145mrad	0.1mrad	8.0mrad	1.6mrad	8.4mrad	3.5mrad

Table 2 Parameters of the injection bumper

Number of cells	1
Strength	3~5mrad
Maximum magnetic density	250G
Effective cell length	350mm
Overall cell length	500mm
Aperture	(150+150)mm×70mm
Ramping time	<100μs
Flat-top time	100~250μs

With this design, a simulation of the beam injection was performed. Fig. 5 plots the orbit of the injected ion beam. Fig. 6 demonstrates the horizontal and vertical beam envelopes of the molecule ions injected into the CSRe. The parameters of the various injection components and the new bumper during the injection are listed in Table 1 and Table 2.

Combining the existing magnetic septum and the four auxiliary coils with the new bumper, the molecular ions injection can be realized without the influence of the present injection system of the CSRe.

2.3 Ultra-cold electron target

An ultra-cold electron target will be equipped in CSRe, and will be located at the first quadrant (See Fig. 2). It consists of an acceleration section (including the electron gun), an interaction section, a collector, solenoid and toroid sections, high voltage system, vacuum system and cooling system. The main parameters of electron target are listed in table 3.

To improve the resolution, an electron target with ultra-low temperature electron beam is necessary. In order to reduce the transverse temperature of electron beam, the maximum magnetic field generated by superconductor solenoid at the cathode position is up to 3.5T, and the guiding magnetic intensity amounts to 0.01-0.05T. Hence, magnetic expansion ratios up to 100 are possible. For strongly magnetized electron beam, the transverse temperature is suppressed. However, the transverse temperature of electron beam could be reduced to 2meV by further development of the photocathode electron source.

Table 3 The main parameters of electron target

Max. energy of electron beam	10keV
Length of E-target	~ 2.0m
Superconducting solenoid (E-gun)	~ 3.5T
Magnetic intensity	<500Gs
α	<100
Photocathode diameter	2 mm

2.4 Diagnosis system for weak ion beams

An accurate current measurement of stored molecular ion beam is essential for experimental study of the DR. For high-mass molecular ion beams, the beam current from the source may range from several nA to tens of nA. The existing diagnostic system of CSRe has a sensitivity of only ~1μA. Therefore, a new diagnostic system capable of measuring the current of weak beams needs to be developed. The new beam diagnostics system should be able to measure continuous or pulsed beams in nA range.

For the bunched beam current measurement, a high sensitivity technique has recently been developed by the Manne Siegbahn Laboratory, which is a combination of a Bergoz Integrating Current Transformer (ICT) and a capacitive pick-up (PU), has a reported lower limit of 100 pA[21]. The ICT apparatus can measure an absolute value of the ion current with a sensitivity of $1nA_{rms}$, and the PU device can obtain a relative current value with sensitivity down to $100pA_{rms}$. The signals of PU are calibrated with that of ICT at relatively high ion current. This technique will be implemented in the HIRFL-MIRFL project.

For the coasting beam current measurement, a set of micro-channel-plate (MCP) will be used to collect the neutral particles from the interaction of ions and residual gas in the ring. A multi-channel-scaler (MCS) with small time interval can be applied to record the output signals of MCP. Assuming that the signal of MCP is proportional to the absolution of the ion current, the ion current is thus obtain by the MCS data multiplying with a constant scaling factor[22].

3. Project schedule

The physical design of HIRFL-MIRP was finished in spring 2010. The main components will be finished in the middle of 2012. The preliminary experiments are planned in 2013.

4. Summary

Exploiting the advantage of the high magnetic rigidity (9.4Tm) of the HIRFL-CSRe, the

HIRFL-MIRFL project can open a new window for dissociative recombination research by extending the mass range of molecular ions up to 150 amu. The design of HIRFL-MIRFL as well as the relevant technical problems has been described. A new bumper with a 100~250μs flat-top time has been designed to allow injection of 320 kV molecular ions into the CSRe ring. An ultra-cold electron target with 2meV transversal temperature has been designed for use in the MIRFL. The diagnostic system for weak ion beam will be built for CSRe in collaboration with the Stockholm group.

Acknowledgements

We are very grateful for the help and useful advice from colleagues at Stockholm University, and the support from National Natural Science Foundation of China (No. 10820101050), National Basic Research Program of China (973 Program, No. 2010 CB832901), and Lanzhou University. We also acknowledge travel support from the US National Science Foundation and NASA.

Reference

[1] W D Geppert, M Larsson 2008 *Mol. Phys.* **106** 2199
[2] M Larsson, A E Orel 2008 *Dissociative Recombination of Molecular Ions* Cambridge University Press New York
[3] J B A Mitchell, S L Guberman (Eds.) 1989 *Dissociative Recombination: Theory, Experiment and Applications* World Scientific Singapore
[4] J B A Mitchell 1990 *Phys. Rep.* **186** 215
[5] B R Rowe, J B A Mitchell, A Canosa (Eds.) 1993 *Dissociative Recombination: Theory, Experiment and Applications* Plenum Press NewYork
[6] D Zajfman, J B A Mitchell, D Schwalm, B R Rowe (Eds.) 1996 *Dissociative Recombination: Theory, Experiment and Applications: III* World Scientific Singapore
[7] M Larsson, J B A Mitchell, I F Schneider (Eds.) 2000 *Dissociative Recombination: Theory, Experiment and Applications: IV* World Scientific Singapore
[8] S L Guberman (Ed.) 2003 *Dissociative Recombination: Theory, Experiment and Applications*, Kluwer Academic/Plenum Publishers NewYork
[9] A Wolf, L Lammich, P Schmelcher (Eds.) 2005 *Sixth International Conference on Dissociative Recombination: Theory, Experiment and Applications DR2004 J. Physics: Conference Series 4*
[10] A I Florescu-Mitchell, J B A Mitchell 2006 *Phys. Rep.* **430** 277
[11] N G Adams *et al.* 2006 *Mass Spectrom. Rev.* **25** 798
[12] R D Thomas 2008 *Mass Spectrom. Rev.* **27** 485
[13] J N Bardsley, M A Biondi 1970 *Adv. At. Mol. Phys.* **6** 1
[14]A Neau *et al.* 2000 *J Chem. Phys.* **113** 1762
[15]M Larsson *et al.*1933 *Phys. Rev. Lett.* **70** 430
[16] B J McCall *et al.* 1998 *Science* **279** 1910 & 2003 *Nature* **422** 500; B J McCall 2001 Ph. D. Thesis University of Chicago
[17]D Strasser *et al.*2002 *Phys. Rev. A* **66** 032719
[18] H Kreckel *et al.*2005 *Phys. Rev. Lett.* **95** 263201
[19] Samantha Fonseca dos Santos *et al.* 2007 *J. Chem. Phys.* **127** 124309
[20] B J McCall *et al.* 2000 *Science* **287**1941
[21] A Paal *et al.* 2006 Proceedings of EPAC 2006 196

[22] M Hamberg 2008 Ph. D. Thesis Stockholm University

Journal of Physics: Conference Series **300** (2011) 012010 doi:10.1088/1742-6596/300/1/012010

The Cryogenic Storage Ring and its application to molecular ion recombination physics

C Krantz, F Berg, K Blaum, F Fellenberger, M Froese, M Grieser, R von Hahn, M Lange, F Laux, S Menk, R Repnow, A Shornikov and A Wolf

Max-Planck-Institut für Kernphysik, Saupfercheckweg 1, 69117 Heidelberg, Germany

E-mail: claude.krantz@mpi-hd.mpg.de

Abstract. The Cryogenic Storage Ring (CSR), presently under construction at the Max-Planck-Institute for Nuclear Physics, will allow the storage of large ionic molecules under optimum experimental conditions. The electrostatic beam optics and the presence of a low-energy electron cooler will allow highly-precise recombination experiments with molecular ions of 160 atomic mass units per charge state. The all-cryogenic design of the storage ring will provide unprecedented vacuum conditions and assure long storage times even for very heavy ion beams. Suppression of the black body radiation background of the beam pipe in combination with electron cooling will give access to internal ion temperatures of 10 K and allow state-selective experiments on infrared-active species, impossible in present-day room-temperature storage rings. We give an overview of the CSR project, point out the scientific opportunities arising from its unique design, and give an outlook on possible first molecular recombination experiments after commissioning of the storage ring.

1. Introduction

Since the 1990s, electron cooler heavy-ion storage rings, besides their applications in other fields of molecular, atomic and nuclear physics, have contributed some of the most important experimental data on the dissociative recombination (DR) of molecular ions with electrons. Using relatively simple systematics, the merged-beams technique provides an energy resolution in the order of 1 meV in slow electron-ion collision reactions [1]. Storage ring experiments are thus a particularly reliable source of low-temperature DR rate coefficients, as needed e.g. for the understanding of reaction kinematics in the chemistry of cold interstellar media, star forming regions, or circumstellar disks [2]. Moreover, the well-defined initial-state conditions that can be achieved by radiative and electron cooling of molecular ions in the storage ring allow precision studies of the DR process itself. Due to the high velocity of the merged electron and ion beams in the laboratory frame, DR product detectors cover an effective 4π solid angle. Setups with neutral fragment imaging and timing capabilities allow complete reconstruction of the DR breakup momentum distributions [3, 4, 5]. In studies of low-energy neutral-product DR, these systems thus fulfil a role similar to that of COLTRIMS in the case of high-velocity charged-product reactions.

The value of the experimental input that magnetic heavy ion storage rings [6], like the Test Storage Ring (TSR) in Heidelberg or CRYRING in Stockholm, have provided to the field of dissociative recombination physics [7] is undisputed. Designed in the 1980s, these machines rely on the principle of magnetic deflection and focussing of the stored ions. This storage technology was originally developed to fulfil the demands of nuclear and particle physics, i.e. handling of species with high charge-to-mass ratios. As research in the field of molecular ions progresses towards more and more complex but still only singly (or weakly) charged systems, the design of classic storage rings becomes a limitation with regard to the nature of the species that can be successfully studied. In addition, the rotational transition energies of molecules are so low that rotations can be populated by the black body radiation from the room-temperature vacuum chamber walls of existing storage rings, which complicates the experimental initial-state conditions. The Cryogenic Storage Ring (CSR) presently under construction [8, 9] at the Max-Planck-Institute for Nuclear Physics (MPIK) will address these limitations and provide the experimental platform for the next generation of precision merged-beam DR measurements.

2. Limitations of the present electron cooler storage ring TSR

We will outline the limitations imposed on DR experiments by magnetic storage rings using as an example the Test Storage Ring TSR of the MPIK in Heidelberg, Germany.

2.1. Limits of the ion beam optics

Magnetic storage rings are momentum selective. The storage condition for particles of charge-to-mass ratio q/m is fulfilled if the ion momentum p divided by the charge state q matches the magnetic rigidity ρB of the ring:

$$\frac{p}{q} = \frac{\gamma m v}{q} = \rho B \ , \tag{1}$$

where γ is the Lorentz factor. The maximum rigidity ρB_{max} is defined by the maximum induction B_{max} of the dipole magnets and the curvature radius ρ of the closed orbit in the storage ring. For the TSR, which was designed for experiments on highly charged atomic ions (HCI), $\rho B_{\text{max}} \approx 1.4\,\text{Tm}$. This value is adapted to the range of HCI beam energies that can be produced at MPIK: with typical charge-to-mass ratios of $0.33\,\text{u}^{-1}$ to $0.5\,\text{u}^{-1}$, HCI can be stored in the TSR at velocities $v \sim 0.1\ c$ which conveniently match the typical output velocity of the accelerator facilities at MPIK. Storage times for HCI in the TSR can be very long, as ion loss cross sections due to reactions with residual gas molecules become small at elevated collision velocities.

For molecular ions, the (positive) charge state is normally limited to unity, as stripping of more than one valence electron from the parent molecule is normally not possible without disrupting the chemical bonds. Even for relatively small and simple molecules, the charge-to-mass ratios of the corresponding ions become very small compared to HCI. E.g. for D_3O^+, $DCND^+$ and CF^+, three of the most heavy systems that have been studied at the TSR, q/m reduces to $0.045\,\text{u}^{-1}$, $0.033\,\text{u}^{-1}$, and $0.032\,\text{u}^{-1}$, respectively. As can be seen from eq. (1), the maximum storage velocity at given maximum magnetic rigidity behaves like

$$v_{\text{max}} \sim m^{-1} \ . \tag{2}$$

Given this decrease of the beam velocity with rising ion mass, it is not surprising that storage times of heavy molecular species are short compared to HCI, typically of a few seconds only, as the residual gas induced ion loss rate rises at lower velocities. With respect to DR experiments, there is consequently an *effective* maximum ion mass. It is reached at the point where the ion storage time τ_{ion} becomes shorter than the internal and phase-space cooling times τ_{cool}. Beyond this point, control of the initial-state conditions of the experiment is no longer assured, strongly affecting the precision of possible DR measurements.

2.2. Limitations of the room-temperature environment

At a given ion velocity, the beam lifetime in a storage ring is defined by the mean free path length of the ion in the residual gas of the beam pipe. The TSR features a vacuum pressure of a few 10^{-11} mbar all along its 55-m-long vacuum chamber. This value is close to the optimum pressure that can be achieved in a room-temperature setup, as the residual gas is in this case dominated by H_2, the main degassing component of the stainless steel chamber walls.

A more fundamental limitation of the 300-K environment is related to molecular physics itself. While vibrational excitations are too high-energetic to be excited by far-infrared black body radiation, so that molecular ions can be safely assumed to relax to their vibrational ground state within the storage time in the TSR, this is not true for rotational excitations. Although rotational single-state experiments have been possible in the TSR in the special case of molecules with zero electric dipole moment using a state-selective ion source [10], the rotational states of stored ions generally have to be expected to assume a population that is in thermal equilibrium with the background radiation of 300 K. Due to this initial-state multiplicity, theoretical predictions of DR rate coefficients have to be convolved with a relatively broad range of rotational states, which fundamentally limits the quality of experimental data as theory benchmarks.

3. The Cryogenic Storage Ring project

The Cryogenic Storage Ring CSR is designed to provide the platform for the next generation of precision experiments with molecular ions. The CSR is an all-cryogenic electrostatic electron cooler storage ring [8, 9] and will feature experimental facilities for studies of ion-neutral, ion-photon and ion-electron interactions.

3.1. Beamline

Unlike in the case of magnetic deflection, the electrostatic beam optics is energy-selective. At a given charge state, q, the maximum kinetic energy, E, is defined by the bending radius, ρ, and the maximum electric field, $\mathcal{E}_{\max}$, that can be applied within the deflector elements:

$$\frac{E}{q} \leq \rho\,\mathcal{E}_{\max}\,. \tag{3}$$

For the CSR, $\rho\,\mathcal{E}_{\max} = 300\,\text{kV}$, hence the ring can store particles at maximum kinetic energies of 300 keV per charge state independent of their mass. In contrast to the Lorentz force, the electric deflecting force does not depend on the ion momentum. As a consequence, the dependence of the maximum storage velocity on the particle mass is weaker in comparison to (2):

$$v_{\max} \sim m^{-1/2}\,. \tag{4}$$

Thus, although in an electrostatic ring the velocity of light ions is limited compared to a magnetic ring by the typical value of $\rho\,\mathcal{E}_{\max}$, very heavy species can be stored at higher velocity. While several electrostatic storage rings have already been commissioned [11, 12] or are currently under construction [13], the CSR will be by far the largest device of this type, with a closed-orbit circumference of approximately 35 m and, to-date, the highest value of $\rho\,\mathcal{E}_{\max}$.

In order to achieve long storage times even for very heavy species, the vacuum system of the CSR is designed to provide very low residual gas pressures. Using a combination of ion-getter pumps, non-evaporative getter (NEG) coatings and UHV-bakeable cryopumps, the residual gas pressure will be reduced to $\sim 10^{-11}$ mbar during room-temperature operation. Further reduction of the residual gas density is achieved by cooling the entire vacuum chamber and ion optics to below 10 K using a superfluid helium cryocooler [8]. Therefore, the beam line has to be enclosed in a cryostat consisting of two layers of thermal radiation shields, several layers of superinsulation

Figure 1. Photographic view of the CSR under construction in the experimental hall of the MPIK accelerator building. 1 – Electrostatic bending and focussing section, 2 – straight experimental section, 3 – connection box to superfluid He cooler.

and a large isolation vacuum chamber. In cold operation, the dominating residual gas component H_2 can no longer be pumped by NEG coatings. Instead, it will be removed by cryocondensation on dedicated copper surfaces in the beam line which are cooled to below 2 K by direct contact with the superfluid helium circuit [14]. This reduces the pressure in the ring to approximately 10^{-13} mbar (room temperature equivalent pressure yielding identical residual gas density), thus improving ion storage times by more than 2 orders of magnitude compared to warm operation. The cryogenic and vacuum concepts of the CSR have been tested in a prototype setup and have met their design goals, as reported recently [15].

A 300 kV injector platform is presently under construction at MPIK. Several types of ion sources dedicated to CSR experiments are planned, including, but not limited to, an EBIT for HCI production and specialised sources for production of cold molecular ion beams. The latter can be supersonic expansion jet [16] sources, electrospray ion sources for generation of large molecular ions [17], or buffer-gas cooled radiofrequency traps which allow relaxation of initially hot ion bunches prior to injection into the CSR [18]. The 300 kV high-voltage platform is sufficiently spacious for several sources to be installed simultaneously, which ensures flexibility in the planning of CSR experiments.

3.2. Experimental facilities

The CSR has a quadratic layout with four straight beamline sections enclosed by four identical corner sections. The latter contain all of the beam bending and focussing optics. Furthermore, the ion beam injection will be performed in a corner section by fast switching of one of the bending elements. Thus, four ~ 2-m-long linear sections are available for experimental instrumentation.

A first straight section is dedicated to ion beam diagnostics, featuring a capacitive pickup electrode for highly-sensitive current measurements. A second section houses a transverse gas-jet target backed by a reaction microscope for studies of fast ion-neutral interactions. The linear chamber following the injection corner will be available for future experimental equipment, but will also include a collinear merging region with a neutral beam. The latter will be fed through the injection beamline at variable velocity and allow experiments on slow ion-neutral collisions.

The fourth experimental section contains the electron cooler and thus the key ingredient to recombination experiments. Electron cooling of slow, massive ions is a challenging task. While pioneering work with slow merged electron beams has been done at the KEK electrostatic storage ring [19], CSR is the first machine of this kind to aim at strong phase-space compression by electron cooling, using long ion storage times and specialised sources for cold, slow electrons. For a wide range of molecular species, this combination of a cooled ion beam and a very slow merged electron beam make the CSR the only next-generation storage ring suited for experiments on low-energy DR.

3.3. Electron cooler

The electron cooler is designed to operate at electron energies E_e from 1 eV to 1 keV. Considering the velocity-matching condition $E_e = E_i \, m_e/m_i$ (with e and i denoting the energy and mass of electron and ion, respectively), the energy range $E_e > 160$ eV is needed only for the generation of non-zero collision energies in merged-beams experiments. For cooling of singly charged, massive molecular ions, electron energies of only a few eV or less are necessary. The electrons are overlapped with the stored ion beam using a novel magnetic beam-merging scheme, which minimises disturbance of the ion trajectory [20].

Well-controlled manipulation of very slow electron beams is difficult, as effects such as the beam space-charge or workfunction differences between the cathode and the beam pipe materials, which are only minor corrections in existing coolers, become dominant contributions to the final electron energy. In addition, the available electron density at low energy is severely limited by the electron gun perveance, as the extraction voltage at the cathode cannot be arbitrarily higher than the final acceleration potential. In the absence of high electron densities, the electron temperature has to be as low as possible in order to assure an acceptable cooling force [21]. Hence, the CSR electron cooler will employ a cryogenic GaAs(Cs,O) photocathode as an electron source, similar to the existing high-resolution electron target of the TSR, which has already demonstrated its ability to cool slow ion beams within a few seconds of storage time [22].

Recent experiments performed at the existing photocathode electron beam setup of the TSR have established a procedure that allows reliable production of very slow electron beams of well-defined energy and density, taking space-charge and contact potential effects into account [23]. The measurements have shown that stable beams with a narrow velocity-spread can be obtained at mean kinetic electron energies down to 1 eV or lower. Given the CSR's ion energy limit of 300 keV, the mass range of singly-charged ion species for which velocity matching can be obtained in the electron cooler extends to at least 160 u.

3.4. Detectors

Similarly to the other experimental facilities, the electron cooler will be backed by particle detectors for rigidity-changed or neutralised products in the bending section located after it in the CSR. The low ion beam energy in combination with the cryogenic extremely high vacuum environment severely limits the spectrum of available detector technologies that can be employed.

Rigidity-changed products, such as those arising from dissociative excitation or ion-pair formation events, will be recorded using counting detectors based on charge multiplication by secondary electron emission from a converter electrode. The secondaries are then detected using an MCP backed by a collector anode. Secondary electron emitters have the advantage of not presenting any dead layer which the impinging products have to traverse for successful detection, so that also low-energetic particles can be counted. Detectors based on this principle [24] have already been successfully used at the TSR for several years.

Neutral DR fragment imaging will be performed using an MCP backed by delay-line readout anodes, which provide multi-hit time and position sensitivity and thus allow measurement of kinetic energy releases in breakup events, in addition to the pure recombination rate counting capability. Successful operation of such a detector in a cryogenic environment has already been reported in [25] and is currently being tested in the CSR prototype (described in [15]).

The explosion cone aperture for detectable DR fragments is limited to $\pm 1°$ by the geometry of the CSR bending elements. For symmetric breakups of the parent molecule into neutral fragments of approximately equal mass, this limitation is insignificant, as only unrealistically large kinetic energy releases of several tens of eV could cause the recombination products to miss the imaging detector. For strongly asymmetric breakups, where the mass of one or several fragments is much smaller than that of the parent molecule, products can miss the detector if the energy release of the recombination reaction exceeds a critical value. The maximum allowed kinetic energy release is in this case proportional to the mass of the light fragment and inversely proportional to the mass of the parent molecule. If, e.g., a single hydrogen atom is emitted from a parent of mass 100 u, stored in the CSR at 300 keV, the geometrical efficiency will be limited already for kinetic energy releases exceeding ~ 1 eV. However, for those events where both particles do hit the MCP, the large velocity difference between the fragments will in this case allow inclusion of time-of-flight measurements in the analysis of the kinetic energy release. Thanks to the much slower ion beam velocity of the CSR, this will be possible with much higher precision compared to present fast-beam setups.

While an MCP-based fragment detector lacks the product mass sensitivity of the present setup at TSR [4], fragmentation branching ratios can be obtained from statistical analysis of the fragment velocity distributions if the emittance of the parent ion beam is small, i.e. if the centre-of-mass of the molecular breakup is sufficiently well-defined by electron cooling [5]. Furthermore, the cryogenic environment of the ring offers realistic opportunities for future fragment imaging setups involving low-temperature micro-calorimeters [26], which can offer even higher fragment mass resolution than present surface-barrier semiconductor detectors.

4. Opportunities for molecular ion recombination experiments at the CSR

CSR is currently the only next-generation electrostatic storage ring for which also DR experiments are planned. Its high mass acceptance combined with electron cooling and the cryogenic environment opens the way to a range of experiments not feasible in present setups.

4.1. DR in a cold environment

With its cold vacuum chamber, the CSR will be the first electron cooler storage ring that truly simulates the environmental conditions found in the interstellar medium (ISM). It will thus allow the study of DR reactions relevant to the chemistry of the ISM without interference by an unrealistic 300-K black-body background radiation.

Aided by superelastic electron collision (SEC) cooling [27], by the quasi radiation-free storage volume, and by the long storage times in the extremely high vacuum of the CSR, initially hot molecules produced in the ion sources will be allowed to relax down to their rovibrational ground-state. Experiments seeking to benchmark theory will greatly profit from this possibility to prepare and maintain pure internal state populations. Possible candidates for first state-selective DR experiments at the CSR are HD^+ and HeH^+, ions which are easy to produce and store at high current intensity and which are known to efficiently relax radiatively. Theory of the simple molecule HD^+ is in fairly good agreement with present room-temperature convolved DR rate coefficients and could thus benefit most from single-state recombination data. For HeH^+, updated theoretical DR rate coefficients have been published recently, but can, up to now, only be compared to rotational-state convolved experimental datasets [28].

4.2. DR of massive molecular ions

The CSR will extend the mass range of molecular ions accessible for precision DR studies. It will be able to store heavy, slow ion species long enough to render electron cooling possible.

Internal relaxation of these ions will be achieved by combination of the cold ion sources on the injector platform (Sec. 3.1), which pre-cool all internal molecular modes including those of low or vanishing infrared activity, and the low-temperature environment in the storage ring, which prevents the re-excitation of modes with stronger infrared activity. Candidates for experiments taking advantage of the high mass range of the CSR include e.g. the carbon fluorine ions CF_n^+, which play an important role in industrial plasmas and of which only the lightest specimen CF^+ could be studied at a high energy resolution in the TSR up to now. Another promising field is the DR of large hydrocarbon or small biomolecule ions. Finally, even electron collision studies of cluster ions will be an option at the CSR, which will allow, e.g., to study the dependence of the recombination process on the internal cluster temperature.

5. Present status and outlook

We have introduced the CSR project as the future platform for next-generation precision DR experiments. The CSR and its ion source array are presently being built at the MPIK (see fig. 1) with the first corner section to be completely assembled in early 2011. The remaining parts of the ring lattice as well as the electron cooler and the detector systems will be constructed during the next two years, such that first stored beams are expected at the beginning of 2013.

References

[1] Lestinsky M, Lindroth E, Orlov D A, Schmidt E W, Schippers S, Böhm S, Brandau C, Sprenger F, Terekhov A S, Müller A and Wolf A 2008 *Phys. Rev. Lett.* **100** 033001

[2] Petrie S and Bohme D K 2007 *Mass Spectrom. Reviews* **26** 258–280

[3] Strasser D, Lammich L, Kreckel H, Lange M, Krohn M, Schwalm D, Wolf A and Zajfman D 2004 *Phys. Rev.* A **69** 064702

[4] Buhr H, Stützel J, Mendes M B, Novotný, Schwalm D, Berg M H, Bing D, Grieser M, Heber O, Krantz C, Menk S, Novotny S, Orlov D A, Petrignani A, Rappaport M L, Repnow R, Zajfman D and Wolf A 2010 *Phys. Rev. Lett.* **105** 103202

[5] Nevo I, Novotny S, Buhr H, Andrianarijaona V, Altevogt S, Heber O, Hoffmenn J, Kreckel H, Lammich L, Lestinsky M, Pedersen H B, Schwalm D, Wolf A and Zajfman D 2007 *Phys. Rev.* A **76** 022713

[6] Wolf A 1999 *Atomic Physics with Heavy Ions* ed Beyer H F and Shevelvo V P (Berlin: Springer) pp 1–20

[7] Larsson M and Orel A E 2008 *Dissociative Recombination of Molecular Ions* (Cambridge University Press, Cambridge)

[8] Hahn R von, Blaum K, Crespo López-Urrutia J R, Froese M, Grieser M, Lange M, Laux F, Menk S, Varju J, Orlov D A, Repnow R, Schröter C D, Schwalm D, Sieber T, Ullrich J, Rappaport M, Zajfman D, Urbain X, Quack H and Wolf A 2008 *Proc. 11th European Particle Accelerator Conference (EPAC08)* pp 394–396 URL `http://jacow.org/e08/papers/mopc137.pdf`

[9] Hahn R von, Blaum K, Fellenberger F, Froese M, Grieser M, Krantz C, Lange M, Menk S, Laux F, Orlov D A, Repnow R, Shornikov A, Sieber T and Wolf A 2011 *Nucl. Instrum. Methods Phys. Res., Sect. B*, accepted

[10] Novotny S et al., *in preparation*, and Novotny S 2008 *Fragmentation of molecular ions in slow electron collisions* (PhD thesis, University of Heidelberg) URL `http://www.ub.uni-heidelberg.de/archiv/8531/`

[11] Møller S P 1997 *Nucl. Instrum. Methods Phys. Res., Sect. A* **394** 281–286.

[12] Jinno S, Takao T, Omata Y, Satou A, Tanuma H, Azuma T, Shiromaru H, Okuno K, Kobayashi N and Watanabe I 2004, *Nucl. Instrum. Methods Phys. Res. A* **532** 477–482.

[13] Rensfeld K-G, Andler G, Bagge L, Blom M, Danared H, Källberg A, Leontein S, Liljeby Löfgren P, Paál A, Simonsson A and Skeppstedt Ö 2004 *Proc. 9th European Particle Accelerator Conference (EPAC04)* pp 1425–1427 URL `http://jacow.org/e08/papers/mopc137.pdf`

[14] Orlov D A, Lange M, Froese M, Hahn R von, Grieser M, Mallinger V, Rappaport M, Sieber T, Weber T and Wolf A 2008 *Advances in Cryogenic Engineering* vol 53 (AIP Conference Proceedings vol 985) ed Weisend II J G *et al.* (American Institute of Physics, Melville NY) pp 1233–1239

[15] Lange M, Froese M, Menk S, Varju J, Bastert R, Blaum K, Crespo López-Urrutia J R, Fellenberger F, Grieser M, Hahn R von, Heber O, Kühnel K-U, Laux F, Orlov D A, Rappaport M L, Repnow R, Schröter C D, Schwalm D, Shornikov A, Sieber T, Toker Y, Ullrich J, Wolf A and Zajfman D 2010 *Rev. Sci. Instrum.* **81** 055105

[16] Proch D and Trickl T 1989 *Rev. Sci. Instrum.* **60** 713–716

[17] Fenn J B, Mann M, Meng C K, Wong S F and Whitehouse C M 1989 *Science* **246** 64–71

[18] Kreckel H, Motsch M, Mikosch J, Glosík J, Plašil R, Altevogt S, Andrianarijaona V, Buhr H, Hoffmann J, Lammich L, Lestinsky M, Nevo I, Novotny S, Orlov D A, Pedersen H B, Sprenger F, Terekhov A S, Toker J, Wester R, Gerlich D, Schwalm D, Wolf A and Zajfman D 2005 *Phys. Rev. Lett.* **95** 263201
[19] Tanabe T, Noda K and Syresin E 2004, *Nucl. Instrum. Methods Phys. Res., Sect. A* **532**, 105–110
[20] Shornikov A, Orlov D A, Grieser M, Blaum K and Wolf A 2009 *Proc. Workshop on Beam Cooling and Related Topics (COOL09)* pp 119–122 URL `http://jacow.org/cool2009/papers/thm2mcco03.pdf`
[21] Orlov D A, Fadil H, Grieser M, Krantz C, Hoffmann J, Novotný O, Novotny S and Wolf A 2007 *Proc. Workshop on Beam Cooling and Related Topics (COOL07)* pp 230–233 URL `http://jacow.org/cl07/papers/frm1c03.pdf`
[22] Krantz C, Orlov D A, Hoffmann J, Lestinsky M, Novotný O, Jaroshevich A S, Terekhov A S and Wolf A 2009, *J. Phys.: Conf. Ser.* **192** 012025
[23] Shornikov A, Orlov D A, Jaroshevich A S, Krantz C and Wolf A, in preparation
[24] Rinn K, Müller A, Eichenauer H and Salzborn E 1982 *Rev. Sci. Instrum.* **53** 829–837
[25] Kuehnel K-U, Schröter C D and Ullrich J 2008 *Proc. 11th European Particle Accelerator Conference (EPAC08)* pp 1179–1181 URL `http://jacow.org/e08/papers/tupc95.pdf`
[26] Burck A, Kempf S, Schäfer S, Rotzinger H, Rodriges M, Wolf T, Gastaldo L, Fleischmann A and Enss C 2008 *J. Low Temp. Phys.* **151** 337–344
[27] Shafir D, Novotny S, Buhr H, Altevogt S, Faure A, Grieser M, Harvey A G, Heber O, Hoffmann J, Kreckel H, Lammich L, Nevo I, Pedersen H B, Rubinstein H, Schneider I F, Schwalm D, Tennyson J, Wolf A and Zajfman D 2009 *Phys. Rev. Lett.* **102** 223202
[28] Haxton D J and Greene C H 2009 *Phys. Rev. A* **79** 022701

Eighth International Conference on Dissociative Recombination (DR2010) IOP Publishing
Journal of Physics: Conference Series **300** (2011) 012011 doi:10.1088/1742-6596/300/1/012011

DESIREE: a unique cryogenic electrostatic storage ring for merged ion-beams studies

R. D. Thomas[a], H. T. Schmidt[a], J. Gurell[a], N. Haag[a], A. I. S. Holm[a], H. A. B. Johansson[a], G. Källersjö[a], M. Larsson[a], S. Mannervik[a], D. Misra[a], A. Orban[a], P. Reinhed[a], K.-G. Rensfelt[a], S. Rosén[a], F. Seitz[a], J. Weimer[a], H. Zettergren[a], G. Andler[b], H. Danared[b], S. Das[b], L. Liljeby[b], M. Björkhage[b], M. Blom[b], L. Brännholm[b], P. Halldén[b], F. Hellberg[b], A. Källberg[b], S. Leontein[b], P. Löfgren[b], B. Malm[b], A. Paál[b], A. Simonsson[b], and H. Cederquist[a]

[a]Department of Physics, Stockholm University, SE-106 91 Stockholm, Sweden.
[b]Manne Siegbahn Laboratory, Stockholm University SE-104 05 Stockholm, Sweden

In this proceedings I will describe the design of a new storage device currently under construction at Stockholm University, Sweden. This device uses purely electrostatic focussing and deflection elements and allows ion beams of opposite charge to be confined under extreme high vacuum and cryogenic conditions in separate "rings" and then merged over a common straight section. This Double ElectroStatic Ion Ring ExpEriment (DESIREE) apparatus allows for studies of interactions between cations and anions at low and well-defined centre-of-mass energies. I discuss the design of the DESIREE facility, highlighting some of the technical advantages of using purely electrostatic over magnetic elements, as well as the issues that have arisen during its development and construction. Finally, the advantages of this design are a boon to fundamental experimental studies and I finish by discussing an example of such potential research.

1 Introduction

One of the driving forces for the initial development of ion storage rings was from high-energy particle physics, and the size of these rings covers a wide range with circumferences ranging from tens to thousands of meters. In recent decades, these devices have also proven successful in addressing the needs of the atomic and molecular physics communities into low-energy processes. The majority of these devices are modeled after the LEAR device at CERN [17], in which magnetic elements such as dipole, quadrupole and higher multipole magnets are used to define and control the ion orbits in the device. Examples of such devices are CRYRING (Stockholm, Sweden [1]), ASTRID (Aarhus, Denmark [6]), TSR (Heidelberg, Germany [4]), TARN II (Tokyo, Japan [14]) and the newly constructed HIRFL-CSR (Lanzhou, China [19]).

Importantly, electrostatic elements are rarely used in these devices, and when they are it is primarily for electrostatic septa for the injection/extraction of ions. This aspect is relevant when it comes to comparing devices based on electrostatic storage as opposed to magnetic storage. Using electrostatic elements clearly removes the problems associated with effects such as remanent fields and hysteresis in the magnets and the requirements for water cooling. Furthermore, magnets are expensive, big and heavy, but with electrostatic elements a compact and less expensive solution can be realized. The fact that all elements are electrostatic means that, for an injector on a given potential, the mass and charge of the injected ions can be changed without needing to change the settings of the ion-optical elements. Another important advantage, especially related to the general direction of research in atomic and molecular physics, is that an electrostatic device can store heavy ions in low charge states, something which is often a problem in magnetic rings where the maximum mass of the stored ions is limited by the bending power of the magnets, and this plays a large role in the experimental uncertainties [16]. Finally, also from a purely scientific perspective, the lack of magnetic fields also plays a role in studied of fundamental physics since such fields can give rise to mixing and transitions between quantum states in the ions being stored in the device.

These advantages were realised through the pioneering work of Møller in Aarhus, Denmark, with the construction of the purely electrostatic storage ting ELISA [7], and through Zajfman and Schmidt and their co-workers through their developments of linear electrostatic ion traps for storage of fast ion beams [20, 12]. Since its construction, several other such devices have been built based on ELISA's design, [15, 3], while construction on several larger electrostatic storage devices is also under way [18, 21, 10, 5, 2]. Two of these represent significant advances on ELISA since they will operate at crogenic temperatures: the first, CSR, is under construction in Heidelberg, Germany, and has a design similar to the "traditional" magnetic storage rings listed earlier, especially it's sister ring TSR which is located in the same laboratory [21]. The second, which is the facility presented in these proceedings, is under construction at Stockholm University, adds a unique twist to storage-ring design: it consists of two separate storage rings with a common section in which, for example, the interaction between oppositely charged ions can be studied in a merged beams configuration [10, 5, 2]. The unique position of the DESIREE facility is further highlighted through its design. Instead of the usual approach, in which the two rings would be mounted in separate beam pipes, the ion optics for the whole device are fully open and the whole "storage volume" is enclosed in a double-walled vacuum vessel. This has many advantages from an experimental perspective: it is easier to create extreme high vacuum conditions, leading to longer storage times; and since all the ion-optical elements are mounted on a common baseplate, which then also serves as the floor for the inner vacuum chamber, everything thermally shrinks together during cooling, allowing the inner vessel to be cryogenically cooled to approximately 10 Kelvin through the use of cryogenerators without significantly affecting the environment.

2 Technical Design

The cryogenic environment, open construction, and the double ring structure all make DESIREE truly unique, though also a considerable technical challenge. In the following section, I will briefly discuss some of the technical highlights of DESIREE.

2.1 Rings and Injectors

The DESIREE facility consists of two high-voltage ion source platforms, one at 100 kV and the other at 25 kV, which service the two storage rings and the common straight merging section that are located in the common vacuum vessel. Each of the rings has two 160° cylindrical bends and four 10° parallel-plate deflectors, where the two 10° deflectors located on either end of the merger section are common for each ring. This immediately means that only ions with opposite charge can be stored in the two rings at the same time. Figure 1 shows a technical schematic of the heart of the DESIREE facility. As can be seen in Fig. 1, the lower ring (ring 2) has a somewhat different layout compared to the upper ring (ring 1). Ring 2 can store ions with different energy or charge compared to the ring 1 and the bending angle will therefore differ from 10° in the two common deflectors. To compensate for this effect, and to make the ion beams collinear in the merging region, additional deflector plates are added to the second ring. Four quadrupole doublets, for focusing of the beams, and a few smaller deflectors for minor corrections of the beam position are located in each of the rings. Ion-injection into the two rings is achieved by rapid switching of the relevant 10° deflectors.

In addition to the "traditional" filament and cold-cathode ion sources which have been used to create the singly charged atomic and molecular ions studied in molecular storage rings such

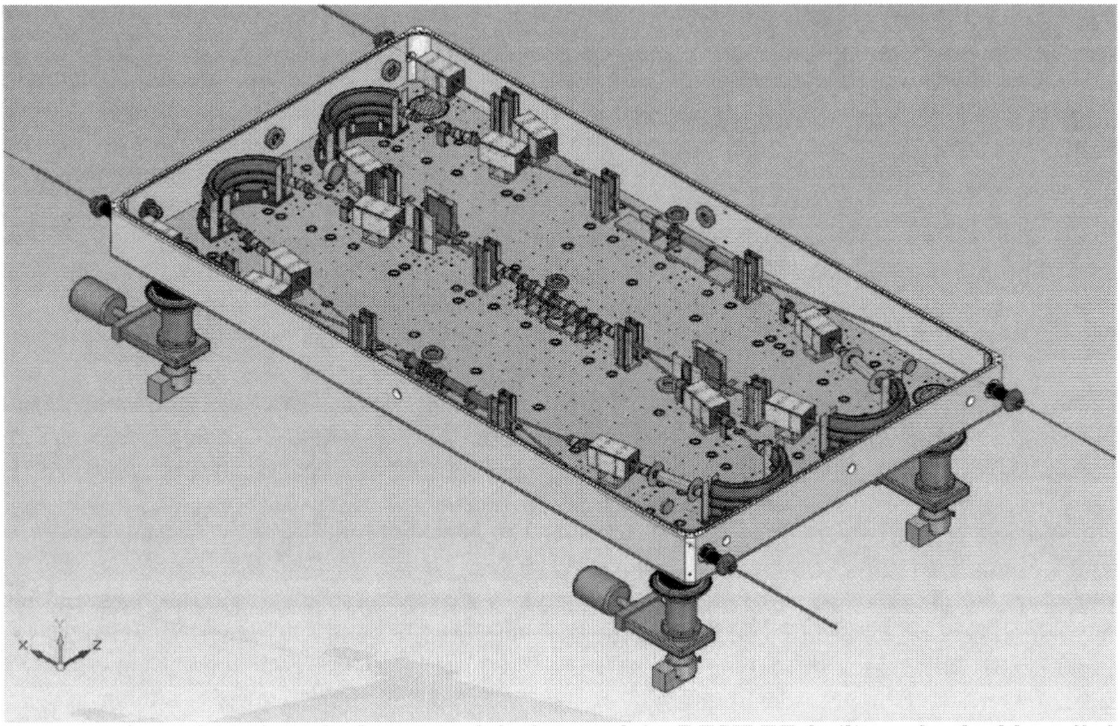

Figure 1: An early technical schematic of the heart of the DESIREE facility - the double walled vacuum vessel - showing the open ion optics, the two separate rings, and the unique merging region. In the text, Ring 2 is the ring to the lower left while Ring 1 is the upper right ring.

as CRYRING [16], the platforms have been designed to accommodate a suite of different ion sources ranging from an expansion source for producing rotationally cold molecular ions, an electrospray source for creating biomolecules and large molecular cluster ions, and a sputter source for generating negative ions. All of these ion sources are available, having been either purchased or constructed in-house, and are currently undergoing testing on a dedicated ion-source experimental platform - though this is beyond the scope of these proceedings.

2.1.1 Mechanical Design

The two rings are housed in a common double-walled vacuum chamber, where the outer chamber is constructed of steel. The external dimensions of the outermost chamber are approximately 5.0 m x 2.0 m x 0.7 m. In order to stand the pressure difference the outer vessel is reinforced by an external framework of I-beams on all sides. The inner chamber, for technical reasons, is made from aluminium. All optical elements and detectors are mounted directly on the bottom of this chamber. The inner chamber is stood-off from the outer chamber via thin-walled stainless steel tubes. Separating the two vessels is a copper thermal screen and several layers of super insulation are placed between this screen and the outer chamber. All components, such as pumps, the cryogenerators, and the electrical feedthroughs are mounted on the bottom of the vessels. This facilitates access to the two vessels from the top. After initial baking and testing, DESIREE will normally be operated over a range of cryogenic temperatures.

During cryogenic operation, the inner chamber will be cooled with four cryogenerators (Sumitomo RDK-415D) where the first stage of each cryogenerator is connected to the thermal screen and the second stage to the bottom of the inner chamber. These connections are made with heavy-duty copper braids to give the best thermal conductivity and to isolate from vibrations. The total heat load on the thermal screen and the inner vessel is estimated to be 60 W and 3 W, respectively, where the main contribution to the heat load on the inner vessel is from blackbody radiation from the screen. If this low heat load can be realized, the final temperature of the inner vessel and the rings should be around 5 K. At cryogenic temperatures the inner vessel will be pumped by the cold walls of the chamber. It is expected that the main contribution to the background pressure will be H_2, which will be reduced by Ti sublimation pumps. After the initial baking at $\approx$ 400 K of the aluminium vessel, a pressure of 1×10^{-10} mbar is expected to be reached at room temperature. The outer chamber, whose main purpose is to thermally insulate the inner vessel, will be pumped with turbo pumps. All feedthroughs and pump-ports located on the inner chamber are sealed with a combination of Helicoflex and Astraseal gaskets while those through the outer chamber are sealed with standard Viton gaskets.

3 Test System

During the initial development of the project, it became quickly clear that in order to meet the technical challenge of designing a complex system like DESIREE a smaller test and training system needed to be constructed. This was necessary to answer such fundamental issues as: the cryogenic properties of the detectors which will be used in DESIREE, and the vacuum and cryogenic properties of the different materials which will be used in its construction. In each of these cases, data often proved to hard to find or was even unknown. Furthermore, questions such as the optimum way to mount the cryogenerators, laser windows, as well as the operation of hitherto standard components such as vacuum gaskets, all needed to be investigated.

The test chamber constructed to address these issues was built as a cryostat with an inner and outer chamber and a copper screen in between, and figure 2 shows a schematic of the test chamber. This chamber then represents a copy of the DESIREE device, though certain differences are worth mentioning here. Although the inner chamber is still made from aluminium, the outer chamber is made from stainless steel. Furthermore, no super insulation is used in the test chamber. In order to minimize the heat conductance between the two vessels, the inner chamber rests only on three thin walled stainless steel tubes located at the bottom of the outer chamber. The test chamber is pumped with two turbo pumps, one for each chamber. Furthermore, the inner chamber can also be pumped by a Ti sublimation pump. The system is equipped with a residual gas analyzer (RGA) for analysis of the background gas.

All components which connect with the inner vessel, such as turbo pumps, pressure gauges, the RGA and all the electrical feedthroughs are mounted on the bottom of the outer vessel. The turbo pump and RGA are interfaced with the inner chamber via thin walled bellows with cooled baffles in order to minimize the heat load on the inner chamber. All flanges on the inner chamber are sealed with only Helicoflex gaskets while standard Viton o-rings are used on the outer chamber where the vacuum requirements are not that high. Cryogenic cooling is achieved by the same type of two-stage cryogenerator that will be used on DESIREE, with a specified cooling capacity of 35 W at 50 K on the first stage and 1.5 W at 4.2 K on the second stage. The first stage is connected to the copper screen and the second to the bottom of the aluminium vessel via a flexible copper braid.

In order to test the cooling, temperature sensors are mounted on the aluminium chamber and the screen. Prior to mounting the cryogenerator, tests were performed to verify its specified cooling power and to measure cooling power at temperatures where it is not specified. The results agreed with the those stated by the manufacturer. During these tests the aluminium chamber bottom reached 9 K while the second stage of the cryogenertor was at 4.5 K. From

Journal of Physics: Conference Series **300** (2011) 012011 doi:10.1088/1742-6596/300/1/012011

Figure 2: Schematic drawing of the test chamber with the outer chamber (light blue) inner aluminium chamber (blue) and thermal copper screen (red).

the cooling power test it can be concluded that the total heat load on the aluminium chamber is 1.5 W. The somewhat higher temperature observed at the inner vessel is due to the limited thermal conductivity of the flexible copper braid.

As mentioned, one of the important tasks for this chamber was to test the cryogenic properties of the different kinds of detectors that were planned for use in DESIREE. One such family of detectors are microchannel-plate based detectors (MCPs), in which these plates are coupled to different types of anode depending on the purpose of the detector. Both resistive- and phosphor screen-anodes have been tested, and the results from the MCP-phosphor screen anode have been reported [11], and showed that these detectors successfully operate under cryogenic conditions.

One of the most significant scientific highlights to come out of these investigations arose from the placing of an electrostatic ion trap in the test chamber, i.e, to mimic the ion-storage properties of DESIREE. Here, a linear ion-trap, ConeTrap [12], was placed inside the inner vacuum chamber. A pulse of ions were injected into the trap, which was then closed, and the lifetime of the ions in the trap studied as function of their storage time. The ions chosen for this particular test were the metastable helium anion, $He^-(1s2s2p\ ^4P_{5/2})$, with the motivation that results from earlier lifetime measurements were limited in the accuracy of systematic effects due to the photo-detachment of the loosely bound 2p electron by 300 K blackbody radiation photons emitted from the surrounding vacuum chamber. In performing this experiment in a suitable cryogenic environment, in practice under 80 K, this effect is eliminated. The results from these tests gave the most accurate measurement for the lifetime of these ions [9], as well as providing valuable information on the ion-beam storage lifetimes as a function of the residual gas pressure in the chamber [8]

4 Proposed experiments

DESIREE has been planned and constructed so that any given experiment can be undertaken in either of the two rings or a single experiment can utilise both rings. The most unique feature is the merging region, and so I briefly discuss an example of the type of experiment which now can be undertaken at DESIREE and which otherwise have proven extremely difficult or impossible. The possibility to perform merged-beams experiments with positive and negative ions that are stored and cooled to low temperatures by temperature equilibrium with the surroundings is the most clearly unique feature of DESIREE.

One of the most successful applications of CRYRING and other similar-sized heavy-ion storage rings over the last decades has been the investigation into reaction of molecular ions with a free electron [16]. Here we consider the mutual neutralisation between small molecular cations and anions, which is related to the molecular recombination in the following way. The formation of polyatomic ions in the interstellar medium is considered to be through chemistry involving ionized species. The degree of ionization in the interstellar clouds is determined by a balance between ionization through cosmic and stellar radiation and neutralisation processes. In the gas phase two such neutralisation processes exist: Electron-cation recombination and cation/anion mutual neutralisation. One of the primary motivations for the studies at CRYRING of dissociative recombination for astrophysically abundant cations (e.g. H_2^+ (HD^+), H_3^+, OH^+, H_3O^+, CO^+, HCO^+, CH_2^+, CH_3^+, and CH_5^+ [[16] and references therein]) has been the role of this process in interstellar chemistry in regions where the negative charge is primarily in the form of free electrons. One of the most recent observations in the interstellar medium has been that of long carbon-chain anions. If present in the amounts required to explain the observed absorption, it is possible that in these regions negative charge is more often in the form of anions than free electrons. In such an environment the role of mutual neutralisation in the ion chemistry is more important than dissociative recombination and the possibility offered by DESIREE to study these processes highly relevant to this field [13].

5 Current Status

The inner and outer vacuum chambers have been machined and fully assembled and are currently undergoing leak testing, including all of the feed-throughs and pumping/injection/optical ports. The copper screen has also been made as have all of the electrostatic/dynamic elements that will be located inside the main double walled chamber - these have also been electrically tested. The 25-keV platform has also been constructed and assembled, as have all parts for both the injection beam lines and their associated ion optics. The next stages include completion of the leak testing of the inner chamber, followed by testing the cryogenic cooling of the whole inner and outer chamber, with all elements in place, as well as a full electrical test.

References

[1] K. Abrahamsson, G. Andler, L. Bagge, E. Beebe, P. Carlé, H. Danared, S. Egnell, K. Ehrnstén, M. Engström, C.J. Herrlander, J. Hilke, J. Jeansson, A. Källberg, S. Leontein, L. Liljeby, A. Nilsson, A. Paal, K.-G. Rensfelt, U. Rosengård, A. Simonsson, A. Soltan, J. Starker, M. af Ugglas, and A. Filevich. Cryring a synchrotron, cooler and storage ring. *Nucl. Instrum. Methods Phys. Res., Sect. B*, 79:269, 1993.

[2] H. Danared, L. Liljeby, G. Andler, L. Bagge, M. Blom, A. Källberg, S. Leontein, P. Löfgren, A. Paál, K.-G. Rensfelt, A. Simonsson, H. T. Schmidt, H. Cederquist, M. Larsson, S. Rosén,

, and K. Schmidt. Desiree - a double electrostatic storage ring for merged-beam experiments. *AIP Conf. Proc.*, 821:465, 2006.

[3] S. Jinno, T. Takao, Y. Omata, A. Satou, H. Tanuma, T. Azuma, H. Shiromaru, K. Okuno, N. Kobayashi, and I. Watanabe. Tmu electrostatic ion storage ring designed for operation at liquid nitrogen temperature. *Nucl. Instr. Methods Phys. Res. A*, 532:477, 2004.

[4] D. Krämer, G. Bisoffi, M. Blum, A. Friedrich, Ch. Geyer, B. Holzer, H. W. Heyng, D. Habs, E. Jaeschke, M. Jung, W. Ott, R. E. Pollock, R. Repnow, F. Schmitt, and M. Steck. One year of operation at the heidelberg tsr. *Nucl. Instrum. Methods Phys. Res. A*, 287:268, 1989.

[5] P. Löfgren, G. Andler, L. Bagge, M. Blom, H. Danared, A. Källberg, S. Leontein, L. Liljeby, A. Paál, K.-G. Rensfelt, A. Simonsson, H. Cederquist, M. Larsson, S. Rosén, H. T. Schmidt, , and K. Schmidt. Design of the double electrostatic storage ring, desiree. In *Proc. EPAC 2006*, page 252, 2006.

[6] S. P. Møller. Astrid-a storage ring for ions and electrons. In L. Lizama and J. Chew, editors, *Conference Record of the 1991 IEEE Particle Accelerator Conference*, page 2811. New York, IEEE, 1991.

[7] S. P. Møller. Elisa, an electrostatic storage ring for atomic physics. *Nucl. Instrum. Methods Phys. Res. A*, 394:281, 1997.

[8] P. Reinhed, A. Orbán, S. Rosén, R. D. Thomas, I. Kashperka, H. A. B. Johansson, D. Misra, A. Fardi, L. Brännholm, M. Björkhage, H. Cederquist, and H. T. Schmidt. Cryogenic kev ion-beam storage in conetrap - a tool for ion-temperature control. *Nucl. Instrum. Methods Phys. Res. A*, 621:83, 2010.

[9] P. Reinhed, A. Orbán, J. Werner, S. Rosén, R. D. Thomas, I. Kashperka, H. A. B. Johansson, D. Misra, L. Brännholm, M. Björkhage, H. Cederquist, and H. T. Schmidt. Precision lifetime measurements of he^- in a cryogenic electrostatic ion-beam trap. *Phys. Rev. Lett.*, 103:213002, 2009.

[10] K.-G. Rensfelt, G. Andler, L. Bagge, M. Blom, H. Danared, A. Källberg, S. Leontein, L. Liljeby, P. Löfgren, A. Paál, A. Simonsson, Ö. Skeppstedt, H. T. Schmidt, H. Cederquist, M. Larsson, and K. Schmidt. Desiree - a double electrostatic storage ring. In *Proc. EPAC 2004*, page 1425, 2004.

[11] S. Rosén, H. T. Schmidt, P. Reinhed, D. Fischer, R. D. Thomas, H. Cederquist, L. Liljeby, L. Bagge, S. Leontein, and M. Blom. Operating a triple stack microchannel plate-phosphor assembly for single particle counting in the 12-300 k temperature range. *Rev. Sci. Instrum.*, 78:113301, 2007.

[12] H. T. Schmidt, H. Cederquist, J. Jensen, and A. Fardi. Conetrap: A compact electrostatic ion trap. *Nucl. Instr. Methods Phys. Res. B*, 173:523, 2001.

[13] H. T. Schmidt, H. A. B. Johansson, R. D. Thomas, W. D. Geppert, N. Haag, P. Reinhed, S. Rosén, M. Larsson, H. Danared, K.-G. Rensfelt, L. Liljeby, L. Bagge, M. Björkhage, M. Blom, P. Löfgren, A. Källberg, A. Simonsson, A. Paál, H. Zettergren, and H. Cederquist. Desiree as a new tool for interstellar ion chemistry. *Int. J. Astrobiol.*, 7:205, 2008.

[14] T. Tanabe, K. Noda, T. Honma, M. Kodaira, K. Chida, T. Watanabe, A. Noda, S. Watanabe, A. Mizobuchi, M. Yoshizawa, T. Katayama, H. Muto, and A. Ando. Electron cooling experiments at ins. *Nucl. Instrum. Methods Phys. Res., Sect. A*, 307:7, 1991.

[15] T. Tanabe, K. Noda, M. Saito, S. Lee, Y. Ito, and H. Takagi. Resonant neutral-particle emission in collisions of electrons with peptide ions in a storage ring. *Phys. Rev. Lett.*, 90:193201, 2003.

[16] R. D. Thomas. When electrons meet molecular ions and what happens next: Dissociative recombinaction from interstellar clouds to internal combustion engines. *Mass. Spec. Rev.*, 27:485, 2008.

[17] T. Walcher. Experiments at the low-energy antiproton ring (lear). *Annual Review of Nuclear and Particle Science*, 38(1):67–95, 1988.

[18] C.P. Welsch, J. Ullrich, C. Gläßner, A. Schempp, R. Dörner, and H. Schmidt-Böcking. Fire-the frankfurt ion storage experiments. *Nucl. Instr. Methods Phys. Res. A*, 527:284, 2004.

[19] J. W. Xia, W. L. Zhan, B. W. Wei, Y. J. Yuan, M. T. Song, W. Z. Zhang, X. D. Yang, P. Yuan, D. Q. Gao, H. W. Zhao, X. T. Yang, G. Q. Xiao, K. T. Man, J. R. Dang, X. H. Cai, Y. F. Wang, J. Y. Tang, W. M. Qiao, Y. N. Rao, Y. He, L. Z. Mao, and Z. Z. Zhou. The heavy ion cooler-storage-ring project (hirfl-csr) at lanzhou. *Nuclear Instruments and Methods in Physics Research Section A: Accelerators, Spectrometers, Detectors and Associated Equipment*, 488:11, 2002.

[20] D. Zajfman, O. Heber, L. Vejby-Christensen, I. Ben-Itzhak, M. Rappaport, R. Fishman, and M. Dahan. Electrostatic bottle for long-time storage of fast ion beams. *Phys. Rev. A*, 55:R1577, 1997.

[21] D. Zajfman, A. Wolf, D. Schwalm, D. A. Orlov, M. Grieser, R. von Hahn, C. P. Welsch, J. R. Crespo Lopez-Urrutia, C. D. Schröter, X. Urbain, and J. Ullrich. The cryogenic storage ring project at heidelberg. *J. Phys.: Conf. Ser.*, 4:296, 2005.

Molecular Ion Dissociation Processes in a Partially Cooled Electrostatic Trap

C. R. Vane[1], M. E. Bannister[1], and S. Deng[1,*], M. Fogle[2], and R. D. Thomas[3]

[1]Physics Division, Oak Ridge National Laboratory, Oak Ridge, TN 37831, USA
[2]Department of Physics, Auburn University, Auburn, AL
[3]Department of Physics, AlbaNova, Stockholm University, SE-106 91, Stockholm, Sweden
E-mail: vanecr@ornl.gov

Abstract. An experimental capability being developed at the Oak Ridge National Laboratory Multi-Charged Ion Research Facility (ORNL MIRF) is described that enables stored cooling and state characterization of atomic and molecular ions of essentially any mass. Ions selected from a variety of available (hot and cold) sources are injected into a 1.5 meter long electrostatic mirror trap, in which excited internal states decay by radiative cooling. Ion lifetime studies or direct measurements of 'in-trap' interactions of molecular ions with electrons or atomic targets are accomplished with either a crossed electron beam or atomic beam (with recoil ion detection) located in the field-free region of the ion-trap, coupled with product velocity imaging. Similarly, extraction of trap-cooled ions can be used 'out-of-trap' for near state-specific electron capture (EC) and dissociative charge exchange (DCE) studies employing higher resolution COLTRIMS techniques coupled with fragment imaging.

1. Introduction

Dissociative interactions of molecular ions with electrons, atoms, and other molecules are important in a wide variety of environments, generating reactive fragments (e.g., free radicals) that often play significant roles in initiating and driving the dominant chemistries. Examples range from cool plasma interactions at fusion near-edge regions, to enhanced chemical vapor deposition plasmas, to planetary upper atmospheres and interstellar clouds, and to molecular breakup phenomena used in mass spectroscopic techniques for investigating the structures of heavy (e.g., biological) molecules. The interactions involved are complex and any hope of untangling the underlying many-body processes and of developing a fundamental understanding of observable charge and energy evolution will require reliable detailed data arising from extreme control of the interacting species' quantum states and environments. One method for achieving such significant control involves storage of molecular ions in an innocuous environment (e.g., low EM field, ultrahigh vacuum, low wall temperature) until radiative decay provides a near ground-state population for direct interaction measurements with a similarly identified target species, or for subsequent excited-state preparation preceding such measurements. Often heavy molecular ions are stored as energetic beams to benefit from kinematic advantages in detection and analysis of products. A number of laboratories are centering their research on enhanced capabilities of intrinsically mass independent all-electrostatic storage traps and rings that are relatively small and therefore cheaper to construct and operate (especially at cryogenic temperatures) than magnetic rings that have been employed so successfully, for example, in exploring and elucidating details of the dissociative recombination processes.

At ORNL we are developing a broad potential experimental capability for detailed, near state-specific studies of dissociation processes occurring for interactions of molecular ions with free electrons and atomic targets, based on a linear electrostatic mirror trap, similar to the design of Zajfman et al [1]. Our research is concentrated on dissociation of O_2^+, O_3^+, and XH_2^+, to complement our dissociative recombination (DR) studies at CRYRING. The MIRF Ion Cooling and Characterization Endstation Trap (ICCE TRAP) has been designed to store atomic and molecular species from a number of ion sources (ranging from internally hot ions produced in a permanent magnet ECR ion source, to heavy biomolecular ions from an electrospray ion source), permit them to radiatively cool, interact them with crossed electron and atomic targets, and measure dissociation fragments using imaging techniques. The linear trap (with 40 mm mirror element aperture) is similar in scale to the Cryogenic Trap for Fast ion beams being developed Max-Planck-Institut für Kernphysik at Heidelberg [2], except that the ORNL trap is designed to operate primarily at room temperature. Unique features of the ICCE TRAP's design enabling a variety of both ion-electron and ion-atom interaction studies include use of side-injection of ions to avoid fast switching of high voltages permitting operation over a wide range of stored ion energies (up to 16 keV per charge), and instrumentation with an intense crossed-beam (5-100 eV) electron target and an 'in-trap' atomic beam target with recoil ion detection.

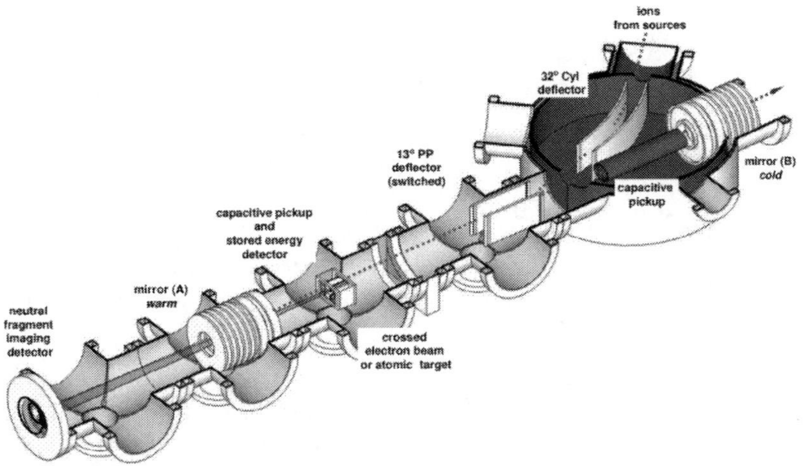

Figure 1. Layout of the ICCE TRAP at the ORNL MIRF. Ions enter from the top right port and are injected along the trap axis to Mirror (A) at the bottom left using pulsed electrostatic deflectors.

A schematic of the apparatus is shown in Figure 1. Atomic and molecular ions enter from the top right and are injected onto the linear trap axis through combination of a 32° cylindrical deflector and a pulsed 13° parallel plate deflector. The ions oscillate between two electrostatic mirrors (A) and (B), the latter designed to operate inside a copper shroud cryo-cooled to ~ 20 K. The trap is pumped by three cryo-pumps, one titanium sublimation pump, and one getter pump. Typical background pressure over the 1.5 m long mirror to mirror trap length is $< 1 \times 10^{-10}$ Torr. Transverse magnetic fields along the axis are reduced to less than 100 mG using a pair of Helmholtz-like coils mounted above and below the trap axis. Stored ions interact with either a low-energy (5-100 eV) crossed electron beam target or an atomic or molecular beam target, LN cooled to ~ 80 K, mounted between the mirrors. IR laser reheating of trapped, cooled molecular ions is also planned through an access port located at the ion turn around point in mirror (A). A 78 mm diameter imaging detector based on a triple-MCP, Z-stack followed by a P46 phosphor screen is located at one end approximately 1 m beyond mirror (A) for measurements of neutral product fragments from molecular ions interacting with the targets, while the other mirror (B) will be instrumented with an a pulsed, field-modifying electrode for extended slow extraction of stored ions to 'out-of-trap' measurements. The imaging detector phosphor light pulses are also focused onto a segmented PMT for timing and particle multiplicity gating. Two dimensional images of multi-fragment hit data are stored on a fast, multi-frame capable camera and the data analyzed real-time. The imaging detector can also function as a position-sensitive detector of charge-reduced atomic ions formed by electron capture (EC) in the atomic target when multi-charged atomic ions are stored in the trap. These ions readily escape the trap along its axis, and a secondary electrostatic deflector (not shown in Figure 1) mounted between mirror (A) and the detector is used to spatially separate ions and neutrals.

Ion loading of the trap is accomplished from the side by switching on the 13° parallel plate deflector, and then rapidly switching off the deflector steering potential prior to return the ion pulse as it is reflected from the first mirror (A). The trap is then filled to a maximum of about 40% of its length by an ion bunch that oscillates between the mirrors. Long term (seconds) trapping is accomplished by tuning the potentials on elements of the mirrors and on Einzel lenses at the entrance to each mirror. Optimization is obtained by tuning two sets of horizontal and vertical 'trim' plates, also located at the entrance to each mirror. The mirror potentials, the highest of which must exceed the energy of the trapped ions divided their charge, are then held static during trap operation, avoiding any fast switching of potentials greater than a few kV even for ion energies up to tens of keV. Exact timing and pulse width for the 13° injection deflector depends on the velocity of the ions being injected, but is typically about 10 μs wide for light ions, permitting loading of approximately 6×10^4 ions per particle nA of beam current.

The number of stored ions can be monitored as a function of time after injection via three separate methods, two non-destructive and one destructive. In the first, a capacitive pickup (19 mm dia x 19 mm long) mounted inside the atomic target grounded shield can be used to measure the relative intensity of beam pulses when the ions are bunched and to infer stored ion mass distribution from Fourier analysis.

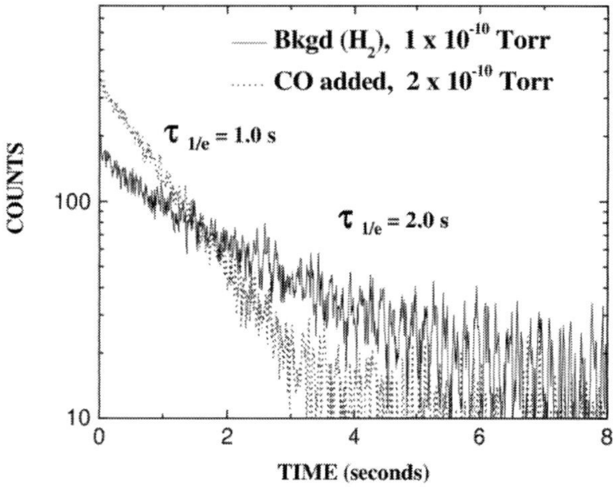

Figure 2. Lifetime curves from measured neutrals for 10 keV CO^+ ions, for background and with approximately 1×10^{-10} Torr (indicated ion gauge pressure) added CO gas.

In the second method, neutrals formed from electron capture, dissociative charge exchange, collisional dissociation (CD), or other neutral fragment-producing dissociation processes – such as dissociative excitation (DE), escape the trap and are detected in the imaging detector. Instantaneous neutrals rates are proportional to the number of stored ions. Figure 2 displays neutral counts (from C, O, and CO) versus trap time (after injection) for 10 keV CO^+ ions stored at 1×10^{-10} Torr background chamber pressure (~90% H_2), and the effect of adding CO target gas.

An atomic beam target is located 22 cm from one of the mirrors and 139 cm from a 78 mm diameter imaging detector. Electron capture generated target recoil ions are extracted by a weak electric field (< 50 V/cm) perpendicular to the trap axis and transported to a fast discrete dynode electron multiplier detector to provide the 'start' signal for time-of-flight (TOF) measurement. Arrival timing of the neutrals made with respect to trap injection, especially when gated with coincidently detected target recoil ions, can be used to represent the time evolution of spatial distribution of the ions in the trap. Under certain mirror element biasing conditions, trapped ions can be automatically retained in narrow (few μs) bunches for hundreds of milliseconds [3]. Example data from a storage oscilloscope for stored 10 keV O^+ ions interacting with an O_2 gas target are displayed in Figure 3 indicating detection timing of neutral fragments and recoil target ions after injection. Signals with a period of 9.7 μs arriving from a 1.4 μs wide ion bunch passing through the target toward the neutral detector are clearly visible, as well as correlated signals from O_2^+ recoil ions created in charge exchange events in each direction of the ion's motion.

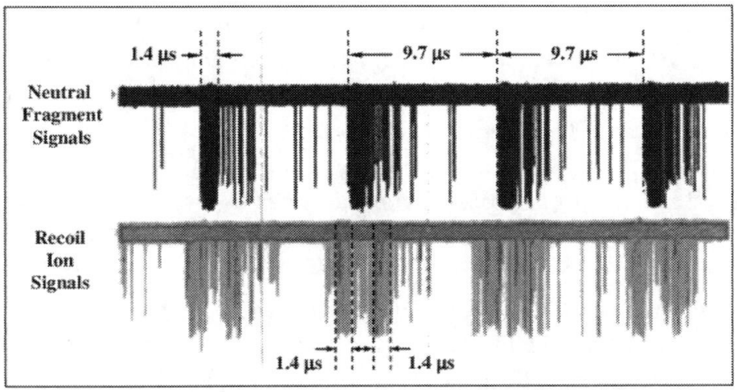

Figure 3. Storage oscilloscope screen shot of signals from neutral O atoms (black) and O_2^+ recoil ions (gray) for trapped 10 keV O^+ ions. The oscilloscope was triggered 1 ms after beam injection and signal traces were accumulated over many trapping cycles to produce this image.

The time-of-flight TOF delay spectral data – [recoil ion = 'TAC start', neutral = 'TAC stop'], can be used to uniquely identify EC or DCE from a particular target atom at the atomic beam target position on an event-by-event basis. An example is shown in Figure 4 for 10 keV O_2^+ ions undergoing EC and DCE with background H_2 and H_2O, and with a small amount of added Ar. By gating data on a selected recoil-

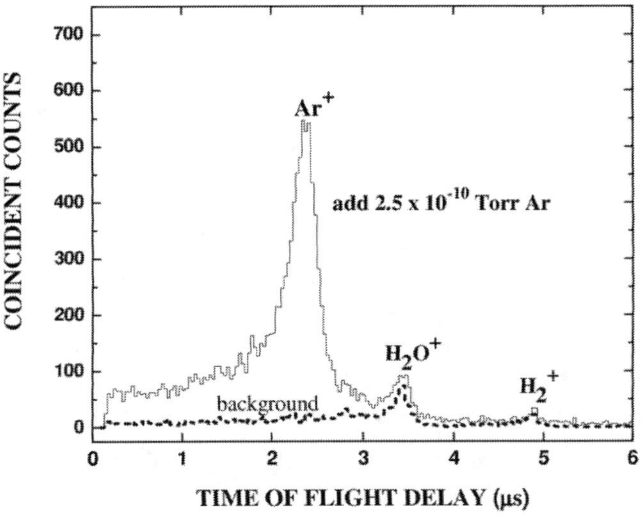

Figure 4. Measured TOF spectra for O and O_2 neutrals from DCE and EC by trapped 10 keV O_2^+ ions detected in coincidence with correlated recoil target ions. Events arising from EC from background gas (mainly H_2O and H_2) are indicated by the dotted line, data from added argon are indicated by the solid line.

Journal of Physics: Conference Series **300** (2011) 012012 doi:10.1088/1742-6596/300/1/012012

neutral TOF delay, measurements can be accumulated for specific ion-atom combinations and as a function of cooling time for ions stored in the trap. For example, imaging of DCE-generated fragments can be simultaneously selected according to trap cooling time of the molecular ion and the target atom element contributing the active electron. Even more specific selection of events permitting identification of initial and recoil final states can be accomplished using high resolution COLTRIMS techniques. But these are presently restricted to external or 'out-of-trap' measurements using extracted ions due to the complexity of the COLTRIMS apparatus required for high resolution.

A third, destructive method being developed for determining the trap lifetime curve for a given ion, energy, and trap tune is based on measurement of total stored ion energy. Ions are injected and stored for a given variable time, then dumped in one cycle onto a calibrated solid-state detector (International Radiation Detectors Inc., IRD) by appropriate synchronized pulsing of the 13° deflector. For bunched ions where the deflector can be switched on during a period in which the ion bunch is completely located within a 'turn-around' in mirror (B), the signal directly represents the total number of ions multiplied by the energy per ion. For un-bunched ions more nearly filling the trap, the IRD detector signal represents a fraction of the stored ions. The advantage of this method is its potential absolute nature (IRD available absolute detector energy calibration and on-line calibration using ion pulses directly from the source) and its relative independence from ion-cooling effects on the details of the measurement.

The ICCE Trap is also used as a metastable atomic ion fraction diagnostic employing stored-ion decay curve measurements for selected singly- and multiply-charged atomic ion beams from the ECR ion source. These are especially of interest for ion species used in merged ion-neutral beams experiments in which significant metastable fractions are expected, e.g., He-like, Be-like, and Mg-like ions. Lifetimes of interesting atomic ions with electric dipole forbidden transitions (mainly M1) range from microseconds to tens of seconds. ICCE Trap ion storage lifetime measurements permit determinations of populations of some metastable ions with lifetimes ranging from milliseconds to seconds [4,5]. These 'in-trap' metastable fraction determinations are possible for stored-ion species for which electron transfer rates from specific targets (e.g., He) depend strongly on the ion's initial internal electronic state, leading to trapped ion overall decay rates that systematically change with the evolving metastable state population. For example, C^{2+} ions from the ORNL MIRF permanent magnet ECR ion source are expected to have a substantial metastable $C^{2+}(1s^2 2s2p\ ^3P)$ population, with component lifetimes of 9.7 and 190 ms [6]. Electron capture by these metastable ions is known to proceed with cross sections about 3-7 times larger than for ground state C^{2+} ions [7]. Initial analysis of lifetime data versus He pressure yield a $C^{2+}(^3P)$ metastable fraction of approximately 45%. Similar techniques can be applied to molecular ion dissociation studies, for example DCE, if rates are sufficiently state-dependent to significantly affect the pace of neutral fragment production.

More detailed measurements, leading to identification of the particular metastable quantum states involved, can be performed in 'out-of-Trap' charge-changing measurements using timed-extraction of ions from the trap and COLTRIMS-based high resolution experimental techniques as described in references [8,9] for molecular ions. The main disadvantage to this 'out-of-trap' method is unavoidably low data acquisition efficiency due to relatively poor duty cycle. The ORNL COLTRIMS apparatus configured for measurements of DCE of molecular ions on gas targets is displayed in Figure 7. Molecular ions from any of the available ion sources will first be injected and cooled (or prepared via laser or electron interactions) in the ICCE TRAP. After a suitable time, atomic or molecular ions can be extracted from the trap through mirror (B) over an extended number of oscillations using a mirror field-modifying technique we have developed using SIMION ion-optics simulations. The extracted ions are then directed into the COLTRIMS apparatus where they interact with a cold atomic beam target. Standard COLTRIMS

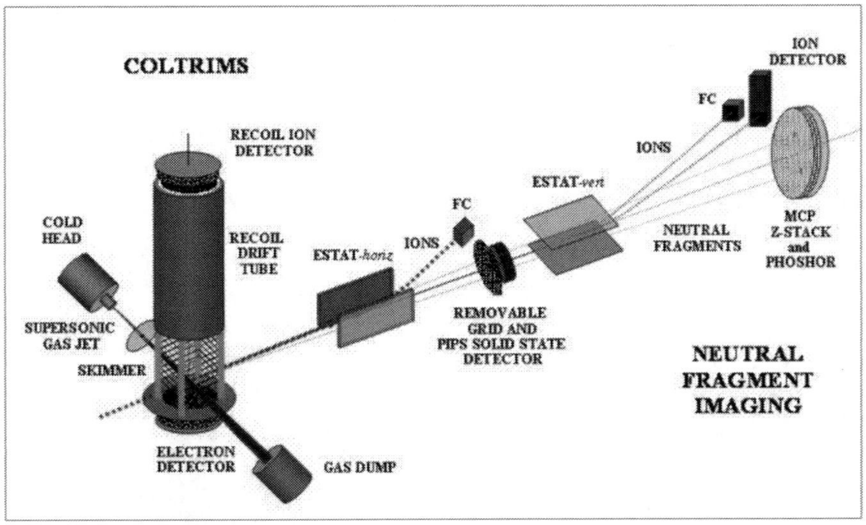

5

Figure 7. COLTRIMS apparatus for near-state specific measurements of EC by atomic ions and DCE by molecular ions using ions cooled and extracted from the ICCE TRAP.

techniques are then used to provide high resolution recoil ion momentum analysis and to derive the collision Q-value for each event. Charge-changed products, including neutral fragments in the case of DCE of molecular ions, are detected in coincidence with recoils using an imaging detector 1.5 m downstream of the COLTRIMS target. Values of Kinetic Energy of Release (KER) are determined from the imaging data, and can be correlated with the recoil data to identify initial and final states [6,7].

Acknowledgements
Research sponsored by the U.S. DOE, Office of Basic Energy Sciences, Division of Chemical Sciences, Geosciences, and Biosciences, under Contract No. DE-AC05-00OR22725 with UT-Battelle, LLC.

*Author SD gratefully acknowledges support from the ORNL Postdoctoral Research Associates Program administered jointly by Oak Ridge Institute for Science Education and ORNL.

References

[1] Zajfman, D., et al, 1997 Phys. Rev. A **55**, R1577 - R1580
[2] Lange, M., et al, Rev. Sci. Instrum. **81**, 055105 (2010)
[3] Pedersen, H. B., et al, Phys. Rev. Lett 87, 055001 (2001).
[4] Trabert, E., Physica Scripta **61**, 257 (2000).
[5] Fardi, A., Cederquist, H., and Schmidt, H. T., Hyperfine Interactions **127**, 247 (2000).
[6] Fogle, M. et al., Astrophys. J. Supp. Ser. **175**, 543 (2008), and references therein.
[7] Ishii, K., Itoh, A., & Okuno, K., Phys. Rev. A, **70**, 042716 (2004).
[8] Schmidt, L. Ph. H., Phys. Rev. Lett. **101**, 173202 (2008).
[9] Johnson, N. G., J.Phys: Conf. Ser. **194,** 102004 and 102019 (2009), and Johnson, N. G., unpublished thesis, Kansas State University 2010.

Eighth International Conference on Dissociative Recombination (DR2010) IOP Publishing
Journal of Physics: Conference Series **300** (2011) 012013 doi:10.1088/1742-6596/300/1/012013

The σ_v reflection symmetry in variational R-matrix theory: Application to Rydberg and doubly-excited states of $^1\Sigma_g^+$ and $^1\Sigma_g^-$ symmetries in diatomic hydrogen

Soumaya Bezzaouia[1], Fatma Argoubi[1], Mourad Telmini[1,2] and Christian Jungen[3,4]

Abstract. The σ_v reflection symmetry for diatomic molecules is discussed in the context of variational R-matrix theory applied to the H_2 molecule ("halfium model"). It is shown that explicit use of the reflection symmetry in the R-matrix calculations improves the results of singly and doubly excited $^1\Sigma_g^+$ states and resonances. R-matrix calculations of non-autoionizing $^1\Sigma_g^-$ doubly-excited states of H_2 situated 27 eV above the ground state are also presented and compared with existing *ab initio* data obtained by other methods.

1 LSAMA, Department of Physics, Faculty of Science of Tunis, University of Tunis El Manar, 2092 Tunis, Tunisia
2 National Centre for Nuclear Science and Technology, Sidi Thabet Technopark 2020 Tunisia
3 Laboratoire Aimé Cotton du CNRS, Université de Paris-Sud, 91405 Orsay, France
4 Department of Physics and Astronomy, University College London, UK

E-mail: mourad.telmini@fst.rnu.tn

1. Introduction

Singly and doubly excited Rydberg states of neutral hydrogen H_2 are known to play a key role in the DR process of H_2^+ ions with slow electrons [1]-[4]. They act as intermediate resonances which promote the conversion of electron energy into nuclear kinetic energy during the reactive collision process. The theoretical description of the collision process requires a good knowledge of the electronic structure of both the hydrogen ion H_2^+ and the neutral molecule H_2, in terms of potential energy curves and/or quantum defects. In particular, the doubly-excited states may autoionize or dissociate into neutral hydrogen atom fragments. Numerous papers [5]-[9] have been published which are dealing with the so-called Q_1 and Q_2 resonance series converging towards the H_2^+ $2p\sigma$ and $2p\pi$ thresholds respectively, and particular attention has been paid to the lowest Q_1 resonance of $^1\Sigma_g^+$ symmetry, because its Born-Oppenheimer potential energy curve crosses the threshold not far from the equilibrium position and therefore dominates the DR process.

In recent years we have developed an *ab initio* R-matrix approach [10] which enables bound states and core-excited scattering states of H_2 to be calculated for fixed nuclei. Our work is based on the pioneering ideas of Greene and Yoo [11][12], which it adapts such as to yield quantum defect matrices that evolve smoothly throughout the bound and continuous energy regions, and which also vary relatively mildly as functions of the molecular geometry (internuclear distance

Journal of Physics: Conference Series **300** (2011) 012013 doi:10.1088/1742-6596/300/1/012013

R). Our "halfium model" combines the variational eigenchannel R-matrix method [13] with the Generalized Multichannel Quantum Defect Theory (GMQDT) implemented using prolate spheroïdal electron coordinates [14]. The approach has been used to investigate singlet ungerade symmetries $(^1\Sigma_u, ^1\Pi_u)$ [10] and gerade symmetries $(^1\Sigma_g, ^1\Pi_g, ^1\Delta_g)$ [15], both for the bound and the continuum (autoionization) regions.

It is interesting to note that even the most recent quantum chemical calculations [16], while following individual potential energy curves out to large internuclear distances, are still limited to the lowest excited states since they do not exceed $n \approx 5$ (where n is the principal quantum number). On the other hand, the halfium model is limited to relatively small $R-$values ($R \approx 10a.u.$), and at this stage is not capable of describing double ionization. One of the advantages of the "halfium model" however is that it produces quantum defects which can be used directly for the calculation of the nuclear-electronic dynamics in the framework of multichannel quantum defect theory (MQDT). The halfium scheme together with the MQDT framework leads to a global analysis of the interactions as no distinction is made between "open" and "closed" channels at the outset, but instead all channels are treated on the same footing irrespective of their corresponding channel thresholds. In this picture doubly excited states are included explicitly as electron-ion collision channels in their own right. It has further been shown [17] how MQDT can describe competing ionization and dissociation processes, and include ro-vibrational dynamics via the frame transformation technique [18]. Therefore MQDT possesses all the ingredients to address the problem of dissociative recombination and, when combined with the halfium model, it is capable to yield *ab initio* predictions of DR processes as well. It thus represents an alternative to the more traditional approach to DR, where the dissociative state (corresponding in H_2 to the lowest $Q_1\ ^1\Sigma_g^+$ resonance) is treated as a valence state rather than as part of an electron-ion channel [1][2].

The previous implementations of the molecular variational R-matrix method [10]-[12] made no distinction between channels of Σ^+ and Σ^- symmetry, in other words, the symmetry property of the electron wavefunction obtained by a reflection of all electron coordinates at a plane containing the two nuclei (symmetry operation σ_v) was not considered explicitly. In this paper, we show how the halfium model may be reformulated such as to include this symmetry property. We show how the convergence in the calculation of $^1\Sigma_g^+$ states is improved by this extension, and we also present initial results on doubly excited states of H_2 of $^1\Sigma_g^-$ symmetry which occur 27 eV above the ground state. These turn out to be non-autoionizing despite their high energy. To our best knowledge, only two previous theoretical studies of Σ^- states in H_2 have been made [19][20], and no experimental evidence for their existence appears to have been found so far.

2. Theory

In the halfium model, the two-electron configuration space is devided into two regions: (i) a reaction volume where the variational R-matrix method is used, and (ii) the remaining space, called asymptotic zone, where GMQDT is used. The connection of the inner zone and outer zone wavefunctions yields the desired reaction matrix or the equivalent quantum defect matrix.

In the original formulation of the halfium model [10][12] each calculation was performed separately for each molecular symmetry defined by the standard quantum numbers Λ (projection of the angular momentum on the internuclear axis, taken positive), electron spin ($S = 0, 1$), and *gerade/ungerade* ($p = 0$ or 1). From elementary molecular theory [21] we know that in a diatomic molecule the symmetry operation σ_v corresponding to a reflection of all electron coordinates at any plane containing the two nuclei also commutes with the electronic Hamiltonian so that an additional quantum number arises: the electronic wavefunction has $-$ symmetry ($q = 1$) or $+$ symmetry ($q = 0$) depending on whether it changes sign or not upon reflection.

The only symmetry for which the q quantum number is relevant at the electronic stage of calculation, is Σ symmetry ($\Lambda = 0$)[22][23]. As far as molecular hydrogen is concerned, the σ_v

symmetry may in practice be disregarded even for Σ states as long as energies below $2p\sigma$ ion threshold are considered. The reason is that only Σ^+ states arise in this range. Indeed, the lowest Σ^- state occurs associated with the $(2p\pi)\epsilon p\pi$ configuration and is expected to lie at around 27 eV above the ground state. The Σ^- states are thus well separated from Σ^+ states, and for many practical purposes need not be taken into consideration.

In what follows we give a brief description of the main modifications required in the halfium model in order to take account of the σ_v symmetry. Each two-electron basis function corresponding to a configuration i, j (cf. [10][12]) is defined in the reaction zone ($\xi_1, \xi_2 \leq \xi_0$) and is specified by the set of quantum numbers $\{S, \Lambda, p, q\}$:

$$\overline{y_{ij}}(\overrightarrow{r_1}, \overrightarrow{r_2}) = \overline{N_{ij}} \left[y_{ij}^{+\,-}(\overrightarrow{r_1}, \overrightarrow{r_2}) + (-1)^q \, y_{ij}^{-\,+}(\overrightarrow{r_1}, \overrightarrow{r_2}) \right] .$$

$\overrightarrow{r_1}$ and $\overrightarrow{r_2}$ are the position vectors of each electron defining the corresponding spheroïdal coordinates (ξ, η, φ). The functions $y_{ij}^{+\,-}$ are products of one-electron orbitals characterized by *signed* orbital angular momentum components:

$$\begin{aligned} y_{ij}^{+\,-}(\overrightarrow{r_1}, \overrightarrow{r_2}) &= N_{ij} \left[\phi_i^{+}(\overrightarrow{r_1}) \phi_j^{-}(\overrightarrow{r_2}) + (-1)^S \phi_j^{-}(\overrightarrow{r_1}) \phi_i^{+}(\overrightarrow{r_2}) \right] \\ y_{ij}^{-\,+}(\overrightarrow{r_1}, \overrightarrow{r_2}) &= N_{ij} \left[\phi_i^{-}(\overrightarrow{r_1}) \phi_j^{+}(\overrightarrow{r_2}) + (-1)^S \phi_j^{+}(\overrightarrow{r_1}) \phi_i^{-}(\overrightarrow{r_2}) \right] , \end{aligned}$$

where $\phi_i^{\pm}(\overrightarrow{r})$ is a one-electron wavefunction associated with the eigenstate $\left|i^{\pm}\right\rangle = \left|n_i \ell_i \lambda_i^{\pm}\right\rangle$ of H_2^+ with $\lambda_i^+ = |\lambda_i|$ and $\lambda_i^- = -|\lambda_i|$. $\overline{N_{ij}}$ and N_{ij} in the preceding equations are normalization factors, and it is understood that ℓ_i is the generalized orbital angular momentum quantum number arising when spheroïdal coordinates are used [10].

The two-electron basis functions are used to solve the Schrödinger equation inside the reaction volume in the framework of the variational R-matrix method. The solutions are denoted Ψ_β and are characterized by stationary logarithmic derivatives b_β on the reaction surface $\max(\xi_1, \xi_2) = \xi_0$. Each solution Ψ_β may be continued in the external zone, i.e. for radii larger than ξ_0, as a linear combination of a regular and irregular two-center Coulomb radial functions. The matching is achieved by expanding each inner-zone variational solution Ψ_β on the reaction surface in terms of so-called "surface harmonics" $\overline{\Phi_k}(\omega)$ (where ω stands for all coordinates except the radial coordinate of the outermost electron) for each asymptotic channel $k \equiv ij$ (see [10]). Taking account of the reflection symmetry we write these surface harmonics as:

$$\overline{\Phi_k}(\omega) = \overline{N_{\Phi_k}} \left[\Phi_k^{+\,-}(\omega) + (-1)^q \Phi_k^{-\,+}(\omega) \right]$$

where

$$\begin{aligned} \Phi_k^{+\,-}(\omega) &= N_{\Phi_k} \phi_i^{+}(\overrightarrow{r}_<) \, Y_{\ell_j \lambda_j^-}(\eta_>, \varphi_>) \\ \Phi_k^{-\,+}(\omega) &= N_{\Phi_k} \phi_i^{-}(\overrightarrow{r}_<) \, Y_{\ell_j \lambda_j^+}(\eta_>, \varphi_>) , \end{aligned}$$

and $\overline{N_{\Phi_k}}$ and $N_{\Phi_k}$are normalization factors. Here the indexes $<$ and $>$ designate the inner and outermost electron respectively; they arise from the indiscernability of the electrons. $Y_{\ell\lambda}(\eta, \varphi)$ are spheroïdal harmonics analogous to the standard spherical harmonics.

The matching of the inner zone solutions to those obtained outside the reaction volume yields short-range reaction matrix $\mathbf{K}$ and equivalent quantum defect matrix $\boldsymbol{\mu}$. The energies of bound and doubly-excited states are finally obtained by standard MQDT procedures as in the standard halfium model [10].

Table 1. $^1\Sigma_g^+$ $ns\sigma$ Rydberg series in H_2. Root mean squares and averaged values of symmetrized (● in Fig 1) and non-symmetrized (○ in Fig. 1) deviations of effective principal quantum numbers with respect to quantum chemical data [24] (cf. the text).

state	$2s\sigma$	$3s\sigma$	$4s\sigma$	average
$\text{rms}_{non-sym}$	0.024	0.025	0.020	0.023
rms_{sym}	0.014	0.014	0.016	0.015
$\overline{\Delta\nu}_{non-sym}$	0.021	0.021	0.016	0.019
$\overline{\Delta\nu}_{sym}$	0.011	0.010	0.010	0.010

3. Results

3.1. *Singly excited bound $^1\Sigma_g^+$ states: the $ns\sigma$ Rydberg series*

Although Σ^+ states in H_2 are generally well separated from Σ^- states, a slight improvement of the bound state energies of the Σ^+ states nevertheless occurs due to improved convergence properties of the calculation when the symmetrized version of the code is used. A Σ symmetry two-electron variational basis typically is composed of $\sigma\sigma$, $\pi\pi$, $\delta\delta$ ... configurations. The two first ion thresholds are of σ−type ($1s\sigma$ and $2p\sigma$), so that the majority of the basis functions will naturally correspond to $\sigma\sigma$ configurations. These have Σ^+ symmetry and consequently are not affected by the basis set symmetrization. By contrast, $\pi\pi$ and $\delta\delta$ configurations yield both Σ^+ and Σ^- symmetry functions. For a given size of the basis set, the symmetrization of the basis function will therefore effectively optimize the basis, pushing down the lowest roots of the Hamiltonian, although the improvement is expected to be small as stated earlier. We have verified that two calculations done, respectively, with and without symmetrization, but including exclusively $\sigma\sigma$ configurations, do give exactly the same results as required. The calculations have been carried out by using reaction boundaries corresponding to $\xi_0 = 25/R + 1$.

Figure 1 displays differences $\Delta\nu = \nu - \nu_{ab\,initio}$ as a function of the internuclear distance R for the $^1\Sigma_g^+$ $ns\sigma$ series of H_2 (principal configuration $[1s\sigma]ns\sigma$, $n = 2-4$)). The effective principal quantum number is defined by the Rydberg equation as $\nu = 1/\sqrt{E_{1s\sigma} - E_n}$, where E_n is the bound state eigen-energy and $E_{1s\sigma}$ the first ionization threshold energy in rydbergs. The effective quantum numbers $\nu_{ab\,initio}$ have been extracted from quasi-exact state-of-the-art quantum chemical calculations [24]. Open symbols in Figure 1 show the differences obtained in a non-symmetrized R-matrix calculation, whereas full symbols designate the differences obtained when a calculation with a symmetrized basis of the same size was carried out. The improvement obtained with the symmetrized code is obvious, particularly for small R values.

A more quantitative analysis of these results is presented in Table 1. For each of the three Rydberg states represented in Fig. 1, we have averaged the deviations over R and evaluated the rms value of $\Delta\nu$. The last column contains the average of the rms value for all three states. This quantity is a global measure of the improvement obtained: the value decreases from 0.023 (for non symmetrized calculations) to 0.015 (for symmetrized calculations), showing that the overall improvement amounts to 35% . Furthermore, by averaging the $\Delta\nu$ values for each state separately, we see that the symmetrized energies are systematically lower than non-symmetrized energies, confirming once again the improvement of the calculations. In terms of averages, the deviation is reduced by a factor 2 ($\langle\overline{\Delta\nu}_{non-sym}\rangle = 0.019$ and $\langle\overline{\Delta\nu}_{sym}\rangle = 0.010$).

3.2. *Doubly excited $^1\Sigma_g^+$ resonances: the Q_1 resonance series*

The Q_1 resonance series are situated above threshold and therefore have the character of autoionizing series. It must therefore be stressed that the Ritz variational principle does not

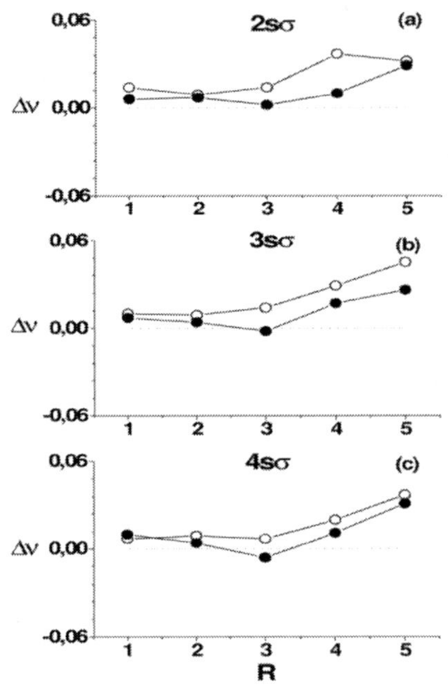

Figure 1. $^1\Sigma_g^+$ $ns\sigma$ Rydberg series in H_2. Deviations $\Delta\nu = \nu - \nu_{ab\ initio}$ of calculated effective principal quantum numbers obtained without (○) and with (●) symmetrization, from the corresponding quantum chemical results [24], plotted as function of the internuclear distance R (in bohr) for (a) $2s\sigma$, (b) $3s\sigma$ and (c) $4s\sigma$. ν is the effective principal quantum number.

apply in this situation as a consequence of the fact that the core excited states are embedded in the open ionization continua. This means that we have no unique criterion at our disposal in order to asses the quality of the calculations. Of course we may expect that the results obtained with symmetrized and with non-symmetrized bases should be basically comparable, and as a rule of thumb we can hope that a lowering of the resonance positions is a sign that the calculation has been improved. The calculations have been carried out as described in Ref. [15].

Tables 2, 3 and 4 list effective principal quantum numbers for the core excited $^1\Sigma_g^+$ states nQ_1 ($n = 1, 2, 3$) resonances. The notations have the same meaning as in Table 1, except that $\nu_{ab\ initio}$ here refers to the R-matrix calculations from Ref. [6]. As before, the last row of each Table displays the averages of the principal quantum number differences. We can see a significant improvement for the lowest resonance (in the sense defined above), whereas the two higher resonances are hardly affected.

3.3. Doubly excited non-autoionizing states of $^1\Sigma_g^-$ symmetry

We have evaluated $^1\Sigma_g^-$ states corresponding to the doubly-excited Rydberg series $(2p\pi)(n'p\pi)$ $n' \geq 3$ and $(2p\pi)(n''f\pi)$ $n'' \geq 4$ converging to $2p\pi$ ion threshold. This kind of calculation had not been possible with the old version of our R-matrix code and is conducted here for the first time. To underline the non-autoionizing character of these metastable states, we denote them as nP_2 $^1\Sigma_g^-$. This notation is inspired by Bottcher's notation nQ_2 for autoionizing doubly-excited states lying in the same energy range [25].

Table 2. Effective principal quantum numbers of the $1Q_1$ $^1\Sigma_g^+$ core excited resonance. Values ν_{sym} and $\nu_{non-sym}$ correspond to symmetrized and non-symmetrized calculations, respectively, and $\nu_{ab\ initio}$ are derived from Ref. [6]. The internuclear distance R is in bohr.

R	$\nu_{non-sym}$	ν_{sym}	$\nu_{ab\ initio}$	$\Delta\nu_{non-sym}$	$\Delta\nu_{sym}$
1.0	1.594	1.578	1.630	−0.036	−0.052
2.0	1.473	1.465	1.467	0.007	−0.002
average				−0.014	−0.027

Table 3. Same as Table 2 for the $2Q_1$ $^1\Sigma_g^+$ core excited resonance.

R	$\nu_{non-sym}$	ν_{sym}	$\nu_{ab\ initio}$	$\Delta\nu_{non-sym}$	$\Delta\nu_{sym}$
1.0	2.783	2.775	2.821	−0.038	−0.046
2.0	2.675	2.680	2.681	−0.006	−0.001
3.0	2.633	2.632	2.649	−0.016	−0.017
4.0	2.724	2.723	2.705	0.019	0.018
average				−0.010	−0.011

Table 4. Same as Table 2 for the $3Q_1$ $^1\Sigma_g^+$ core excited resonance.

R	$\nu_{non-sym}$	ν_{sym}	$\nu_{ab\ initio}$	$\Delta\nu_{non-sym}$	$\Delta\nu_{sym}$
1.0	3.800	3.794	3.817	−0.017	−0.023
2.0	3.693	3.696	3.702	−0.009	−0.006
3.0	3.663	3.662	3.672	−0.009	−0.010
4.0	3.740	3.737	3.731	0.009	0.006
5.0	3.885	3.818	3.822	0.063	−0.004
average				−0.007	−0.007

Table 5. Energies of the non-autoionizing $1P_2$ $^1\Sigma_g^-$ doubly-excited state in H_2 for different internuclear distances R. E_S, this work. E_K, Ref. [20]. E_M, Ref. [19]. Energies and R are in atomic units. See the text for details.

	previous		present		
R	E_K	E_M	E_S	ΔE_{SK}	ΔE_{SM}
1.0	−0.5523	−0.5520	−0.5522	0.0001	−0.0002
2.0	−0.5031	−0.5029	−0.5030	0.0001	−0.0001
3.0	−0.4568	−0.4566	−0.4567	0.0001	−0.0001
4.0	−0.4176	−0.4174	−0.4175	0.0001	−0.0001
5.0	−0.3850	−0.3848	−0.3848	0.0002	0.0000

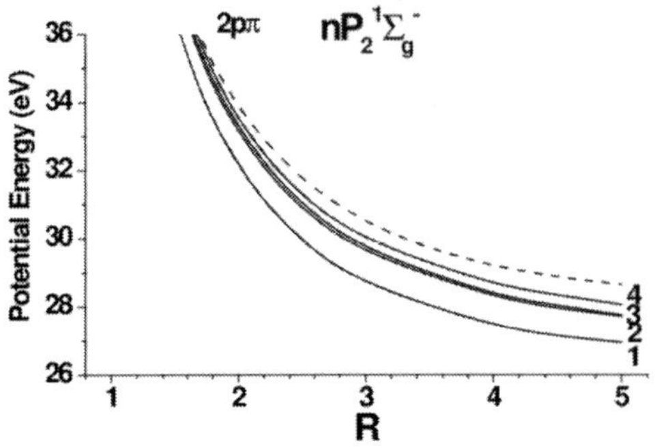

Figure 2. Potential energy curves of $nP_2\ ^1\Sigma_g^-$ states. The $2p\pi$ threshold is represented by a dashed line.

Table 6. Energies of non-auto-ionizing states $nP_2\ ^1\Sigma_g^-$ ($n = 2 - 4$) obtained here, E_S, and by Martin [19], E_M, and their differences for different internuclear distances R. Energies and R are in atomic units.

	$2P_2$			$3P_2$			$4P_2$		
R	E_S	E_M	ΔE_{SM}	E_S	E_M	ΔE_{SM}	E_S	E_M	ΔE_{SM}
1.0	−0.5132	−0.5133	0.0001	−0.5060	−0.5062	0.0002	−0.4978	−0.4978	0.0000
2.0	−0.4664	−0.4665	0.0001	−0.4610	−0.4613	0.0003	−0.4518	−0.4518	0.0000
3.0	−0.4226	−0.4227	0.0001	−0.4194	−0.4195	0.0001	−0.4088	−0.4087	−0.0001
4.0	−0.3857	−0.3857	0.0000	−0.3844	−0.3847	0.0003	−0.3725	−0.3724	−0.0001
5.0	−0.3563	−0.3564	0.0001	−0.3548	−0.3550	0.0002	−0.3432	−0.3432	0.0000

Table 5 lists the present results (indexed by S in the Table) obtained for the $1P_2\ ^1\Sigma_g^-$ state for different internuclear distances R, and compares them with those of Martin [19] (indexed by M) and Komasa [20] (indexed by K). To the best of our knowledge, these two papers are the only existing publications reporting *ab initio* calculations of Σ^- states of H_2. The Table indicates that there is overall agreement between the different calculations. As expected, the quantum chemical energies of Ref. [20] are lowest ($\Delta E > 0$), but the two other sets of results fall to within a reasonable margin of the order of 50 cm^{-1}. Note that the calculations of Ref. [20] are no doubt of benchmark quality. They extend to large R values (not shown in Table 5), but are limited to the lowest state of $^1\Sigma_g^-$ symmetry. Ref. [19] on the other hand reports five excited states, but only up to $R \leq 5$ a.u.

The R-matrix method in principle is able to provide *ab initio* data for arbitrarily high principal quantum numbers as well as for internuclear distances up to $R = 10$ a.u. at least. Figure 2 illustrates preliminary calculations of a few higher potential energy curves of metastable non-autoionizing $nP_2\ ^1\Sigma_g^-$ ($n = 1 - 4$) states. These curves are seen to run parallel to the $2p\pi$ ionization threshold as is characteristic for Rydberg states. In Table 6 we list the presently obtained energies E_S along with those of Martin [19], E_M, and the differences ($\Delta E_{SM} = E_S - E_M$) for the states $nP_2\ ^1\Sigma_g^-$ ($n = 2 - 4$). The overall agreement is similar as obtained for the lowest $^1\Sigma_g^-$ state.

4. Conclusion

We have implemented the σ_v molecular reflection symmetry in the framework of the "halfium" variational R-matrix code and investigated its effect on the calculation of H_2 $^1\Sigma_g^+$ states in the discrete range and in the ionization continuum. Preliminary results on $^1\Sigma_g^-$ non-autoionizing doubly-excited states have also been presented, and have been compared with published data from other authors. These states cannot autoionize by electronic interaction (configuration mixing), but they may be weakly autoionized and/or predissociated by rotational-electronic non-adiabatic coupling. Systematic calculations of singlet and triplet H_2 Σ^- states are underway and will be published in a forthcoming paper [26].

Acknowledgements

We thank Drs. S. L. Guberman, I. F. Schneider, C. H. Greene and X. Urbain for stimulating discussions concerning this work. We are grateful to Dr. F. Martin for providing unpublished data on doubly-excited states of H_2. M. T. and Ch. J. acknowledge financial support from NSF and NASA. Ch. J. was supported in part by the ANR (France) under Contract No. 09-BLAN-020901. F. A. and Ch. J. have also benefitted from support by the E. Miescher Foundation (Basel, Switzerland).

References

[1] Giusti A 1980 *J. Phys. B: At. Mol. Phys.* **13** 3867
[2] Giusti-Suzor, A Bardsley J N and Derkits C 1983 *Phys. Rev. A* **28** 682-691
[3] Schneider I F, Dulieu O and Giusti-Suzor A 1991 *J. Phys. B: At. Mol. Phys.* **24** L289-97
[4] Florescu A I, Ngassam V, Schneider I F and Suzor-Weiner A 2003 *J. Phys. B: At. Mol. Opt. Phys.* **36** 1205-1215
[5] Kouchi N, Ukai M and Hatano Y 1997 *J. Phys. B: At. Mol. Opt. Phys.* **30** 2319-2344
[6] Tennyson J 1996 *At. Data. Nuc.* Data Tables **64** 253-277
[7] Guberman S L 1983 *J. Chem. Phys.* **78** 1404-1413
[8] Sanchez I and Martin F 1997 *J. Chem. Phys.* **106(18)** 7720-7730
[9] Sanchez I and Martin F 1999 *J. Chem. Phys.* **110 (14)** 6702-6713; AIP Document No. PAPS JCPSA6-110-007914
[10] Telmini M and Jungen C 2003 *Phys. Rev. A* **68** 062704
[11] Greene C H and Yoo B 1995 *J. Chem. Phys.* **99** 1771
[12] Yoo B 1990 *Ph.D. Thesis* Louisiana State University, Baton Rouge LA
[13] Greene C H 1983 *Phys. Rev. A* **28** 2209
[14] Jungen C and Texier F 2000 *J. Phys. B: At. Mol. Opt. Phys.* **33** 2495
[15] Bezzaouia S, Telmini M and Jungen C 2004 *Phys. Rev. A* **70** 012713
[16] Corongiu G and Clementi E 2010 *Int. J. Quant. Chem.*
[17] Matzkin A, Jungen C and Ross S C 2000, *Phys. Rev A* **62** 062511
[18] Ross S and Jungen C 1994 *Phys. Rev. A* **49** 4353
[19] Martin F 1999 *J. Phys. B: At. Mol. Opt. Phys.* **32** L181-L187
[20] Komasa J 2008 *Phys. Chem. Chem. Phys.* **10** 3383-3389
[21] Landau L and Lifschitz E, 1966 *Quantum mechanics*, Mir editions, Moscow
[22] Xie J and Zare R N 1990 *J. Chem. Phys.* **93** 3033
[23] Raseev G and Jungen Ch 1998 *Phys. Rev. A* **57** 2407
[24] Kolos W and Wolniewicz L 1969 *J. Chem. Phys.* **50** 3228. Wolniewicz L and Dressler K 1994 *J. Chem. Phys.* **100** 444. Orlikowski T, Staszewska G and Wolniewicz L 1999 *Mol. Phys.* **96** 1445. Staszewska G and Wolniewicz L 2002 *J. Mol. Spectrosc.* **212** 208 http://www.phys.uni.torun.pl/publications/ifiz/luwo/
[25] Bottcher C 1976 *J. Phys. B: Atom. Molec. Phys.* **9**, 2899
[26] Argoubi F et al (in preparation)

Dissociative Electron Attachment to HCN, HCCH and HCCCN

S. T. Chourou and A. E. Orel

Department of Applied Science, University of California, Davis, Davis CA 95616 USA

E-mail: stchourou@ucdavis.edu

Abstract.
Studies of polyatomic molecules show that more than one degree of freedom comes into play in the dissociative process of certain molecules upon low-energy electron collision. In many of these cases, bond breaking may only proceed if the initial symmetry of the target molecule is distorted. We have performed a multi-dimensional computational treatment of the dissociative electron attachment (DEA) to three polyatomic systems, HCN, HCCH, and HCCCN below 10 eV and compared their nuclear motion and dissociation cross section. We further investigated the isotope effect in the case of HCN and HCCH. We have adopted the same computational approach in the study of these systems. This approach consists of time-dependent wavepacket propagation where the Hamiltonian operator involves the multi-dimensional complex potential energy surfaces (PESs) of the resonant anion.

1. Introduction

Treatment of electron driven processes for polyatomics introduces challenges in comparison to diatomic molecules due to the rapid rise of computational complexity as a result of the increased number of internal degrees of freedom of the {electron, molecule} system. A number of theoretical studies aimed at gaining insight into the dissociative paths for polyatomic systems such as dissociative electron attachment to formic acid [1] and water [2, 3] as well dissociative recombination (DR) to HCO^+ [4] H_3^+ [5] and $HCNH^+$ [6]. In many of these calculations, approximations have been applied to reduce the system dimensionality by restricting the nuclear motion to a subset of internal coordinates and limiting the number of electronic states involved in the computation. In this work, a comparative summary of theoretical investigations of polyatomic systems is presented. A discussion of the theoretical approaches and computational methods used in DEA to polyatomics is further provided to extend these techniques in future investigations of DEA and DR to polyatomics. This proceeding is organized as follows. First, a brief overview of the theory of DEA to polyatomics applied in the present studies is outlined. Secondly, these principles are applied toward the study of specific molecules HCN, HCCH and HCCCN systems in terms of the study of dissociation mechanism and the computation of DEA cross sections.

2. Theory

Based on the early formalism put forward by O'Malley [7] and in the framework of the local complex potential approximation [8], the nuclear wave equation is given by

Journal of Physics: Conference Series **300** (2011) 012014 doi:10.1088/1742-6596/300/1/012014

$$\left[E - \mathbf{K_q} - E_{res}(\mathbf{q}) + \frac{1}{2}i\Gamma(\mathbf{q})\right]\xi_{E\nu}(\mathbf{q}) = \sqrt{\frac{\Gamma(\mathbf{q})}{2\pi}}\chi_\nu(\mathbf{q}) \tag{1}$$

where $\mathbf{K_q}$ is the nuclear kinetic energy operator, E_{res} is the resonance energy, Γ is the autoionization width, χ_ν stands for the initial vibrational state of the molecular target, $\xi_{E\nu}$ represents the DEA wavefunction of the system E is the total energy of the {electron, molecule} system. The quantities found in Eq.(1) are expressed as a function of the variable $\mathbf{q}$ which represents the nuclear internal coordinates of the molecule. We further derive a time-dependent expression of this motion equation along the lines of the work of Turner and McCurdy [9]:

$$\begin{cases} \left[\mathbf{K_q} + E_{res}(\mathbf{q}) - \frac{1}{2}i\Gamma(\mathbf{q})\right]\xi_\nu(\mathbf{q},t) = i\partial_t\xi_\nu(\mathbf{q},t); \\ \xi_\nu(\mathbf{q},0) = \sqrt{\frac{\Gamma(\mathbf{q})}{2\pi}}\chi_\nu(\mathbf{q}) \end{cases} \tag{2}$$

The details of this derivation are introduced elsewhere [10]. This formulation therefore, suggests a computational approach based on 2 main steps.

2.1. Resonant PESs

First, we construct the multi-dimensional PESs of the neutral target molecule and of the resonant anion formed by attachment of the incident electron. The neutral potential energy surfaces are computed using standard structure calculation methods at the multi-configuration self-consistent field and the multi-reference configuration interaction levels. The resonances parameters are computed using the Complex Kohn Variational method [11] at the electron collision energy range surrounding the DEA peaks observed experimentally. All these surfaces are expressed as a function of a suitable system of internal coordinates.

2.2. Dissociation dynamics and cross section

The Multiconfiguration Time-Dependent Hartree (MCTDH) [12] method was used to solve the time-dependent Schrödinger equation (2). We ran the nuclear dynamics using the Heidelberg MCTDH Package [13] to compute the evolution of the wavepacket representing the transient resonant state and calculate the corresponding outgoing flux. These steps leading to the DEA cross section in the case of a triatomic system are discussed in [14] and references therein.

3. Systems of interest

We focused on three polyatomic systems: HCN, HCCH and HCCCN. These molecules have linear equilibrium geometry and have an unsaturated Π electronic system. A low-energy (below 10 eV) incident electron is trapped by the centrifugal barrier giving rise to a π^*-shape resonance. This resonant state may decay via bond breaking. These molecules have been studied experimentally [15, 16, 17, 18] and authors have predicted that the DEA process involves inherent polyatomic effects.

3.1. HCN

We studied the DEA channel given by

$$\mathrm{HCN}\ (X^1\Sigma^+,\nu) + \mathrm{e}^-(E) \rightarrow (\mathrm{HCN})^{-*}\ (^2\Pi_g) \rightarrow \mathrm{CN}^-(^1\Sigma,\nu') + \mathrm{H}\ (^2S),$$

The DEA process for HCN below 3 eV is shown to be dominated by a "tunneling" effect through a potential barrier in the lowest resonant surface. Inspection of the propagating wavepacket shows a slight spread along the bending degree of freedom (a broadening of about

Journal of Physics: Conference Series **300** (2011) 012014 doi:10.1088/1742-6596/300/1/012014

20% at FWHM at 6 fs). A more recent calculation [19] using a more accurate resonant PES derived form a relaxed-SCF (RSCF) calculation shows a better agreement with the experiment of Allan *et al.* [15] compared to the previous static exchange calculation [20]. At the RSCF level of calculation, the trial electronic scattering wavefunction takes account of the polarization and correlation effects; the reader may refer to [19] for more details about this computation.

Furthermore, the improved surface topology of HCN (due to the inclusion of the polarization and correlation effects) results in a significantly better agreement with the isotope ratio found experimentally by the same authors. Since the RSCF calculation leads to a shift of the resonance position to lower energies and to a narrowing of the autoionization widths, this results suggests that initial wavepacket undergoes a significant scaling of its amplitude by the $\sqrt{\frac{\Gamma(\mathbf{q})}{2\pi}}$ factor in Eq.(1). The resulting smaller "entry amplitude" leads to an overall weaker flux of the propagating wavepacket.

Recent calculation of the DEA cross section for HCN and DCN is shown in Fig.(1) together with the experimental plots in [15].

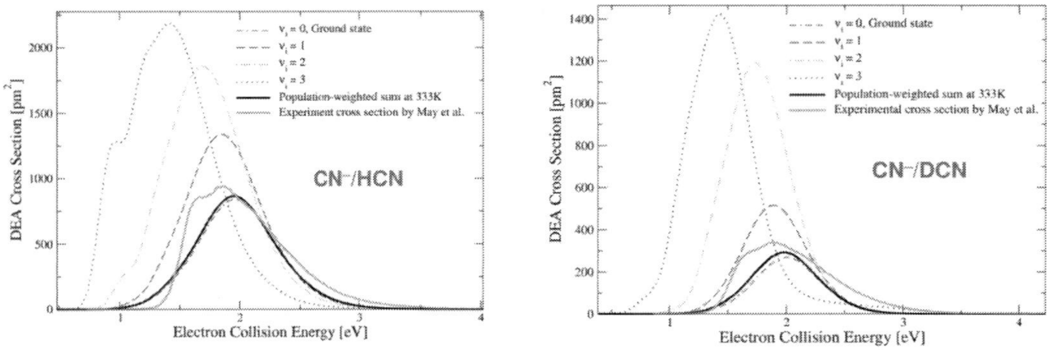

Figure 1. Absolute DEA cross sections for the initial angular vibrational states $\nu_i = 0$, 1, 2 and 3 and population-weighted sum at 333 K compared with experimental findings in [15] for HCN (left panel) and DCN (right panel).

3.2. HCCH

We treated the reaction channel given by

$$C_2H_2\ (X^1\Sigma_g^+,\, \nu) + e^-(E) \rightarrow (C_2H_2)^{-*}\ (^2\Pi_g) \rightarrow C_2H^-(^1\Sigma^+, \nu') + H(^2S)$$

The dissociation mechanism for acetylene shows a more pronounced bending effect as the wavepacket profile along the bending dimension presents a broadening of about 45% at FWHM at 6 fs. This indicates that the majority of the flux collected asymptotically comes from the wavepacket portion that circumvents the potential energy barrier. By calculating the population-weighted sum of the contribution of the excited angular vibrational states of acetylene, a better agreement with the experiment is obtained at 300 K in terms of isotope ratio $\sigma_{0\rightarrow DEA}^{(C_2D^-)}/\sigma_{0\rightarrow DEA}^{(C_2H^-)}$ at 3 eV. Despite an improved value of this isotope ratio (17.9 versus the experimental value of 14.4 [21]), the absolute values of the population-weighted sum lies about 1.87 higher than the DEA absolute cross sections measured at 300 K.

Similarly to the HCN case, we considered a RSCF scattering calculation to evaluate the effect of the polarization and correlation effects on Γ and hence the DEA cross section. We performed a RSCF calculation for 3 points around the equilibrium geometry of HCCH and determined the average ratio between the resulting autoionization widths and those computed at the static exchange level. The average ratio was found to be approximately $\gamma = 0.92$ indicating, as expected, a narrowing of the electron-HCCH scattering cross sections. Moreover, an average decrease of the resonance positions of approximately $\varepsilon = 0.3$ eV is observed. By applying the ratio γ as a scaling factor to the function $\Gamma(\mathbf{q})$ and the value ε as a shift to the potential $E_{res}(\mathbf{q})$ in Eq.(2), the population-weighted sum reduces to about 4.9 pm^2 for HCN and to 0.27pm^2 for DCN. In Fig.(2), we summarize the DEA cross section results in [21] and compare with experimental finding in [16].

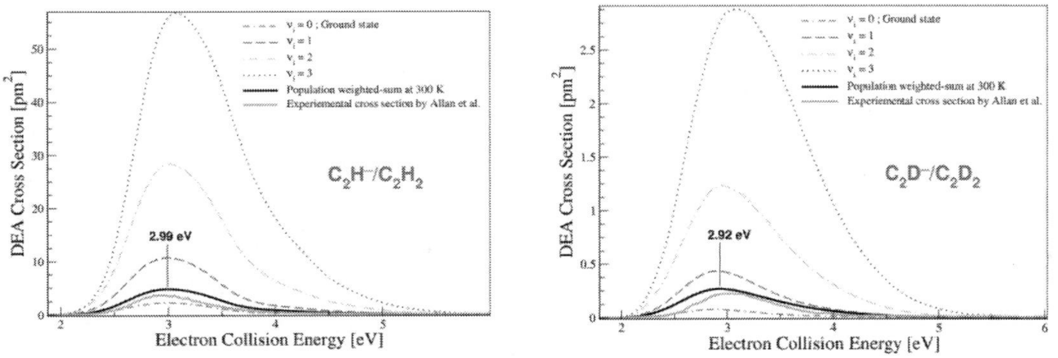

Figure 2. Absolute DEA cross sections for the initial angular vibrational states ν_i = 0, 1, 2 and 3 and population-weighted sum at 300 K compared with experimental findings in [15] for C_2H_2 (left panel) and C_2D_2 (right panel).

One must note that this only constitutes a rough estimate of the resonance parameters, since as seen in the HCN case, a more complete (and computationally costly) description of the complex anion surface is needed to have a correct account of dissociation dynamics.

3.3. HCCCN

We took into consideration the following channels for DEA to HCCCN:

$$\mathrm{HCCCN}\ (\mathrm{X}^1\Sigma^+,\nu) + \mathrm{e}^-(E) \rightarrow \mathrm{HCCCN}^{-*} \rightarrow \begin{cases} \mathrm{CCCN}^-(^2\Sigma^+,\nu_I'^-)+\mathrm{H}(^2S)\text{: (I);} \\ \mathrm{CN}^-(^1\Sigma^+,\nu_{II}'^-)+\mathrm{HCC}(^2\Sigma^+,\nu_{II}')\text{: (II);} \\ \mathrm{HCC}^-(^1\Sigma^+,\nu_{III}'^-)+\mathrm{CN}(^2\Sigma^+,\nu_{III}')\text{: (III);} \end{cases}$$

In order to account for these dissociation channels, the coupling between the two lowest resonant PESs considered for this calculation must be introduced. In fact, these complex surfaces present a conical intersection and a diabatization procedure is hence employed to construct the two diabatic states to be used in the expression of the nuclear Schrödinger equation. As such, the MCTDH approach is set to solve a system of two coupled equations of motion as detailed in [22].

Furthermore, the HCCCN molecule has been modeled as three-body system: H, (CC) and (CN) and the internal degrees of freedom considered in this calculation are given by $\mathbf{q} = (r_1, r_2, \theta_1, \theta_2)$ as depicted in Fig.(3). The resulting DEA flux from a 3-D calculation taking into account the coordinates r_1, θ_1 and θ_2 (r_2, θ_1 and θ_2) is normalized with the experimental yield of $CCCN^-$ (CN^- and CCH^-) anions measured by Field *et al.* [18] and plotted in Fig.(4).

This calculation seems to have an acceptable agreement with these measurements. The major part of the observed DEA spectrum seems to cluster in the 2-9 eV energy band with peaks in the

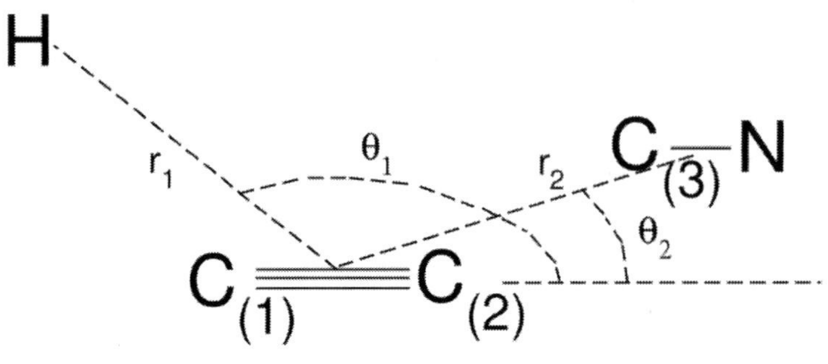

Figure 3. Coordinate system used in DEA dynamics calculation. In this system of internal coordinates we include the distance between the center of mass of $C_{(1)}C_{(2)}$ and the hydrogen r_1, the distance between the center of masses of $C_{(1)}C_{(2)}$ and CN r_2 and respective bending angles

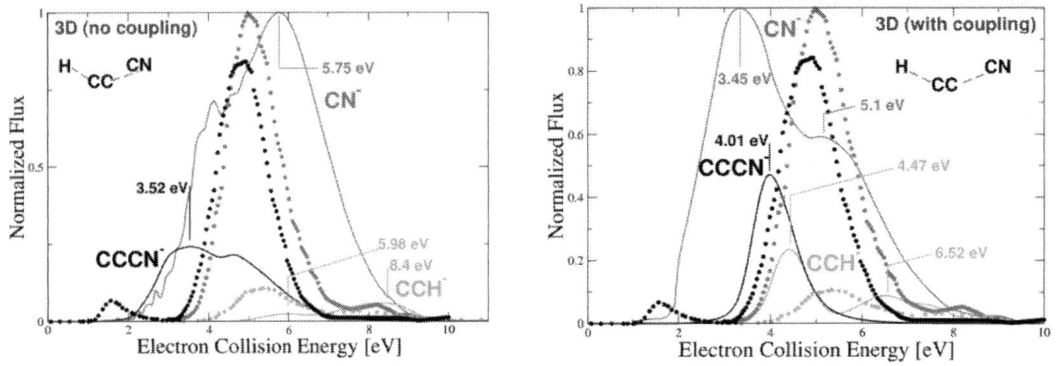

Figure 4. Normalized flux from 3-D calculations allowing the bending of the (H)-(CC)-(CN) system. Left panel: we propagate on the A' resonant states (non-coupled). The dominant yields for CCH^- (orange curve) and $CCCN^-$ (black curve) are about 8% and 25% of that of CN^- (red curve). Other secondary peaks for each of the fragments is also observed. Right panel: we include the coupling between the A' resonant states. The dominant yields for CCH^- (orange curve) and $CCCN^-$ (black curve) are about 23% and 49% of that of CN- (red curve). A secondary CN^- peak at 5.1 eV has a height of approximately 60% while a secondary CCH^- at 6.52 eV has a height of about 10%. Both cases Experimental results of [18] are in dotted lines and follow the same color coding as the computed curves with regard to the fragment they represent.

vicinity of the main CN^- maximum peak. Our calculations show that this collection of peaks is shifted to lower electron energies as the dimensionality of the system is increased, eventually approaching the experimental central value of 5.28±0.15 eV.

By introducing the coupling, the general distribution structure and the peak height ratios of the DEA flux is closer the experimental spectra; however, the positions and widths of the peaks are altered. Hence, one can infer the existence of a trade-off between the inclusion of addition degrees of freedom (enhancing the positions and widths of the peaks with little effect on the fragment ratios) on one hand and the introduction of the coupling (enhancing the fragments ratios and worsening the peaks' positions and widths) on the other hand.

4. Discussion and Conclusion

These calculations demonstrate that the resonant anions show a propensity for deviating from their linear geometry upon electron attachment. Molecular bending appears to drive the dissociation dynamics and therefore, the DEA process cannot be described accurately using a diatomic (1-D) model.

This work was organized to address incrementally larger systems. Treating the dissociative channel in HCN leading to formation of hydrogen and the cyanide anion CN^- followed by the DEA to acetylene to explore the breakup into hydrogen and HCC^- served as a basis for investigation of the larger cyanoacetylene molecule and its fragmentation into $CCCN^-$ (i.e. production of hydrogen), CN^- and CCH^-. In principle, this approach can be generalized to other polyatomics. The practical benefit of utilizing the qualitative and quantitative dynamical characteristics of the sub-groups is to make a well-founded choice of the reduced system of internal coordinates appropriate for the study of the larger system (i.e. do not consider the degrees of freedom that are unlikely to contribute significantly to the dynamics).

Therefore, relatively simple molecules (3-5 atoms) can serve as a starting point of systems exhibiting inherent polyatomic effects. Tackling DEA and DR problems for larger polyatomic molecules may be achieved by methodically splitting the system into groups and adding in degrees of freedom as needed guided by information gathered from the prior studies of the subsystems.

Acknowledgments

We acknowledge support from the National Science Foundation under Grant No.PHY-05-55401 and the U. S. DOE Office of Basic Energy Science, Division of Chemical Science.

References

[1] T. N. Rescigno, C. S. Trevisan, and A. E. Orel, Phys. Rev. Lett. **96**, 213201 (2006)
[2] D.J. Haxton, T.N. Rescigno, and C.W. McCurdy, Phys. Rev. A **78**, 040702 (2008)
[3] D.J. Haxton, C.W. McCurdy, and T.N. Rescigno, Phys. Rev. A **75**, 012710 (2007)
[4] Åsa Larson, Stefano Tonzani, Robin Santra and Chris H. Greene, Journal of Physics: Conference Series 4, Sixth International Conference on Dissociative Recombination, 148-154 (2005)
[5] Dos Santos Samantha Fonseca; Kokoouline Viatcheslav; Greene Chris H., Journal of Chemical Physics, Volume 127, Issue 12, pp. 124309-124309-8 (2007)
[6] V. Ngassam, A. E. Orel and A. Suzor-Weiner, Journal of Physics: Conference Series 4, Sixth International Conference on Dissociative Recombination, 224-228 (2005)
[7] T. F. O'Malley, Phys. Rev. **150**, 14 (1966)
[8] A.U. Hazi, T.N. Rescigno, and M. Kurilla, Phys. Rev. A **23**, 1089 (1981)
[9] C. W. Mcurdy and J. L. Turner, J. Chem. Phys. **78** No. 11 (1983)
[10] D. J. Haxton, C. W. McCurdy and T. N. Rescigno Phys. Rev. A **73**, 062724 (2006)
[11] T.N. Rescigno, C.W. McCurdy, A.E. Orel and B.H. Lengsfield,III, "The Complex Kohn Variational Method", Proceedings of the Workshop on Comparative Study of Current Methodologies in Electron-Molecule Scattering, ISBN 0-306-44911-0 (1993)
[12] Beck M. H., Meyer H-D., Z. Phys. D **42**, 113 (1997)
[13] G. A. Worth, M. H. Beck, A. Jäckle, and H.-D. Meyer, The MCTDH Package, Version 8.2 (2000); H.-D. Meyer, Version 8.3; See *http://www.pci.uni-heidelberg.de/tc/usr/mctdh/*
[14] D. J. Haxton, Zhiyong Zhang, H-D. Meyer, T. N. Rescigno, and C. W. McCurdy Phys. Rev. A **69**, 062714 (2004)
[15] O. May, D. Kubala, and M. Allan, Phys. Rev. A **82**, 010701 (2010)
[16] O. May, J. Fedor, B. C. Ibanescu, and M. Allan Phys. Rev. A **77**, 040701(R) (2008)

[17] P. D. Burrow, A. E. Howard, A. R. Johnston and K. D. Jordan, J. Phys. Chem. **96** No. 19 (1992) 7570-1578
[18] K. Graupner, T. L. Merrigan, T. A. Field, T. G. A. Youngs and P. C. Marr, New Journal of Physics **8**, 117 (2006)
[19] S. T. Chourou and A. E. Orel, "Isotope effect in dissociative electron attachment to HCN" (Accepted)
[20] S. T. Chourou and A. E. Orel, Phys. Rev. A **80**, 032709 (2009)
[21] S. T. Chourou and A. E. Orel, Phys. Rev. A **80**, 034701 (2009)
[22] S. T. Chourou and A. E. Orel, "Dissociative electron attachment to HCCCN" (Submitted)

Eighth International Conference on Dissociative Recombination (DR2010) IOP Publishing
Journal of Physics: Conference Series **300** (2011) 012015 doi:10.1088/1742-6596/300/1/012015

The role of the Jahn-Teller coupling in dissociative recombination of H_3O^+ and H_3^+ ions

N Douguet[1,4], A Orel[2], I Mikhailov[1], I F Schneider[3,4], C H Greene[5] and V Kokoouline[1,4]

[1]Department of Physics, University of Central Florida, Orlando, FL 32816, USA
[2]Department of Applied Science, University of California at Davis, Davis, CA 95616, USA
[3]LOMC-FRE-CNRS-3102 Université du Havre, Le Havre, 76600 France
[4]Laboratoire Aimé Cotton CNRS-UPR-3321, Université Paris-Sud, 91405 Orsay, France
[5]Department of Physics and JILA, University of Colorado, Boulder, CO 80309, USA

E-mail: douguet@physics.ucf.edu

Abstract. We discuss the importance of the non Born-Oppenheimer Jahn-Teller coupling in the dissociative recombination of H_3O^+ and H_3^+ ions with low energy electrons. The coupling is known to be important for dissociative recombination of H_3^+, for which detailed and simplified theoretical models have been previously developed. In this study we discuss how the detailed theoretical model reproduces individual resonances in the experimental spectrum of H_3^+. We present a simplified theoretical model of the dissociative recombination of H_3O^+: Similarly to H_3^+, we consider the Jahn-Teller coupling as a key factor responsible for the capture of the electron into a Rydberg state associated with excited vibrational levels of the ion. We use a simplified Jahn-Teller model for highly symmetrical ions (not only for H_3O^+ and H_3^+) and adopt the normal mode approximation for the vibrational states of H_3O^+. Using the multi-channel quantum defect formalism and accurate *ab initio* calculations, we derive the cross sections for electron capture by the ion that is initially in the ground vibrational level. In our approximation, once the electron is captured by the ion, the autoionization probability is neglected compared to the predissociation probability. This allows us to evaluate the cross section for the dissociative recombination of H_3O^+ using just a few parameters obtained from *ab initio* calculation. The total cross section obtained is in good agreement with data from storage ring experiments.

1. Introduction

Reactive collisions of ionized species play an important role in the overall chemical evolution of the cold interstellar medium (ISM). One of the important charged species in the ISM, the hydronium ion H_3O^+ is formed by a chain of chemical reactions, starting with H_3^+ [1]

$$\begin{aligned} O + H_3^+ &\rightarrow OH^+ + H_2 \\ OH^+ + H_2 &\rightarrow H_2O^+ + H \\ H_2O^+ + H_2 &\rightarrow H_3O^+ + H\,. \end{aligned} \tag{1}$$

The produced H_3O^+ molecular ion has a closed electronic shell and does not react with atomic or molecular hydrogen, but can be depleted by the dissociative recombination (DR) with electrons, which are also abundant in diffuse interstellar clouds. The ion was also observed in tails of comets. For example, observations by the spacecraft Giotto [2] reported the hydronium ions as the most abundant ionic species in the comet Halley. It was suggested that in the inner coma of the comet, DR is the major mechanism of H_3O^+ destruction.

One of the products of dissociative recombination of H_3O^+ with electrons is water: In several experimental studies [1, 3, 4] it was demonstrated that H_2O is produced by the recombination of H_3O^+ with the branching ratio for the water pathway of about 18-33%: 33±8% in the older ASTRID experiment [1], 25±1% in the newer ASTRID experiment [4], and 18±5% in CRYRING experiment [3]. The slightly different results for the branching ratio in three experiments [1, 3, 4] could possibly be explained by a different population of excited rovibrational states of H_3O^+ in the storage rings. The water channel in DR of H_3O^+ is the reason that DR is believed to be the major source of water production in interstellar clouds [5, 6]. Therefore, understanding main mechanisms in DR of H_3O^+ is important for a correct evaluation of water abundance in the ISM. Finally, the dissociative recombination of H_3O^+ ions is thought to play an important role in the chemistry of the lean methane-oxygen flames [7], and in the electrical control of the instability of propane-air flames [8].

A theoretical description of dissociative recombination in H_3O^+ (and other polyatomic ions, in general) is complicated due to the large number of vibrational degrees of freedom and many electronic states of the neutral molecule H_3O that have to be included into the complete theoretical model. Theoretical methods for DR in triatomic molecular ions, H_3^+ [9, 10, 11], HCO^+ [12, 13], and LiH_2^+ [14] have been developed during the last decade, showing relatively good agreement with the experimental data. The structure of the H_3^+, HCO^+, and LiH_2^+ ions are similar in the way they recombine with electrons: They all have a closed electronic shell, so their dissociative resonant states lie above their ground electronic potentials such that the potential energy surfaces of the resonant electronic states do not cross the ionic potential surface in the Franck-Condon region. This makes direct DR impossible for such molecular ions. The dominant DR mechanism for these ions proceeds therefore through capturing an electron into a Rydberg state associated with a vibrationally excited ionic core.

A completely quantum-mechanical treatment that has been employed previously for DR in H_3^+ [10, 11, 15] requires the knowledge of the full electronic potential surfaces in a large configuration space, for both the ion and the neutral state. If used for H_3O^+, the treatment would require accurate *ab initio* calculations performed for the ground electronic state of the ion and for several excited electronic states of the neutral molecule H_3O for all vibrational degrees of freedom. Once the potential energy surfaces are available, one would have to do the vibrational dynamics in several dimensions, ideally along all six vibrational coordinates. Thus, the full completely quantum-mechanical treatment for indirect DR of polyatomic ions requires a massive computational effort in order to compute the DR cross section.

It is desirable to produce a simpler model, which takes into account only the key factors in the DR mechanism with the goal to give a correct estimate for the DR cross section. An important step forward for such simple DR theory for polyatomic ions was made in the previous study of DR in HCO^+ [12, 13] (both studies were based on the ideas suggested in Ref. [9]), for which the dissociative channels are not taken into account explicitly, and autoionization is neglected. Competition between autoionization and dissociation was discussed qualitatively in [12], where it was shown that autoionization should only slightly decrease the probability for dissociation after the electron is captured into a predissociated Rydberg state. On the other hand, that later treatment used *ab initio* calculations on the full configuration space, including different electronic couplings. Moreover, adiabatic vibrational states depending parametrically on the dissociative coordinates were computed, the procedure of channel elimination was performed and energies and widths of resonances were obtained. Finally an averaged cross section for electronic capture in any ionic vibrational state was computed. That simplified treatment was still quite complicated.

Jungen and Pratt [16, 17] made another step further in simplifying the treatments for both

H_3^+ [17] and HCO^+ [16], by considering solely the capture into the first excited doubly degenerate vibrational normal mode (harmonic approximation). This later approximation simplifies greatly the model and their results are in very good agreement with the more detailed approaches. Our study of DR of H_3O^+ will rely on the same basic principles discussed in Refs. [9, 12, 13, 16, 17], except that the H_3O^+ ion has two Jahn-Teller active normal modes. Before introducing the approach in detail, we start with a brief description of the H_3O^+ vibrational modes around the equilibrium position.

The simplified model discussed below provides the overall DR cross-section for H_3O^+. Resonances that would be present in high resolution experimental data can not be reproduced by this model. No high-resolution experimental data is available for DR of H_3O^+, but for H_3^+ such a high-resolution experimental DR spectrum is available. Therefore, the detailed theoretical model [10, 15, 11] developed previously can be used to check how theory can reproduce details in the DR spectrum. Another part of this study is devoted to a comparison of resonance features in the theoretical and experimental DR spectra.

This paper is organized in the following way. In the next section, we describe the vibrational normal modes of H_3O^+ and the Jahn-Teller interaction that couples the vibrational motion of the ion with the motion of the incident electron. In Section 3, we introduce the simplified theoretical model for calculation of the DR cross-section in closed-shell molecular ions of the C_{3v} symmetry. In Section 4, we present new results on DR of H_3^+. Namely, we compare the resonances obtained in the complete theoretical model with resonances observed in the high-resolution DR experiment with H_3^+. In Section, 5 we discuss obtained results and possible improvements of the theoretical treatment.

2. Vibrational modes of H_3O^+ and the Jahn-Teller model

In its equilibrium position, the H_3O^+ ion has a pyramidal shape, described by the C_{3v} symmetry point group. However, the barrier for tunneling for the oxygen atom through the plane of the three hydrogen nuclei is quite weak, therefore D_{3h} symmetry point group can be used for the qualitative analysis of vibrational modes. (For the D_{3h} group, the expressions for vibrational normal modes do not depend on actual force constants of H_3O^+; the C_{3v} normal modes do depend on the force constants). In this picture, the oxygen atom is situated in the middle of the equilateral triangle formed by the three hydrogen nuclei, at a distance R_{OH} =1.039Å. The normal modes in the D_{3h} configuration are given in Ref. [18]. For our purposes we only need the expressions for the two doubly-degenerate modes E': Q_2 and Q_4

$$\begin{aligned}
Q_{2x} &= \sqrt{\frac{1}{12}}(2x_1 - x_2 - x_3 - \sqrt{3}y_2 + \sqrt{3}y_3),\\
Q_{2y} &= \sqrt{\frac{1}{12}}(2y_1 - y_2 - y_3 + \sqrt{3}x_2 - \sqrt{3}x_3),\\
Q_{4x} &= \sqrt{\frac{3m}{M+3m}}x_o - \sqrt{\frac{M}{3(M+3m)}}(x_1 + x_2 + x_3),\\
Q_{4y} &= \sqrt{\frac{3m}{M+3m}}y_o - \sqrt{\frac{M}{3(M+3m)}}(y_1 + y_2 + y_3)\,.
\end{aligned} \tag{2}$$

In the above expressions, we used the Cartesian coordinates (x_i, y_i, z_i) of the three hydrogen nuclei i =1,2,3 with mass m, and the oxygen nucleus with coordinates (x_o, y_o, z_o) and mass M (z-components do not contribute into Q_2 and Q_4). The Q_2 degenerate mode is similar to the asymmetric stretching of the H_3 triangle in H_3^+ with the fixed oxygen atom. The second degenerate normal mode Q_4 represents a motion of the oxygen in the x-y plane.

In the study, we consider the H_3O^+ ion vibrating near its equilibrium configuration, therefore we need the normal coordinates at that specific configuration in the C_{3v} point group symmetry.

Journal of Physics: Conference Series **300** (2011) 012015 doi:10.1088/1742-6596/300/1/012015

The two E modes at the C_{3v} configuration are linear combinations of the modes shown above for D_{3h} with coefficients depending on the force constants of H_3O^+. However, the mixing between Q_{2x} and Q_{4x} on one hand and between Q_{2y} and Q_{4y} on the other hand for H_3O^+ is small. Therefore, the character of the normal modes Q_2 and Q_4 at C_{3v} geometry is mainly determined by the expressions above. It is important to stress here that in this study the H_3O^+ modes are obtained numerically by the standard normal modes analysis made using the MOLPRO package. The expressions above are provided just to give an idea about the character of the modes. For the simplicity of the discussion, we will use the same notations Q_2 and Q_4 for the normal modes at the C_{3v} equilibrium configuration.

The frequencies of the four $2A_1 \oplus 2E$ normal modes of H_3O^+ are given in Table 1. The frequencies have been obtained from *ab initio* calculations using the MOLPRO package [19] and are compared with accurate (beyond the normal mode approximation) calculations by Bowman *et al.* [20], who used the MULTIMODE code. The results agree well, with relative error less than 10%. In order of increasing frequencies, the first symmetric normal mode Q_1 is the so-called *umbrella* mode, the second doubly degenerate mode $Q_2 = (Q_{2x}, Q_{2y})$ represents the deformation, the third mode Q_3 represents the symmetric O-H stretching and finally, the fourth degenerate mode $Q_4 = (Q_{4x}, Q_{4y})$ is the antisymmetric O-H stretching.

	Representation	Present calculations	Bowman *et al.* [20]
ω_1	A_1	900	972
ω_2	E	1704	1639
ω_3	A_1	3609	3405
ω_4	E	3705	3711

Table 1. Frequencies (in cm^{-1}) of H_3O^+ normal modes obtained in the present study compared to the energies for the lowest vibrational states obtained by a direct diagonalization of the vibrational Hamiltonian by Bowman *et al.* [20].

We now introduce the vibronic Jahn-Teller coupling in the H_3O radical. The H_3O electronic Rydberg states with non-zero projection Λ (more precisely, with $\Lambda \neq 3k$ where k is an integer number) of the electronic angular momentum on the symmetry axis of the molecule are doubly degenerate at the equilibrium position of the ion (and any equilateral configuration of the three hydrogen nuclei). Such states with $\Lambda \neq 0 (\mathrm{mod}\, 3)$ belong to the E irreducible representation of the C_{3v} group. For any geometry different from equilateral, energies of the two electronic states split apart. The energy splitting between the two electronic states is linear with respect to Q_{2x}, Q_{2y}, Q_{4x}, and Q_{4y} [21]. Because the splitting between the two states is small or zero near the equilateral geometry, the non-Born-Oppenheimer interaction, Jahn-Teller coupling, between the vibrational and electronic motions is large or infinite. This coupling is responsible for the large probability of capturing the incident electron into a Rydberg state associated with an excited vibrational level. We include the Jahn-Teller coupling into the DR treatment using the parametrized Hamiltonian describing the coupled states suggested by Moffit and Liehr [22] and Longuet-Higgins [21]. The parametrized Hamiltonian is briefly discussed below.

We collectively denote all the nuclear and electronic coordinates by Q and q respectively and consider two degenerate electronic states $\phi_+(q)$ and $\phi_-(q)$ corresponding to two projections $\Lambda = \pm 1$ of the electronic angular momentum on the molecular symmetry axis. We also need two vibrational functions $\chi_v^+(Q)$ and $\chi_v^-(Q)$ corresponding to the vibrational motion along potential surfaces of the two electronic states. A solution of the vibronic equation is sought in the form

$$\Psi = \chi_v^+(Q)\phi_+(q) + \chi_v^-(Q)\phi_-(q) \,. \tag{3}$$

We consider only small displacements of the nuclei from the equilibrium position of the ion and assume at this point that the electronic states do not depend on small nuclear displacements. It is also assumed that there are no other electronic states, which are accidentally close in energy to the degenerate pair of states. This is needed to avoid any accidental perturbation of the two-state parametrized Hamiltonian. The electronic states in Eq. (3) transform according to

$$C_3\phi_\pm = \exp(\mp 2i\pi/3)\phi_\pm\,, \tag{4}$$

$$\sigma_v\phi_\pm = \phi_\mp\,, \tag{5}$$

where C_3 is the rotation by $2\pi/3$ around the molecular symmetry axis and σ_v the reflection through a vertical plane. Instead of the real normal coordinates (Q_{2x}, Q_{2y}) and (Q_{4x}, Q_{4y}), we will use the complex normal modes Q_i^+ and Q_i^- ($Q_i^\pm = Q_{ix} \pm iQ_{iy}$) and polar normal modes ρ_i and φ_i ($Q_{ix} = \rho_i\cos\varphi_i, Q_{ix} = \rho_i\sin\varphi_i$) with $i = 2, 4$. The complex coordinates transform in the same way as the electronic functions in Eq. (4).

We need to solve the equation $\hat{H}_{JT}\Psi = E\Psi$, with $\hat{H}_{JT} = \hat{T}_{vib} + \hat{H}_{elec}$, where the vibrational $\hat{T}_{vib}$ and electronic $\hat{H}_{elec}$ parts of the Hamiltonian depend on the normal coordinates. We obtain the following form for the matrix $\hat{H}_{elec}$ in the basis of the two electronic states

$$\hat{H}_{elec} = W_0\hat{I} + \hat{H}_{elec,Q_2} + \hat{H}_{elec,Q_4} \tag{6}$$

with

$$\hat{H}_{elec,Q_2} = \begin{pmatrix} \frac{1}{2}k_2\rho_2^2 & f_2\rho_2 e^{-i\varphi_2} + g_2\rho_2^2 e^{2i\varphi_2} \\ f_2\rho_2 e^{i\varphi_2} + g_2\rho_2^2 e^{-2i\varphi_2} & \frac{1}{2}k_2\rho_2^2 \end{pmatrix} \tag{7}$$

and with the similar expression for $\hat{H}_{elec,Q_4}$ depending on ρ_4 and φ_4. In Eq. (6), $\hat{I}$ is the 2×2 identity matrix and W_0 is the energy of H_3O at the equilateral geometry. The one-dimensional normal modes Q_1 and Q_3 do not participate in the coupling between the degenerate electronic states. Therefore, the solutions of the two-state vibronic Hamiltonian can be written

$$\Psi = \eta_1(Q_1)\eta_3(Q_3)(\eta^+(Q_2,Q_4)\phi_+(q) + \eta^-(Q_2,Q_4)\phi_-(q))\,. \tag{8}$$

If we set the coordinate ρ_4=0 (thus $\hat{H}_{elec,Q4}$=0) but not ρ_2 and diagonalize the electronic Hamiltonian in Eq.(6) we obtain the electronic Born-Oppenheimer surfaces $U_{a,b}(\rho_2,\varphi_2)$

$$U_{a,b}(\rho_2,\varphi_2) = W_0 + \frac{1}{2}k_2\rho_2^2 \pm [f_2\rho_2 + \frac{1}{2}g_2\rho_2^2\cos(3\varphi_2)], \tag{9}$$

and the Born-Oppenheimer eigenfunctions

$$\phi_a = \frac{1}{\sqrt{2}}[\phi_+ e^{-i\varphi_2/2} + \phi_- e^{i\varphi_2/2}]\,, \tag{10}$$

$$\phi_b = \frac{1}{\sqrt{2}}[-i\phi_+ e^{-i\varphi_2/2} + i\phi_- e^{i\varphi_2/2}]\,. \tag{11}$$

In a similar way, the Born-Oppenheimer energies and eigenstates can be obtained for the Q_4 degenerate mode.

If plotted as functions of two coordinates Q_{2x} and Q_{2y}, the potential surfaces in Eq. (9) exhibit a conical intersection near the equilateral geometry $\rho_2 = 0$ with the minimum of energy situated at $f_2^2/2k_2$ below the energy of the intersection. The electronic potential surfaces obtained by *ab initio* calculations exhibit the same behavior. This allows fitting the *ab initio* data to the analytical behavior of Eq. (9). Therefore, the parameters k_i, f_i, g_i, and W_0 of the Jahn-Teller coupling can be obtained from the *ab initio* energies of excited electronic states of H_3O.

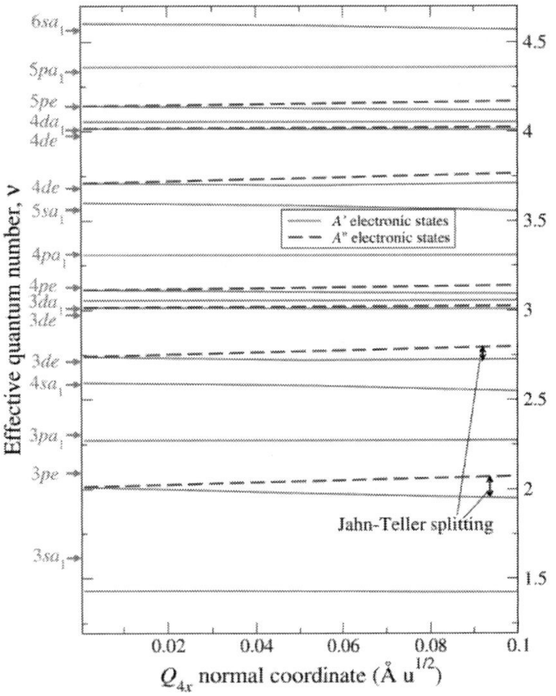

Figure 1. (Color online) Effective quantum numbers of several excited electronic states of H_3O as a function of the Q_{4x} coordinate. The Q_{4x} coordinate is plotted in units Å$\sqrt{\text{u}}$, where u is the atomic mass unit, dalton. The figure gives also by red arrows effective quantum numbers obtained in Ref. [23] for equilibrium geometry of H_3O^+. A detailed comparison of the effective quantum numbers at the equilibrium is provided in Table 2. For higher states, the quantum defect $n - \nu$ is nearly independent on n. The independence of the quantum defect on n is a confirmation of the accuracy of the present *ab initio* calculation.

Table 2. Effective quantum numbers of several excited electronic states of H_3O at the equilibrium of the ion obtained in Ref. [23] and this study.

Electronic state	Ref. [23]	This study
$3sa_1$	1.616	1.432
$3pe$	2.096	2.013
$3pa_1$	2.306	2.274
$4sa_1$	2.587	2.598
$3de$	2.714	2.736
$3de$	2.974	3.015
$3da_1$	3.008	3.055
$4pe$	3.128	3.113
$4pa_1$	3.314	3.310
$5sa_1$	3.562	3.604
$4de$	3.682	3.706
$4de$	3.980	4.016
$4da_1$	4.008	4.016
$5pe$	4.134	4.052
$5pa_1$	4.313	4.143
$6sa_1$	4.542	4.360

We use the quantum defect theory to calculate the DR cross-section. In this approach it is convenient to represent the *ab initio* potential energy surfaces in the form of effective quantum numbers $\nu(Q)$ or quantum defects $\mu(Q)$ that are related to the potential energies $U(Q)$ according to (in atomic units - a.u.)

$$U(Q) = U^+(Q) - \frac{1}{2[\nu(Q)]^2} = U^+(Q) - \frac{1}{2[n - \mu(Q)]^2}, \tag{12}$$

where $U^+(Q)$ denotes the ionic potential and n is the principal quantum number associated with the corresponding electronic state (with energy $U(Q)$). For excited electronic states the quantum defects $\mu(Q)$ are only weakly-dependent on n.

Figure 1 shows the Q_{4x}-dependence of effective quantum numbers obtained from *ab initio* potential energy surfaces calculated in this study. The results at the equilibrium geometry of H_3O^+ are compared with *ab initio* data from Ref. [23]. According to quantum defect theory, the quantum defects and the Jahn-Teller splitting for the excited electronic states should not depend on the principal quantum number. The results shown in the figure are in agreement with that. Because during the DR process, the electron is initially captured into a highly excited Rydberg state, we need to use the Jahn-Teller parameters obtained from the excited Rydberg states.

3. Simplified model for electronic capture and calculated cross section

We assume that the ion is initially in the ground vibrational level v and we neglect the rotation of the ion. The energy of the ground vibrational level is chosen to be the origin of the total energy. The incident electron with kinetic energy E_{el} arrives with the projection Λ on the molecular symmetry. Using the approximation suggested in Ref. [12], we calculate the cross section $\langle\sigma\rangle$ for the electron capture, where $\langle\sigma\rangle$ means that we evaluate the cross section averaged over autoionization Rydberg resonances associated with the vibrational state v'. For this purpose, we use the element $\langle v', \Lambda'|\hat{S}|v, \Lambda\rangle$ of scattering matrix describing the $|v, \Lambda\rangle \to |v', \Lambda'\rangle$ scattering process

$$\langle\sigma\rangle = \frac{\pi}{2E_{el}} \left|\langle v', \Lambda'|\hat{S}|v, \Lambda\rangle\right|^2 \Theta((E_{v'} - E_v) - E_{el}). \tag{13}$$

The Heaviside function in Eq. (13) means that electron can only be captured into a Rydberg state attached to the vibrational state v' with energy $E_{v'}$ if the total energy $E_v + E_{el}$ is smaller than $E_{v'}$. To evaluate the matrix element, we apply the vibrational frame transformation

$$\langle\sigma\rangle = \frac{\pi}{2E_{el}} \left| \int \langle v'|Q\rangle S_{\Lambda',\Lambda}(Q)\langle Q|v\rangle dQ \right|^2 \Theta((E_{v'} - E_v) - E_{el}), \tag{14}$$

where $S_{\Lambda',\Lambda}(Q)$ are elements of the clamped-nuclei scattering matrix $\hat{S}(Q)$. The quantities $S_{\Lambda',\Lambda}(Q)$ can also be considered as matrix elements of the same scattering operator $\hat{S}$ as in Eq. (13) but written in a different representation: $\langle Q', \Lambda'|\hat{S}|Q, \Lambda\rangle = S_{\Lambda',\Lambda}(Q)\delta(Q - Q')$. The matrix $\hat{S}(Q)$ depends on Q and can be written in terms of the quantum defect matrix as $\hat{S}(Q) = \exp[2i\pi\hat{\mu}(Q)]$. For small vibrations, we can expand the quantum defect matrix to first order over small displacements Q_i around $Q = 0$,

$$\hat{\mu}(Q) = \hat{\mu}(0) + \sum_i \left(\frac{\partial\hat{\mu}}{\partial Q_i}\right) Q_i. \tag{15}$$

Here and below, all partial derivatives are assumed to be taken at the equilibrium geometry of H_3O^+. Then, the scattering matrix is expressed as

$$\hat{S}(Q) = \exp[2i\pi\hat{\mu}(0)] \exp\left[2i\pi \sum_i \left(\frac{\partial\hat{\mu}}{\partial Q_i}\right) Q_i\right]. \tag{16}$$

We expand the exponential term of the right side to first order

$$\hat{S}(Q) = \exp\left[2i\pi\hat{\mu}(0)\right]\left(\hat{I} + 2i\pi\sum_i\left(\frac{\partial\hat{\mu}}{\partial Q_i}\right)Q_i\right). \tag{17}$$

In Eq. (14), the ionic vibrational functions in the bracket are orthogonal to each other. Therefore, the constant terms in Eq. (17) for the $\hat{S}(Q)$ matrix do not contribute to the cross section of Eq. (14). Hence, combining Eqs. (14) and (17) we obtain

$$\langle\sigma\rangle = \frac{2\pi^3}{E_{el}}\left|\sum_i\left(\frac{\partial\mu_{\Lambda,\Lambda'}}{\partial Q_i}\right)\langle v'|Q_i|v\rangle\right|^2\Theta(E_v' - E_v - E_{el}), \tag{18}$$

where $\mu_{\Lambda,\Lambda'} \equiv \mu_{\Lambda,\Lambda'}(Q)$ is an element of the quantum defect matrix $\hat{\mu}(Q)$ written in the same representation as $S_{\Lambda,\Lambda'}(Q)$.

Equation (18) is quite general and can be used to calculate the cross section for the electron capture by different molecular ions. We now apply it to DR in H_3O^+ using the parameters of the Jahn-Teller coupling obtained above. These parameters are used below to construct the scattering matrix $\hat{S}(Q)$. Thus, they determine the H_3O^+ DR cross section. We will consider the symmetric normal modes Q_1 and Q_3, as well as the doubly degenerate normal modes $Q_2 = (\rho_2, \varphi_2)$ and $Q_4 = (\rho_4, \varphi_4)$ as the vibrational coordinates.

We start with an electronic state with potential

$$U(Q) = U^+(Q) - \frac{1}{2(n - \mu(Q))^2}, \tag{19}$$

and taking derivative with respect to Q_i, we obtain

$$\left(\frac{\partial\mu}{\partial Q_i}\right) = -\left(\frac{\partial\nu}{\partial Q_i}\right) = -\nu^3\left(\frac{\partial U}{\partial Q_i}\right). \tag{20}$$

The potential does not depend to first order on Q_1 or Q_3, i.e. $\left(\frac{\partial\mu_{\Lambda,\Lambda'}}{\partial Q_1}\right) = \left(\frac{\partial\mu_{\Lambda,\Lambda'}}{\partial Q_3}\right) = 0$, for any electronic states Λ and Λ'. In the same way, using Eq. (9), we find that the potentials do not depend to first order on φ_2 and φ_4 near the equilateral geometries, i.e. $\left(\frac{\partial\mu_{\Lambda,\Lambda'}}{\partial\varphi_2}\right) = \left(\frac{\partial\mu_{\Lambda,\Lambda'}}{\partial\varphi_4}\right) = 0$. Hence, in the first order of approximation, the only non-vanishing terms in Eq. (18) are the ones that depend on ρ_2 and ρ_4. Therefore, the main contribution into the electron capture is due to the Jahn-Teller coupling.

We assume that the incident electron has projection $\Lambda = -1$ on the molecular axis (Eqs. (3) and (6)), the final electronic state has projection $\Lambda' = 1$, once the electron is captured into a Rydberg state attached to an excited vibrational level of the ionic core. For the cross section calculation, we need to evaluate the derivatives $\partial\mu_{\Lambda',\Lambda}/\partial\rho_i$ at the equilibrium of the ion. Because the diagonal elements of $\hat{\mu}(Q)$ are the same, the derivative $\partial\mu_{\Lambda',\Lambda}/\partial\rho_i$ is equal to the derivative of eigenvalues $\mu_a(Q)$ and $\mu_b(Q)$ of $\hat{\mu}$. The derivatives $\partial\mu_a/\partial\rho_i$ and $\partial\mu_b/\partial\rho_i$ are obtained from *ab initio* calculations (shown in Fig. 1) or directly using Eqs. (9) and (12) from the potential energy surfaces of the excited electronic states of H_3O that are coupled by the Jahn-Teller interaction,

$$\left(\frac{\partial\mu_{1,-1}}{\partial\rho_2}\right) = \left|\frac{\partial U_a}{\partial\rho_2}\right|\nu^3 \quad \text{and} \quad \left(\frac{\partial\mu_{1,-1}}{\partial\rho_4}\right) = \left|\frac{\partial U_a}{\partial\rho_4}\right|\nu^3. \tag{21}$$

As the last step of the treatment, we consider the vibrational states of H_3O^+ in the normal mode approximation and label them as $|v_1 v_2^{l_2} v_3 v_4^{l_4}\rangle$, where l_2 and l_4 are the vibrational angular momenta associated with the degenerate normal modes. The initial state of the ion is its ground vibrational state, thus $|v\rangle = |00^0 00^0\rangle$. The only non vanishing terms in the cross section (18) are $\langle 01^1 00^0|\rho_2|00^0 00^0\rangle = 1/\sqrt{2\omega_2}$ and $\langle 00^0 01^1|\rho_4|00^0 00^0\rangle = 1/\sqrt{2\omega_4}$. We use mass-scaled normal coordinates in these expressions, so the masses dropped out. The frequencies ω_i are given in Table 1. So, in this approximation the electronic capture can only occur into these two final vibrational states. Also, the cross section in Eq. (18) has to be multiplied by two because the final vibrational state is doubly degenerate. It gives the final DR cross section

$$\langle\sigma\rangle = \frac{2\pi^3}{E_{el}\omega_2}\left(\frac{\partial\nu_a}{\partial\rho_2}\right)^2 \Theta(\hbar\omega_2 - E_{el}) + \frac{2\pi^3}{E_{el}\omega_4}\left(\frac{\partial\nu_a}{\partial\rho_4}\right)^2 \Theta(\hbar\omega_4 - E_{el})\,. \tag{22}$$

The derivatives $\partial\nu_a/\partial\rho_i$ are obtained from the dependencies $\nu_a(\rho_i)$ such as shown in Fig. 1. There are several electronic e states that are doubly-degenerate at the equilateral configuration and that are split by the Jahn-Teller coupling once the C_{3v} symmetry is lifted. These states can be considered as different superpositions of the hydrogen-like states $p\pi$, $d\pi$, $d\delta$. As it is clear from Fig. 1 the Jahn-Teller coupling is not negligible between $d\pi$ states, for example. Therefore, the electron can be captured into these states. In Eq. (22) we account separately for the contribution from each such Jahn-Teller pair of electronic states. The final cross-section is plotted in Fig. 2. In that figure we also show the experimental DR cross section obtained in the CRYRING [3] and ASTRID [4, 24] storage rings.

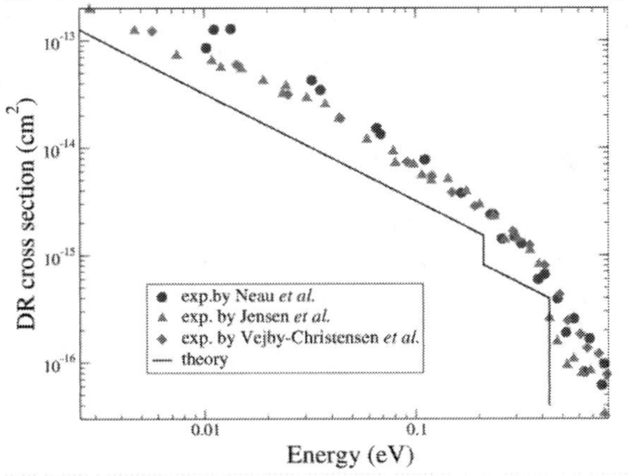

Figure 2. (Color online) Theoretical cross section for dissociative recombination of H_3O^+ with low-energy electrons. The experimental results from Refs. [3, 4, 24] are also given.

4. Resonant H_3^+ dissociative recombination

Tests of the preceding simplified theoretical treatment of DR have been encouraging, in terms of predicting average recombination rates for increasingly complex polyatomic molecules. But it is still of interest to push theoretical and experimental capabilities to the greatest possible level of detail for some fundamental species like H_3^+. To date it has not yet been possible to compare experiment and theory at the level of individual resonances in the spectrum. This is in part because even though this triatomic ion is the simplest, it is nevertheless predicted to have a spectrum dominated by a dense forest of resonances, which would require extremely high experimental resolution in order to test their positions, widths, and intensity profiles.

Some experimental evidence exists for the dense resonances in this energy range for the e+H_3^+ system, in the form of the near-threshold photoionization spectrum of metastable H_3 that was measured years ago by Helm and coworkers (see [25] and references therein). Photoionization is a different observable than dissociative recombination, of course, but it is well known that a resonance seen in one channel will have the same position and width in any other collision channel [26]. However, little experimental evidence exists to support this predicted dominance of resonance pathways for the DR observable. This is presumably in part because of the fact that collision experiments typically have a much greater energy bandwidth than laser photoionization experiments, which makes any given narrow resonance less pronounced in a DR experiment. A second reason is that many more symmetries are present in collision experiments, which are less selective in terms of final state angular momentum and parity channels than are photoionization experiments. It is worth noticing that the same level of theory that has been applied to the DR calculation for H_3^+ has also been shown [27] to describe the Helm photoionization experiments, with excellent agreement in the energy range where rotational autoionization dominates the decay, but somewhat more modest agreement in ranges where vibrational autoionization and predissociation dominate the resonance decay pathways. This concurrence of theory and experiment for high resolution photoionization measurements provides some additional confidence in the resonance properties predicted by theory, but since this concurrence is achieved only for one symmetry of the ortho species, it is not enough to conclude that all symmetries are predicted to have the correct resonance structure in the corresponding DR process occurring in the same energy range.

The level of theory has not been changed since the latest improvements were published in 2007 in [11]. However, some new thermal averages have been computed on a fine energy mesh, to explore the observability of resonance structures and their sensitivity to the ionic rotational temperature. The main results are being reported elsewhere, specifically in a joint experiment-theory publication [28], which includes a careful new analysis of the source ion rotational temperature. We show here one example of the level of agreement that is emerging from that study, plotted in Fig. 3 in the form of the reduced dissociative recombination rate coefficient, which is the DR rate coefficient multiplied by $\sqrt{E_d}$. The preliminary ionic rotational temperature relevant to the experimental results shown in this figure has been estimated to be 380 K. For the energy range shown, the theory has been computed for a rotational ion temperature of 100 K, a transverse temperature in the relative collision energy of 500 μK and a parallel temperature of 25 μK. At this ion temperature and electron beam energy resolution, signatures of resonances begin to emerge that are clearly identifiable. Modulations in the experiment, believed to have an ion temperature of approximately 300 K, do show some correlation with the theoretical modulations. While there remain discrepancies in the overall rate by a factor of 2-4 in some energy ranges, the overall agreement continues to be encouraging, even at this level of comparison, which is far more detailed than has been achieved previously. The energy range below that shown in the figure exhibits poorer agreement between theory and experiment, as is discussed in greater detail in Ref. [28].

Another positive advance in our understanding of H_3^+ dissociative recombination that has been achieved during the past few years has occurred in the context of afterglow experiments and their interpretation. Glosik and collaborators [29, 30, 31, 32, 33] have demonstrated through combined experimental and theoretical analysis that the neutralization process in plasma containing H_3^+, electrons, and ground state helium atoms is predominantly mediated by a three-body process involving He atoms. The three-body recombination rate at temperatures around 300 K tends to be in the range $K_3 \approx 1 - 10 \times 10^{-25}$ cm$^6/s$, which means that for He densities in the range 10^{17} to 10^{18} cm^{-3}, the effective rate of ternary recombination can compete with or even dominate over the binary recombination rate. Nevertheless, by extrapolating results to the limit of zero He density, the binary recombination rate for e + H_3^+ collisions has been obtained and good agreement has been found with both *ab initio* theory and storage ring experiments.

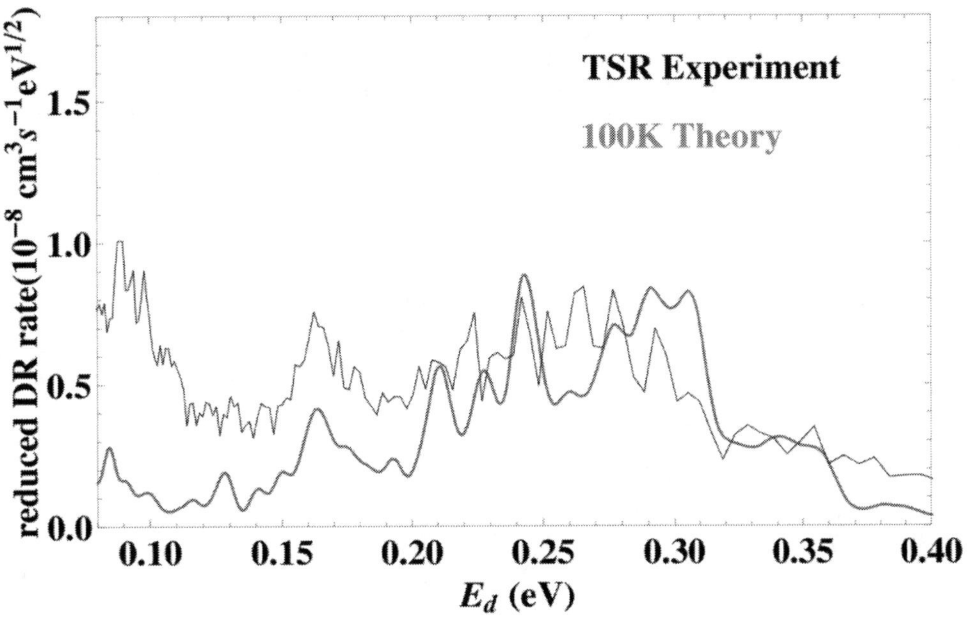

Figure 3. (Color online) Comparison of theoretical and experimental rate coefficients for low energy electron recombination in collisions with H_3^+. The experimental results are adapted from Petrignani *et al.* [28], whose measurements were carried out at the TSR facility in Heidelberg. The analysis of Ref. [28] suggests that the rotational temperature of the ions is approximately 380 K, whereas the theory at the level of Ref. [11] has been computed for a rotational ion temperature of 100 K to make the resonances stand out somewhat more strongly.

5. Discussion of the results and perspectives

We investigated the DR rate of H_3O^+ using the simplified theoretical model. In this model, we assume that once the electron is captured by the ion into a vibrationally excited level, the probability of autoionization is small compared to the predissociation probability. Therefore, we calculate the cross section for the electron capture. The non-Born-Oppenheimer Jahn-Teller interaction between the motion of the incident electron and the vibrational motion of the ion is the key ingredient in the present model. The obtained theoretical DR cross section is in good agreement with the results from the storage ring experiment by Neau *et al.* [3], which indirectly confirms that autoionization probability is indeed small compared to the predissociation probability.

In this model, the incident electron can excite only one quantum of degenerate vibrational modes of the ionic core. The excitation of higher vibrational levels is neglected. One possible simple improvement of the model would be to account for the excitation of higher vibrational levels of E irreducible representation. In the present model it can be made if one considers the fact that the higher vibrational levels of the E irreducible representation have small contributions from the $|01^100^0\rangle$ and/or $|00^001^1\rangle$ vibrational levels due to the anharmonicity of the ionic potential. This improvement will not change the cross section at low energy in a noticeable way, but it will give a non-zero cross section at incident electron energies above $\hbar\omega_4$. The present theoretical cross section is identically zero above $E_{el} = \hbar\omega_4$, although the experimental cross section is small but measurable.

We have also discussed the dissociative recombination of H_3^+, for which a detailed theoretical treatment has been developed. The overall theoretical cross section for the dissociative recombination of H_3^+ is in good agreement with experimental data. Because the state of theory and experiment for DR of H_3^+ is such that they both provide information about positions of individual resonances, we compare detailed features in the experimental and theoretical DR spectra. The overall agreement is satisfactory but there are several regions in the spectrum where theory fails to reproduce experimental features.

Acknowledgments

This work was done with the support of Triangle de la Physique under contract 2008-007T-QCCM and of National Science Foundation under grants PHY-0855622 and PHY-0855092 and by the DOE Office of Science. We thank the experimental group of A. Wolf, A. Petrignani, O. Novotny, and their collaborators for many informative discussions and for providing their experimental data for comparison with theory.

Travel support for C. Greene was provided by the NSF and NASA grants given in the preface to this volume.

References

[1] Andersen L H, Heber O, Kella D, Pedersen H B, Vejbi-Christensen L and Zajfman D 1996 *Phys. Rev. Lett.* **77** 4891
[2] Haider S A and Bhardwaj A 2005 *Icarus* **177** 196
[3] Neau A, Khalili A A, Rosén S, Padellec A L, Derkatch A M, Shi W, Vikor L, Larsson M, Semaniak J, Thomas R, Nâgård M B, Andersson K, Danared H and af Ugglas M 2000 *J. Chem. Phys.* **113** 1762
[4] Jensen M J, Bilodeau R C, Safvan C P, Seiersen K, Andersen L H, Pedersen H B and Heber O 2000 *Astrophys. J.* **543** 764
[5] van Dishoeck E F, Keene J and Phillips T G 1992 *Astrophys. J.* **399** 533
[6] Neufeld D A, Lepp S and Melnick G J 1995 *Astrophys. J. Suppl. Ser.* **100** 132
[7] Prager J, Riedel U and Warnatz J 2007 *Proc. Combust. Inst.* **31** 1129
[8] Wisman D L, Marcum S D and Ganguly B N 2007 *Combust. Flame* **151** 639
[9] Kokoouline V, Greene C H and Esry B D 2001 *Nature* **412** 891
[10] Kokoouline V and Greene C H 2003 *Phys. Rev. Lett.* **90** 133201
[11] Fonseca dos Santos S, Kokoouline V and Greene C H 2007 *J. Chem. Phys.* **127** 124309
[12] Mikhailov I A, Kokoouline V, Larson A, Tonzani S and Greene C H 2006 *Phys. Rev. A* **74** 032707
[13] Douguet N, Kokoouline V and Greene C H 2008 *Phys. Rev. A* **77** 064703
[14] Haxton D J and Greene C H 2008 *Phys. Rev. A* **78** 052704
[15] Kokoouline V and Greene C H 2003 *Phys. Rev. A* **68** 012703
[16] Jungen C and Pratt S T 2008 *J. Chem. Phys.* **129** 164311
[17] Jungen C and Pratt S T 2009 *Phys. Rev. Lett.* **102** 023201
[18] Ogurtsov I, Gorinchoy N and Balan I 2007 *J. Mol. Struct.* **838** 107 proceedings of the Symposium on the Jahn-Teller Effect
[19] Werner H J, Knowles P J, Lindh R, Manby F R, Schütz M *et al.* 2008 Molpro, version 2008.3, a package of *ab initio* programs
[20] Bowman J M and Carter X C H S 2002 *Spectrochim. Acta A* **58** 839
[21] Longuet-Higgins H C 1961 *Adv. Spectrosc.* **II** 429
[22] Moffitt W and Liehr A D 1957 *Phys. Rev.* **106** 1195
[23] Luo M and Jungen M 1997 *Chem. Phys.* **241** 297
[24] Vejbi-Christensen L, Andersen L H, Heber O, Kella D, Pedersen H B, Schmidt H T and Zajfman D 1997 *Astrophys. J.* **483** 531
[25] Mistrík I, Reichle R, Müller U, Helm H, Jungen M and Stephens J 2000 *Phys. Rev. A* **61** 033410
[26] Fano U 1961 *Phys. Rev.* **124** 1866
[27] Kokoouline V and Greene C H 2004 *Phys. Rev. A* **69** 032711
[28] Petrignani A, Altevogt S, Berg M H, Bing D, Grieser M, Hoffmann J, Jordon-Thaden B, Krantz C, Mendes M B, Novotny O, Novotny S, Orlov D A, Repnow R, Sorg T, Stützel J, Wolf A, Buhr H, Kreckel H, Kokoouline V and Greene C H 2011 *Phys. Rev. A (accepted for publication)*
[29] Glosik J, Korolov I, Plasil R, Novotny O, Kotrik T, Hlavenka P, Varju J, Mikhailov I A, Kokoouline V and Greene C H 2008 *J. Phys. B: At. Mol. Opt. Phys.* **41** 191001
[30] Glosík J, Plašil R, Korolov I, Kotrík T, Novotný O, Hlavenka P, Dohnal P, Varju J, Kokoouline V and Greene C H 2009 *Phys. Rev. A* **79** 052707
[31] Glosík J, Korolov I, Plašil R, Kotrík T, Dohnal P, Novotný O, Varju J, Š Roučka, Greene C H and Kokoouline V 2009 *Phys. Rev. A* **80** 042706

Journal of Physics: Conference Series **300** (2011) 012015 doi:10.1088/1742-6596/300/1/012015

[32] Kotrík T, Dohnal P, Korolov I, Plašil R, Š Roučka, Glosík J, Greene C H and Kokoouline V 2010 *J. Chem. Phys.* **133** 034305
[33] Plašil R, Kotrík T, Dohnal P, Varju J, Hejduk M, Korolov I, Roučka S, Kokoouline V and Glosík J 2010 *Molec. Phys.* **108** 2253

Eighth International Conference on Dissociative Recombination (DR2010) IOP Publishing
Journal of Physics: Conference Series **300** (2011) 012016 doi:10.1088/1742-6596/300/1/012016

Calculation of dissociating autoionizing states using the block diagonalization method: Application to N_2H

A. P. Hickman[1], D. O. Kashinski[1,2], R. F. Malenda[1], F. Gatti[3] and D. Talbi[4]

[1] Department of Physics, Lehigh University, 16 Memorial Dr. E, Bethlehem, PA, 18015, USA
[2] Current address: Department of Physics and Nuclear Engineering, United States Military Academy, West Point, NY, 10996 USA
[3] Université Montpellier II - Institut Charles Gerhardt, CTMM - UMR 5253, place Eugène Bataillon, 34095 Montpellier, France
[4] Université Montpellier II - GRAAL, CNRS - UMR 5024, place Eugène Bataillon, 34095 Montpellier, France

E-mail: aph2@lehigh.edu,Dahbia.Talbi@graal.univ-montp2.fr

Abstract. We report the calculation of preliminary potential surfaces necessary to treat dissociative recombination (DR) of electrons with N_2H^+. We performed multi-reference, configuration interaction calculations with a large active space for N_2H^+ and N_2H, using the GAMESS electronic structure code. Rydberg-valence coupling is strong in N_2H, and a systematic procedure is desirable to isolate the appropriate dissociating, autoionizing states. We used the block diagonalization method, which requires only modest additional effort beyond the standard methodology. We treated both linear and bent geometries of the molecules, with N_2 fixed at its equilibrium separation. The results indicate that the crossing between the dissociating neutral curve and the initial ion potential is not favorably located, suggesting that the direct mechanism for DR will be small. Dynamics calculations using the multi-configuration, time-dependent Hartree (MCTDH) method confirm this conclusion.

1. Introduction

Although nitrogen is very important in the chemistry of the interstellar medium, it cannot be directly observed. Its abundance must be inferred indirectly by detecting N_2H^+, which can be produced by the reaction

$$N_2 + H_3^+ \rightarrow N_2H^+ + H_2. \tag{1}$$

It was long asssumed that the N_2 consumed in the reaction above was almost completely recovered by dissociative recombination (DR). The measurements of Adams *et al* [1] in 1990 indicated that the branching ratios in the dissociative recombination process strongly favor the $N_2 + H$ channel:

$$e^- + N_2H^+ \rightarrow \begin{cases} N_2 + H \\ NH + N. \end{cases} \tag{2}$$

However, in 2004 Geppert *et al* [2] reported measurements that the branching ratio to the $NH + N$ channel was dominant. This unexpected result cast doubt on the assumptions mentioned above and motivated additional experimental and theoretical work.

In 2007, Molek *et al* [3] reported a new experimental measurement of the branching ratio and concluded that the upper limit for the branching ratio to the NH + N channel was 5%. Also in 2007, Talbi [4] reported a theoretical investigation of the potential surfaces for linear N_2H and N_2H^+. She concluded that the likely outcome of dissociative recombination would be the N_2 + H channel, with N_2 in the first electronically excited state. There is now general agreement that the branching ratio to the NH + N channel is less than about 5% [5].

Talbi's work [4, 6] has highlighted the need for further calculations to determine appropriate diabatic potential surfaces for both linear and bent geometries of the molecule. This work reports preliminary calculations of that type.

2. What is needed for a DR calculation

A detailed calculation of dissociative recombination must provide potential curves for both the initial molecular ion and the dissociating autoionizing state. The surfaces needed are illustrated in figures 1 and 2. Calculating the potential surface for the initial molecular ion (N_2H^+ in the present case) is straightforward and involves standard techniques of quantum chemistry. However, calculating the dissociating state is much more difficult because that state is embedded in the continuum of scattering states that correspond to $e^- + N_2H^+$.

Recent work by ourselves [7, 8] and others [9, 10] has shown that the block diagonalization method provides a very powerful technique to determine the necessary dissociating, autoionizing states of N_2H. The block diagonalization method also allows us to determine the autoionization width Γ, which is related to the probablity that the electron incident on the molecular ion is captured into the dissociating state.

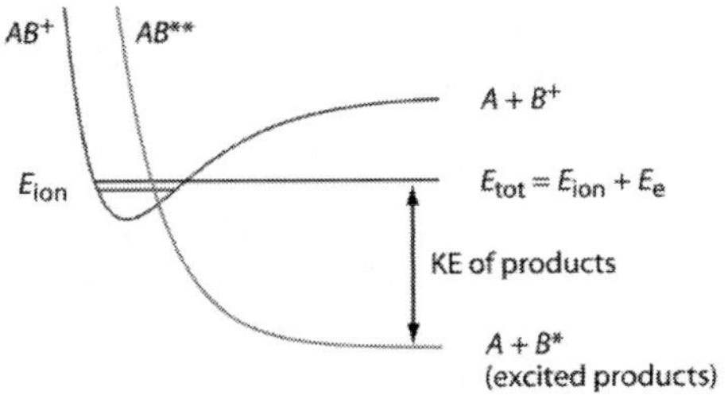

Figure 1. Potential surfaces for the molecular ion and the dissociating neutral state needed for the "direct mechanism" of DR.

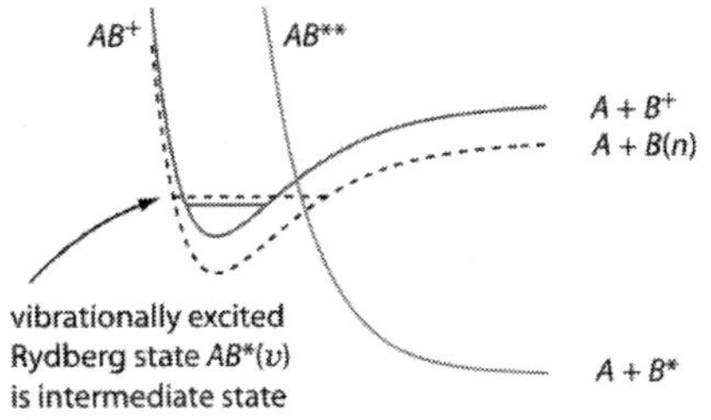

Figure 2. Additional potential surfaces for excited states of the neutral molecule are needed to treat "indirect" process for DR.

The principal difficulty that one encounters when trying to calculate the dissociating surface using the standard techniques of electronic structure is that the states of interest are (1) highly excited and (2) strongly mixed with other states. Multiple curve crossings occur, and it can be very difficult to isolate the desired potential surface. Before going into the details of our calculations, we will present an overview of the results that can be obtained by using the block diagonalization method to separate the desired states from the many other states that appear in a standard calculation.

Figure 3 shows the standard adiabatic curves obtained, and figure 4 shows the corresponding diabatic curves. These calculations were performed in the linear geometry, and R_d is the distance from the hydrogen atom H to the center of mass of the N_2. The adiabatic curves exhibit multiple curve crossings; diabatization allows us to identify the appropriate dissociation channels.

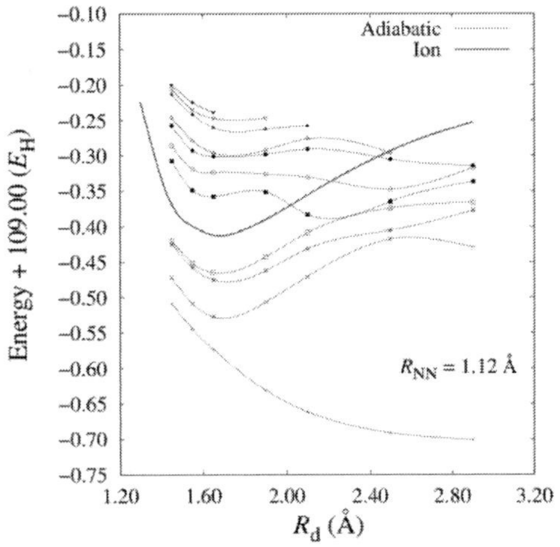

Figure 3. Adiabatic potentials determined from the GAMESS electronic structure calculations without any additional analysis.

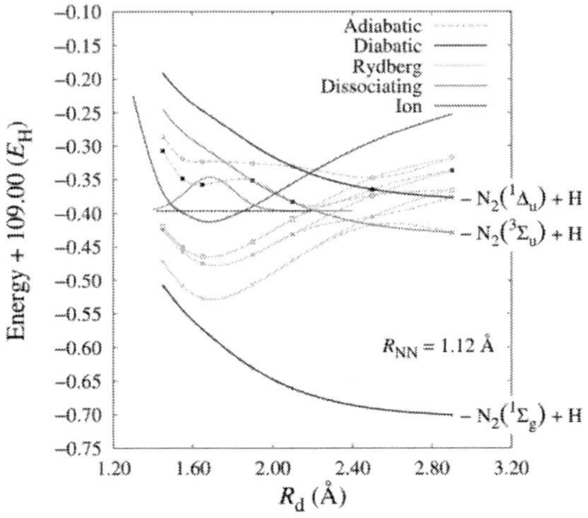

Figure 4. Diabatic potentials determined using the block diagonalization method. A vibrational wave function for the ion is also shown.

3. Block diagonalization method

The block diagonalization method [9] provides an effective technique for transforming the results of a standard electronic structure calculation into diabatic potential curves. An advantage of the method is that one can perform a conventional calculation of the desired size and accuracy and then obtain diabatic curves with comparable reliability. A second advantage is that the numerical effort for the diabatization is quite modest compared to the original calculation (which usually involves diagonalizing a very large, sparse matrix.)

The method requires extra effort in the determination of molecular orbitals (mo's). Briefly, one must insure that the variation of the mo's $\{\phi_i\}$ with molecular geometry is small. This result can be achieved by setting up the calculation in such a way that the adiabatic energies are invariant under a rotation of certain molecular orbitals, and then selecting an appropriate rotation at each geometry. For example, the MCSCF energies are invariant under an arbitary rotation $\mathbf{U}$ of the mo's in the active space [11, 12]. One can replace a set of orbitals $\{\phi'_i\}$ optimized in an MCSCF with a new set

$$\phi_i = \sum_j U_{ij}\phi'_j, \tag{3}$$

and the MCSCF energies will not change. [9] pointed out that one can take advantage of this degree of freedom by choosing $\mathbf{U}$ so that each mo in the set $\{\phi_i\}$ resembles as closely as possible a corresponding orbital in a set of "reference" orbitals. The reference orbitals are defined in a way that makes them easy to interpret and insures that they are slowly varying. We use the algorithm we previously implemented [7] based on a series of Jacobi rotations to maximize the overlap of each ϕ_i with the corresponding reference orbital.

The configuration interaction (CI) electronic wave function Ψ_n for the nth state is represented as the sum of configurations Φ_m, each constructed from the mo's ϕ_i:

$$\Psi_n = \sum_{m=1}^{N} c_{mn}\Phi_m, \tag{4}$$

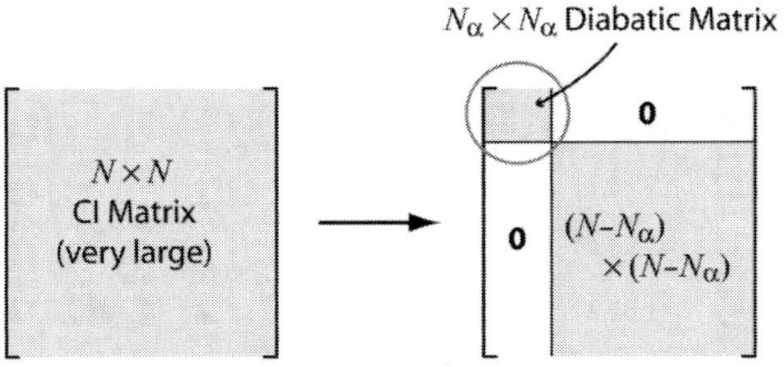

Figure 5. Schematic illustration of block diagonalization. The $N_\alpha \times N_\alpha$ block of interest can be found by applying the transformation **T** to the diagonal matrix **E** according to Eq. (5). The larger block of size $N - N_\alpha$ need not be explicitly determined.

The number of coefficients N in the sum in Eq. 4 may be quite large (of order 10^6). However, one can usually identify a small set of N_α configurations ($N_\alpha \sim 2$–10) that make the dominant contribution to N_α electronic states of interest. Then N_α will be the dimension of the diabatic Hamiltonian, and for the diabatization one only needs the $N_\alpha \times N_\alpha$ matrix of values c_{mn} for the coefficients of the N_α dominant configurations in the N_α states of interest. We denote this matrix by **S**. The diabatic Hamiltonian matrix $\mathbf{H}_{\rm dia}$ can be expressed as a transformation of the diagonal matrix **E** whose nonzero elements are the adiabatic eigenvalues $E_1, \ldots, E_{N_\alpha}$:

$$\mathbf{H}_{\rm dia} = \mathbf{T}^\dagger \mathbf{E} \mathbf{T}, \tag{5}$$

where (†) denotes the adjoint (transpose for a real transformation), and

$$\mathbf{T} = \mathbf{S}^{-1} \left(\mathbf{S}\mathbf{S}^\dagger\right)^{1/2}. \tag{6}$$

This analysis of the adiabatic eigenvalues and eigenvectors is straightforward and only involves matrices of order N_α. Since $\mathbf{H}_{\rm dia}$ is explicitly constructed by a unitary transformation of the matrix **E** of adiabatic eigenvalues, the eigenvalues of the small matrix $\mathbf{H}_{\rm dia}$ will be exactly the same as the chosen N_α eigenvalues of the large matrix determined by the CI calculation. The determination of $\mathbf{H}_{\rm dia}$ is illustrated schematically in figure 5.

4. Electronic structure calculations

The electronic structure calculations reported here were performed using the GAMESS code [13]. The basis set included a 6-311$G(d,p)$ on each N plus diffuse functions and a $10s4p1d$ basis on H; there were 69 total basis functions. Using the notation of Talbi [4] for the molecular orbitals, the orbital occupancy of the ground state of N_2H^+ is

$$1s_a^2\, 1s_b^2\, (NN)^2\, (SP)^2\, (NH)^2\, \pi_x^2\, \pi_y^2 \Big|,$$

where the vertical bar $\left(\Big|\right)$ signifies the highest orbital occupied in the ion core. The orbital occupancy of the N_2H ground state is

$$1s_a^2\, 1s_b^2\, (NN)^2\, (SP)^2\, (NH)^2\, \pi_x^2\, \pi_y^2 \Big| (NH^*)^1.$$

The Rydberg states of N_2H have one electron in a highly excited orbital outside of the ion core:

$$1s_a^2\, 1s_b^2\, (NN)^2\, (SP)^2\, (NH)^2\, \pi_x^2\, \pi_y^2 \Big| (NH^*)\, \pi_x^*\, \pi_y^*\, (NN^*)\, ({\rm Ryd}\,1)^1\, ({\rm Ryd}\,2)\, ({\rm Ryd}\,3) \ldots;$$

we use a light color to indicate that orbitals such as (NH^*) are not occupied. The dissociating valence states of N_2H have an unoccupied core orbital:

$$1s_a^2\, 1s_b^2\, (NN)^2\, (SP)^2\, (NH)^2\, \pi_x\, \pi_y^2 \Big| (NH^*)^1\, \pi_x^{*\,1}\, \pi_y^{*\,1}$$

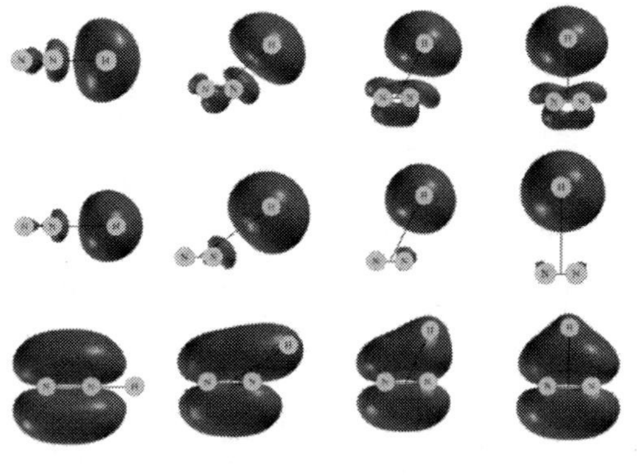

Figure 6. Variation of the molecular orbitals with molecular geometry. The top row shows the antibonding NH* orbital for $R_\mathrm{d} = 1.61$ Å, and $\theta = 0$, 30, 60, and 90 degrees. The second row shows the same orbtial for $R_\mathrm{d} = 2.46$ Å, at the same angles. The third row shows the A' bonding π orbital for $R_\mathrm{d} = 1.61$ Å, and $\theta =$ 0, 30, 60, and 90 degrees. The smooth changes in the molecular orbitals ensure that the block diagonalization process gives reliable diabatic potential curves.

$$1s_a^2\, 1s_b^2\, (NN)^2\, (SP)^2\, (NH)^2\, \pi_x^2\, \pi_y \Big| (NH^*)^1\, \pi_x^{*\,1}\, \pi_y^{*\,1}$$

The dissociating state important for DR is a linear combination of states corresponding to the two orbital occupancies given above.

We have completed multi-reference, configuration interaction (MRCI) calculations with an active space that includes the (NH), π_x, π_y, NH*, π_x^*, and π_y^* orbitals. We included single and double excitations from the active space, leading to 449,892 configurations. We typically calculated 25-85 eigenvalues and eigenvectors of the electronic Hamiltonian. These results are preliminary; final calculations will use a larger active space that includes Rydberg orbitals.

An important requirement for the block diagonal method is that the molecular orbitals vary smoothly with geometry. For most of the calculations reported here, it was sufficient to use natural orbitals determined by a preliminary CI calculation on the N_2H ground state. These orbitals can be determined by the GAMESS code. It was necessary to watch for "orbital swaps." GAMESS orders the molecular orbitals according to an orbital energy; this procedure can lead to an unwanted interchange of two or more orbitals as the geometry changes. By estimating the orbital overlap between corresponding orbitals at similar geometries we were able to reorder the orbitals as necessary. In a few cases, such as small values of R_d, it was necessary to define reference orbitals explicitly as described in section 3. Figure 6 illustrates the smooth variation of the orbitals we used.

We primarily investigated the case in which DR leads to breaking the NH bond, so we froze the NN bond length at the equilibrium value appropriate for N_2H (taken to be $R_\mathrm{NN} = 1.12$ Å). For this case it is convenient to use Jacobi corrdinates, which are illustrated in figure 7. R_d is the distance from the H atom to the center of mass of the N_2, and θ is the bending angle ($\theta = 0$ corresponds to a linear molecule.) One can then expand the potential surface using Legendre polynomials of even order:

$$V(R_\mathrm{NN}, R_\mathrm{d}, \theta) = \sum_{\lambda(\mathrm{even})}^{2N} V_\lambda(R_\mathrm{NN}, R_\mathrm{d}) P_\lambda(\cos\theta) \tag{7}$$

We typically performed calculations for 8–11 different angles at each value of R_d. The calculated points are shown in figure 7. We have plotted the points as functions of $\sin^2\theta = 1 - \cos^2\theta$ in order to facilitate least squares fitting. A Legendre polynomial expansion with only even terms for $\lambda = 0, \ldots, 2N$ is an Nth degree polynomial in $\cos^2\theta$ or, equivalently, in $\sin^2\theta$. By plotting the calculated points vs. $\sin^2\theta$ we could assess the goodness of fit using standard polynomial fitting routines.

The curves shown in figure 7 are typically five-term polynomials (in $\sin^2\theta$). It is clear from these curves that an exact fit to the calculated points would have additional undulations. We

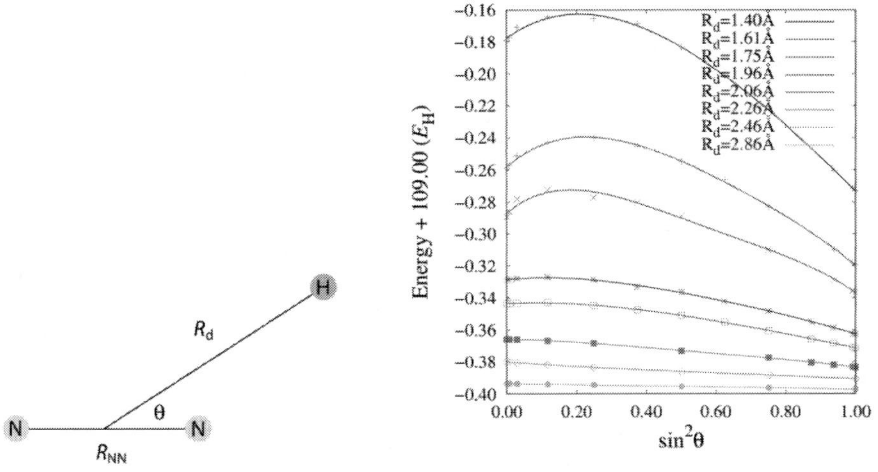

Figure 7. The left panel illustrates Jacobi coordinates. The right panel shows the angular dependence of dissociating potential. Calculated points and a five-term polynomial fit are shown.

prefer to have a somewhat smoother fit, because the additional variation in the calculated points is likely due to the uncertainty is selecting the optimum N_α and the most important configurations for each geometry. Because of the many curve crossings, the set of configurations that make the most important contribution is constantly changing. Note that the points most distant from the fitted curve tend to be at small values of R_d, where the potential is the steepest and small uncertainties are unlikely to be important.

Once the Legendre polynomial expansion coefficients in Eq. (7) have been determined for several values of R_d, they can easily be interpolated using splines. (R_{NN} is fixed.) Then the full surface can be interpolated at arbitrary values of R_d and θ. The interpolated results are shown in figures 8 and 9. For both of these figures, we have plotted the energy vs. the cartesian coordinates of the H atom in the plane of the molecule ($x = R_d \cos\theta$ and $y = R_d \sin\theta$).

5. Dynamics calculations

We have performed preliminary dynamics calculations using the multi-configuration, time-dependent Hartree (MCTDH) method [14]. This method has emerged over the past few years as a very efficient procedure for numerical solutions of the time-dependent Schroedinger

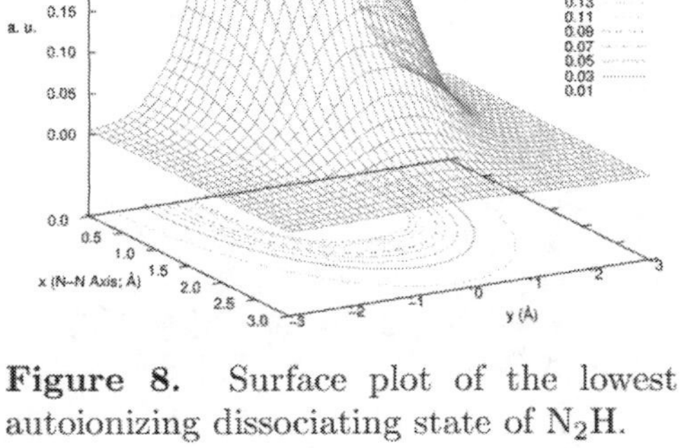

Figure 8. Surface plot of the lowest autoionizing dissociating state of N_2H.

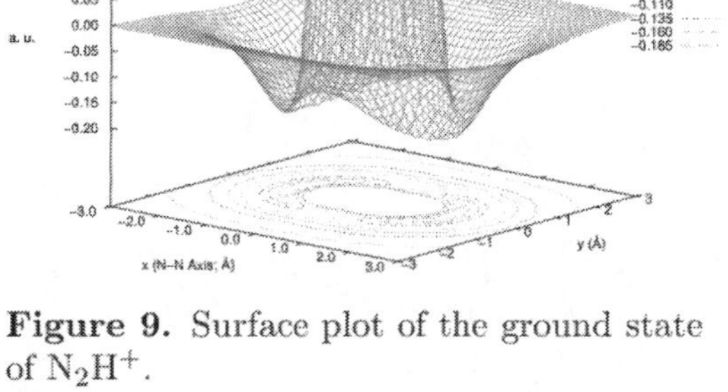

Figure 9. Surface plot of the ground state of N_2H^+.

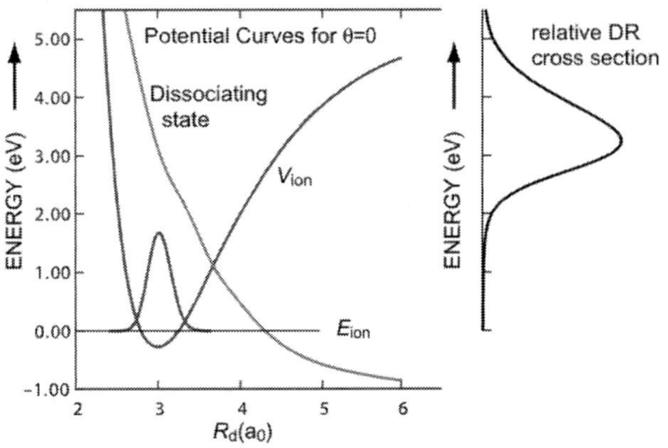

Figure 10. Potential curves for N_2H^+ and N_2H and relative cross sections for DR calculated using the MCTDH method. The dynamics calculations include the angular dependence of the potential surface but only the direct mechanism for DR. The energy scale is the same for both panels; the peak of the DR cross section occurs near an incident electron energy of 3 eV. The cross section from the direct mechanism for very low electron energies is very small.

equation. The MCTDH code permits the solution of the Schroedinger equation at various levels of accuracy; for our relatively small (triatomic) molecules, we were able to use the methods that provide an essentially exact solution.

The potential surfaces shown in figure 10 for the linear molecule ($\theta = 0$) demonstrate that the Franck-Condon factors between the vibrational ground state of N_2H^+ and the N_2–H scattering states are poor. However, as the molecule bends, the crossing between the dissociating potential and the ion potential moves to smaller values of R_d. Since the zero point motion of the molecular ion leads to a distribution of bending angles, it seemed possible that direct DR might occur. We performed dynamics calculations to investigate this possibility

Since we do not yet have the autoionization width, we performed the DR calculations using a model with simplifying assumptions (which can be avoided in future work). We used the built-in capability of the MCTDH code to treat photodissociation; if one assumes that the Γ is small and constant the processes are formally very similar. A small value of Γ corresponds to assuming that the survival factor is unity, a constant Γ is essentially an unknown multiplicative constant, so at this point we can only determine the relative cross section.

Using the full angular dependence of the potential surface we calculated for N_2H^+, we first determined the ground state wave function for N_2H^+. Then we used that wave function as the initial condition (the wave function at $t = 0$) for propagation on the dissociating surface. The relative DR cross sections we determined are presented in figure 10. The peak of the calculated cross sections occurs for an incident electron energy of about 3 eV; the value for low energy electrons (300 K or lower) is clearly very much less. The figure also shows (on the same energy scale) the ionic and dissociating potential curves; it is clear that the peak of the DR cross section occurs near the energy where the Franck-Condon factor between the initial vibrational wave function and the final scattering state is the largest. The conclusion seems clear that for any reasonable value of Γ, the absolute cross section (for direct DR) for low-energy electrons will be very small.

6. Concluding remarks

We have reported preliminary large scale electronic structure calculations for the ground state of N_2H^+ and for several states of N_2H, including the dissociating state important for dissociative recombination. Initial dynamics calculations that include the bending of the molecule but freeze the N_2 bond length suggest that the direct mechanism of DR is not efficient. Further work is needed for a definitive calculation.

Acknowledgments
APH and DOK acknowledge support from NSF, including Teragrid resources provided by the National Center for Supercomputing Applications (NCSA) under grant number TG-PHY090021. DT acknowledges support from the French national PCMI program. NSF and NASA provided additional travel suppport for DT and RFM.

References
[1] Adams N G, Herd C R, Geoghegan M, Smith D, Canosa A, Gomet J C, Rowe B R, Queffelec J L and Moriais M 1991 *J. Chem. Phys.* **34** 4852–4857
[2] Geppert W D, Thomas R, Semaniak J, Ehlerding A, Millar T J, Österdahl F, af Ugglas M, Djurić N, Paál A and Larsson M 2004 *Ap. J.* **609** 459–464
[3] Molek C D, McLain J L, Poterya V and Adams N G 2007 *Phys. Rev. A* **29** 1548
[4] Talbi D 2007 *Chem. Phys.* **332** 298–303
[5] Larsson M and Orel A 2008 *Dissociative Recombination of Molecular Ions* (New York: Cambridge University Press)
[6] Talbi D 2009 *J. Phys. Conf. Ser.* **192** 012015
[7] Spirko J A, Mallis J T and Hickman A P 2000 *J. Phys. B: At. Mol. Opt. Phys* **33** 2395–2407
[8] Hickman A P, Miles R D, Hayden C and Talbi D 2005 *Astron. and Astrophys.* **438** 31–37
[9] Pacher T, Cederbaum L S and Köppel H 1988 *J. Chem. Phys.* **89** 7367–7381
[10] Domcke W and Woywod C 1993 *Chem. Phys. Lett.* **216** 362–368
[11] Roos B O 1987 *Adv. Chem. Phys.* **69** 399–445
[12] Schmidt M W and Gordon M S 1998 *Ann. Rev. Phys. Chem.* **49** 233–266
[13] Schmidt M W, Baldridge K K, Boatz J A, Elbert S T, Gordon M S, Jensen J H, Koseki S, Matsunaga N, Nguyen K A, Su S, Windus T L, Dupuis M and Montgomery Jr J A 1993 *J. Comp. Chem.* **14** 1347–1363
[14] Meyer H D, Gatti F and Worth G A (eds) 2009 *Multidimensional Quantum Mechanics: MCTDH Theory and Applications* (Weinheim: Wiley-VCH)

Journal of Physics: Conference Series **300** (2011) 012017 doi:10.1088/1742-6596/300/1/012017

R-matrix with Pseudo-States (RMPS) method: application to CH^+ resonances curves

Dermot Madden, Jonathan Tennyson and Rui Zhang

Department of Physics and Astronomy, University College London, Gower St., London WC1E 6BT, UK

E-mail: d.madden@ucl.ac.uk, j.tennyson@ucl.ac.uk

Abstract.

In a series of calculations on both electron and positron collisions with small molecules the R-Matrix with Pseudo-States (RMPS) method has been found to recover polarisation effects neglected in other close-coupling methods including the standard R-matrix procedure. The molecular R-Matrix and RMPS methods is being applied to determine low-lying resonance states of CH^+ as a function of internuclear separation. Initial results are presented for both a standard R-matrix close-coupling model and for an RMPS calculation. Eigenphase sums and resonances below the $^3\Pi$ threshold are presented for $^2\Pi$ total symmetry. These resonances are classified by their quantum defects and compared to previous results. Prospects for these and other calculations using the RMPS method are discussed.

1. Introduction

The R-matrix method provides a rigorous *ab initio* procedure for studying electron molecule collisions and, in the context of dissociative recombination (DR), for providing resonance curves and widths ("couplings"). For few-electron targets the method has been shown to give highly accurate results [1, 2, 3]. However for many-electron targets the standard implementation of the R-matrix method can struggle to fully converge polarisation effects; a feature shared with other close-coupling methods, see the comparison given by refs. [4, 5, 6] for example.

R-matrix calculations have been used as the basis for successful studies of DR [7, 8, 9]. However for many-electron targets it has proved necessary to "calibrate" [8] the resonance curves, for example using spectroscopic data. Since the method is essentially variational, these curves are systematically too high in energy. This problem can only be cured using a more complete treatment of polarisation effects.

The R-matrix with pseudo-states (RMPS) method was introduced by Gorfinkiel and Tennyson [6, 10] initially to extend the range of electron-collision calculations above the target ionisation threshold and to treat the ionisation process itself. It was however rapidly realised that the method also gave a significantly improved treatment of target

Journal of Physics: Conference Series **300** (2011) 012017 doi:10.1088/1742-6596/300/1/012017

polarisation. Recently this has been explicitly demonstrated in a systematic study of the polarisability of small molecules [11].

The RMPS method has been used to model and characterise resonance effects in electron impact detachment of C_2^- [12, 13]. It has been shown to give excellent results for low energy collisions with the highly polarisable Li_2 molecule [14]. Perhaps most impressively, the method has been used to compute positron – molecule cross sections which reproduce the measurements, see Figure 1. These studies also yield greatly enhanced positron annihilation rates in line with observations. Both these processes are particularly poorly modelled by conventional R-matrix calculatins [15, 16].

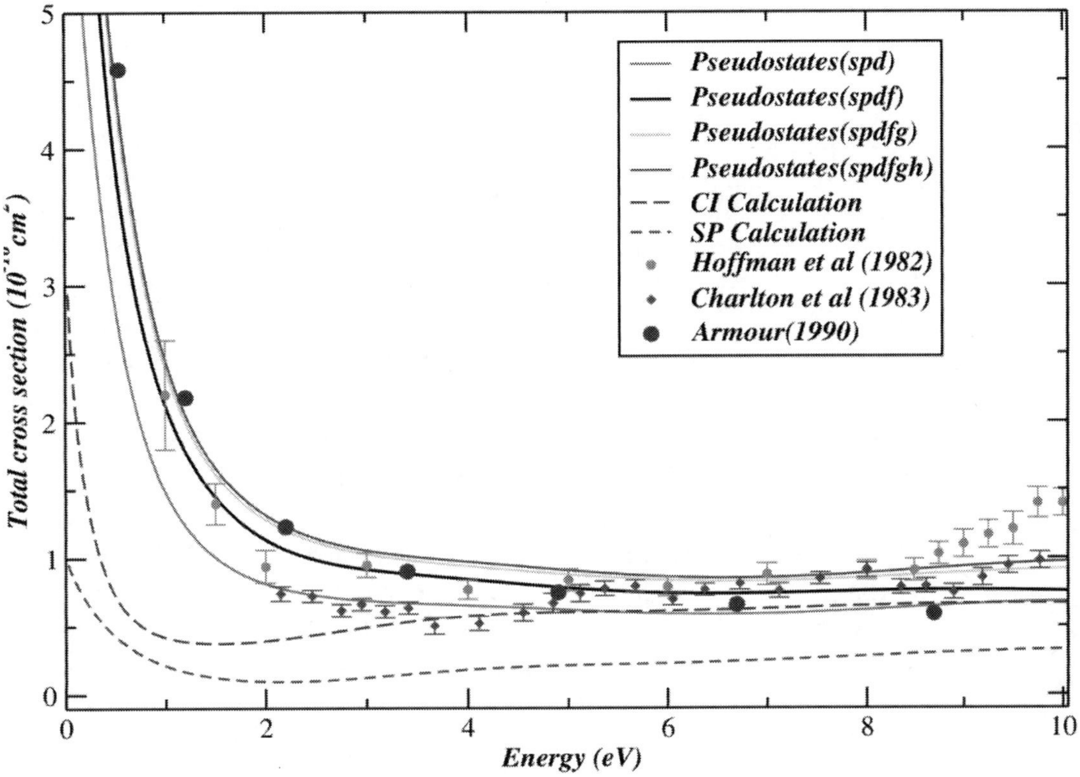

Figure 1. Positron – H_2 elastic cross sections. Solid curves are RMPS calculations with angular functions with $\ell \leq 2$ to $\ell \leq 5$ due to Zhang *et al* [17]; long dash curve is a standard R-matrix close coupling calculation; short dash curve is for a static plus polarisation R-matrix calculation. Hoffmann *et al* [18] and Charlton *et al* [19] are measured cross sections. Armour *et al* [20] are calculations which used wavefunctions that explicitly include the positron-electron coordinate but only consider Σ_g^+ symmetry scattering. The rise in the observed cross section at higher energies is caused by the onset positronium formation at energies above 7 eV.

Given the sucess of the RMPS method, it is natural to ask whether its use can improve predictions of DR resonance curves. To study this we have chosen to look at electron collisions with CH^+.

CH^+ is one of many molecules important for astrophysical observations and is of interest for the formation of large hydrocarbons in the interstellar medium. A key feature of work on CH^+ has being its observed overabundance in the interstellar medium compared to calculated predictions [21]. Our aim is to use *ab initio* methods to look at dissociative recombination of CH^+. There have been a number of previous theoretical [22, 23] and experimental [24, 25, 26] studies of this problem. However there remains issues with the accurate prediction of the rate of dissociative recombination under astrophysical conditions. This could potentially have a large impact on the overall abundance of CH^+ in the interstellar medium. In this report we present preliminary results from our study.

2. Method

The R-Matrix approach is based on dividing configuration space into two regions. The inner region is defined by a sphere of radius a, centred at the centre-of-mass of the molecule. This sphere is chosen to enclose the target charge distribution. In the inner region the interactions are strong and multi-centred and include both exchange and correlation between the scattered electron and the electrons of the target. In the outer region, exchange and correlation effects are neglected as only the long range multi-pole potential is important. For electron-CH^+ scattering, the long-range Coulomb forces are accounted for by the use of Coulomb functions in the asymptotic region; the main issue is therefore to account for the effects of the target dipole moment at long-range.

The R-matrix method, in common with other close-coupling methods, only includes a finite number of states in the close-coupling expansion. It therefore, of necessity, does not account for higher-lying target states or the targer continuum. Intermediate energy processes, where the collision energy lies near to or above the ionisation threshold, cannot be treated correctly without accounting for these states. The RMPS method includes an extra pseudo-continuum basis set which allows for the construction of an extra set of target states. These are refered to as pseudostates as they are not true eigenstates of the target molecule, but they can, if selected correctly, be used to describe the missing electronic target states and the continuum which is discretized within the R-matrix sphere.

The previous R-matrix study of resonance curves in the CH^+ system used Slater Type Orbitals (STOs) to represent the target and numerical functions for the continuum [27]. Since the RMPS method is only implemented in the polyatomic R-matrix code [28], both target, continuum and, indeed, pseudo-continuum orbitals are represented using Gaussian Type Orbitals (GTOs). For continuum orbitals GTOs up to and including g waves were used [29]. The polyatomic R-matrix code cannot treat linear symmetries and all calculations were performed in C_{2v} symmetry. It is reasonably straightforward to reconstruct the full symmetry from these calculations particularly since considerable care was taken to ensure that the calculation preserved the degeneracy structure of the calculations.

Journal of Physics: Conference Series **300** (2011) 012017 doi:10.1088/1742-6596/300/1/012017

One of us has recently written a comprehensive review of molecular R-matrix calculations [30] and the reader is refered there for further details of the method.

3. Target calculations

The CH^+ target was represented in the R-Matrix and RMPS calculations using a cc-pVTZ basis set, a radius of 12 a_0 for the R-matrix box and keeping the two lowest electrons frozen for the calculations. For the standard R-matrix calculation we used a complete active space (CAS) configuration interaction (CI) space of $(1a_1)^2(2\text{-}4a_1,1b_1,1b_2)^4$. For the RMPS calculation we added a pseudo-continuum orbital basis of 10s,10p,6d orbitals, with exponents generated using α=0.17 and β=1.4. The RMPS configuration is $(1a_1)^2(2\text{-}4a_1,1b_1,1b_2)^3$ $(5\text{-}14a_1,2\text{-}7b_1,2\text{-}7b_2,1\text{-}3a_2)^1$.

Potential energy curves for CH^+ for the four lowest states: the X $^1\Sigma$ ground state, a $^3\Pi$, A $^1\Pi$ and b $^3\Sigma^+$ were compared with the high accuracy electronic structure calculations of Barinvos and Van Hemert [31]. The upper and lower panels of Figure 2 show these curves for the R-Matrix and RMPS methods respectively. The RMPS run shows a clear improvement in these energy curves. In the R-Matrix data the $^3\Pi$ and $^1\Pi$ curves dissociate to a slightly lower energy than the $^3\Sigma$ and $^1\Sigma$ curves. In the RMPS run this divergence disappears and the four curves dissociate correctly to the same limit ($C^+(^2P)$ + $H(^2S)$).

The calculations of Barinvos and Van Hemert used a cc-pV6Z basis set with with added diffuse functions and polarisation functions to account for the core polarisation. The comparison of the RMPS calculation with these results confirms that the overall shape of our curves are broadly correct with a smaller basis set (cc-pVTZ) and calculation. The remaining discrepency in the curves appears to be caused by the fact that our RMPS curves slightly underestimate the dissociation energy of CH^+.

4. Scattering calculations

Many models were tested but only results for close-coupling calculations using the standard R-matrix and RMPS target models specified above will be given. The RMPS calculations retained 47 states, $6(^1\Sigma^+)$, $7(^3\Sigma^+)$, $8(^1\Pi)$, $9(^3\Pi)$, $5(^1\Sigma^-)$, $7(^3\Sigma^-)$, $3(^1\Delta)$, $2(^3\Delta)$, cutting off at 43.3 eV above the ground state. The outer region R-matrices were propagated to a radius of 100a_0.

Figure 3 shows eigenphase sums with $^2B_1$ ($^2\Pi$) symmetry for the standard and RMPS calculations. Both methods shows the very complicated resonance structures which converge on the first excitation threshold, a $^3\Pi$. However the lower (and more realistic) threshold for this state in the RMPS study leads to a significant compressed energy scale.

A challenge is presented by the need to resolve the dense collection of resonances just below the $^3\Pi$ threshold. The present calculations use the module in the UK R-Matrix polyatomic code called RESON [32] which automatically fits the eigenphase sums to a

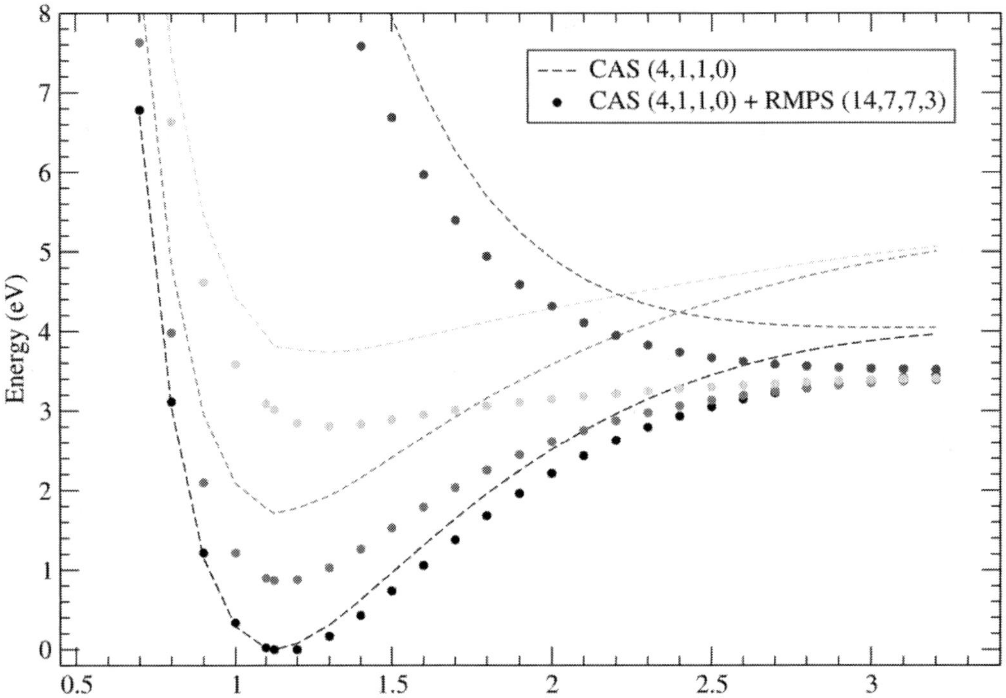

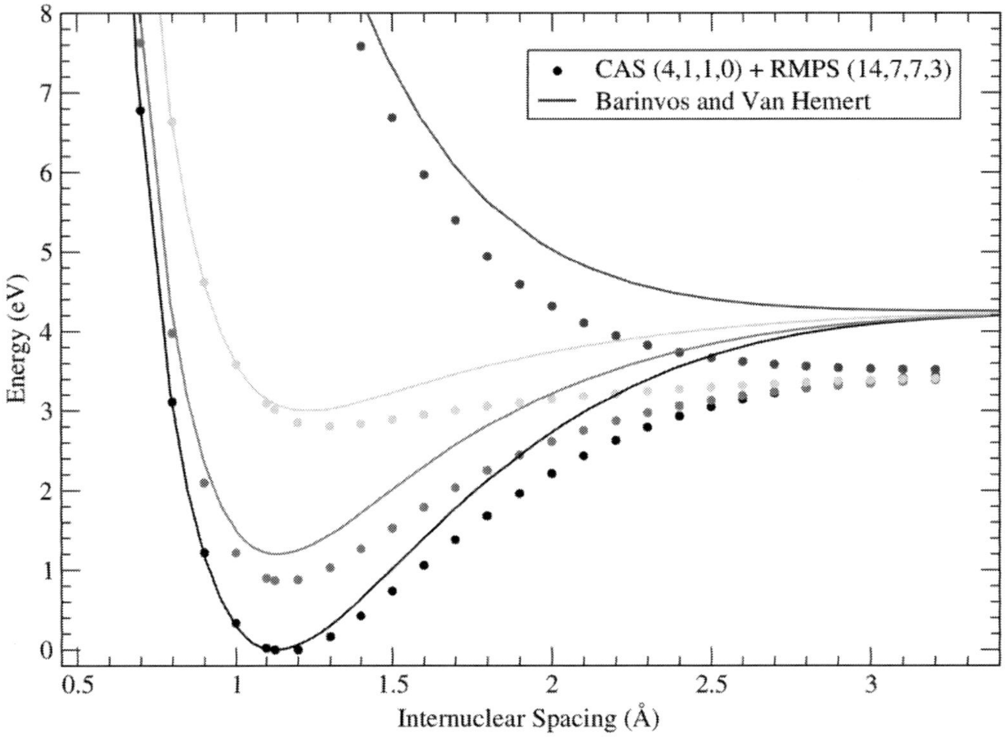

Figure 2. Potential energy curves for the four lowest electronic states of CH^+. Both figures show the following states assending in the order $^1\Sigma^+$, $^3\Pi$, $^1\Pi$ and $^3\Sigma^+$. Upper figure, CAS-CI calculation and RMPS Calculation. Lower figure: RMPS calculation and calculations by Barinvos and Van Hemert [31]. In all cases zero energy has been set to minimum of the ground state.

Breit-Wigner form in a recursive fashion. These resonance parameters are then used to generate complex quantum defects. Future work will examine use of a similar automated implementation of the time-delay method [33, 34] which has been shown to better resolve densely packed resonances. However, use of the time-delay method has a price in terms of speed and efficiency as it requires some hand intervention to produce results. A direct search for the resonances as poles in the complex-energy S-matrix will also be investigated. This method is particularly appropriate for an R-matrix implementation since it is only necessary to consider complex energies in the outer region [35]. Similarly the QB method [36] provides an R-matrix specific resonance characterisation procedure.

Analysis of the resonances allow us to identify four distinct series which we can tentatively assign as being the sσ, pσ, dσ and dδ series. Resonances associated with f and g waves have essentially zero quantum defect which means that their resonances coincide making them difficult to resolve with a Briet-Wigner fit. Figure 4 shows how our calculated quantum defects vary with bondlength in R-Matrix calculations, the results shown in this figure are from a R-Matrix calculation that used the DZP basis set, a larger CAS of $(1a_1)^2(4\text{-}8a_1,1\text{-}3b_1,1\text{-}3b_2,1\text{-}1a_2)^4$ and retained 15 states in the scattering calculation.

An alternative method of matching resonances below the $^3\Pi$ threshold between neighbouring geometries is to plot their quantum defects using Edlén plots. These plots show the quantum defects of each resonance against the resonance energies relative to the threshold energy. Figure 5 shows such plots for the sσ series for the standard R-Matrix and RMPS calculations. This figure clearly shows the increase of the quantum defects obtained with the RMPS calculations.

Table 1 compares our quantum defects for the (a $^3\Pi$)$np\sigma$ those of Carata *et al* [37] at the CH^+ equilibrium bondlength. Table 2 gives our effective quantum numbers of the bound states which form the $(X^1\Sigma^+)np\sigma$ series.

Table 1. Quantum defects for the (a $^3\Pi$)$np\sigma$ series at R=1.127Å from our R-Matrix, RMPS calculations and the multichannel quantum defect of the series from Carata *et al* [37].

n	RMPS	R-Matrix	Carata et al.
5	0.8361		
6	0.8212	0.5069	
7	0.8126	0.5341	
8	0.8078	0.5430	
9	0.8052	0.5477	
10		0.5509	
11	0.8052	0.5534	
MQDT			0.8530

Journal of Physics: Conference Series **300** (2011) 012017 doi:10.1088/1742-6596/300/1/012017

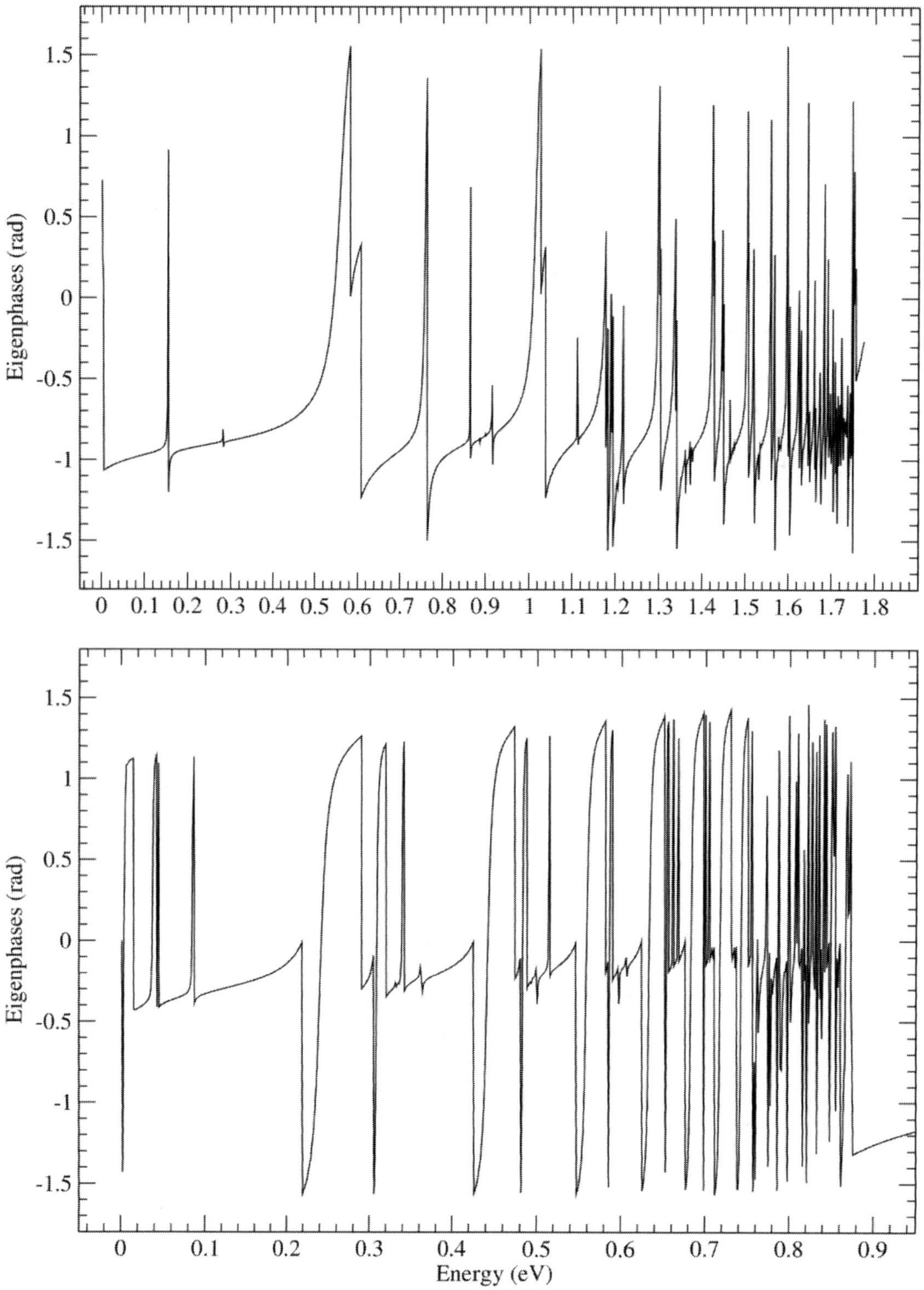

Figure 3. Eigenphases at equilibrium (R = 1.127Å) from the standard R-Matrix calculation (upper figure) and the RMPS calculation (lower figure). Note that in the R-Matrix calculation the $^3\Pi$ threshold is at 1.76 eV and in the RMPS calculation 0.87 eV.

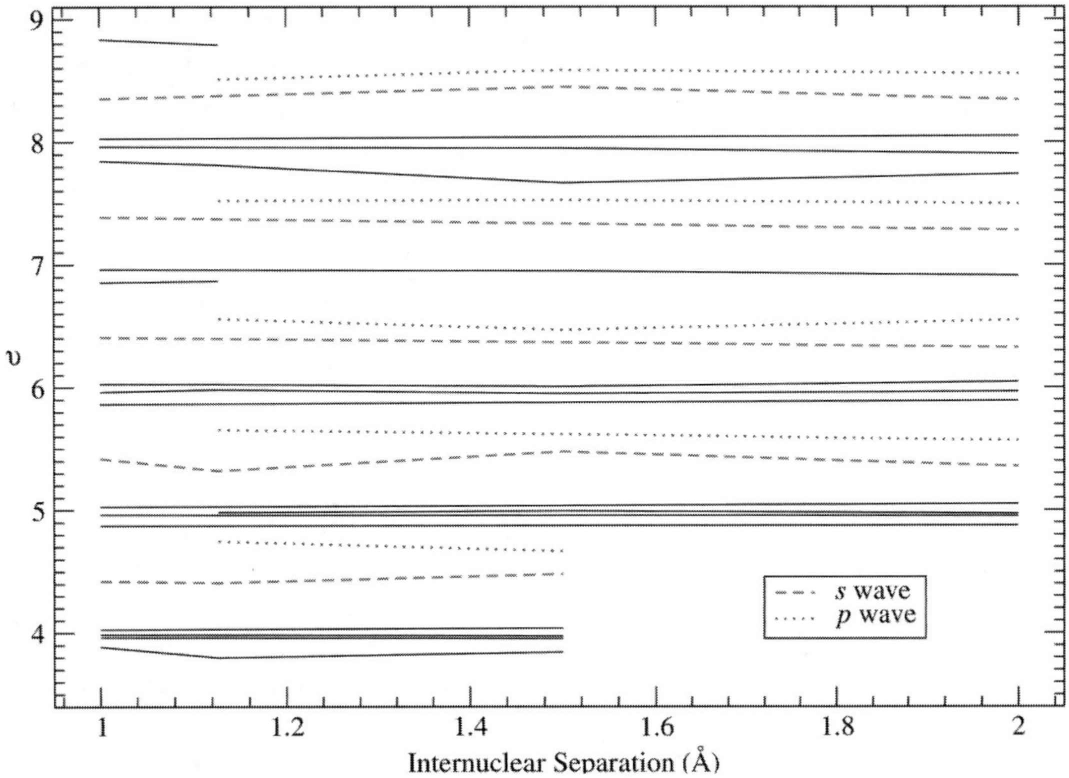

Figure 4. Effectuve quantum number, ν as a function of geometry for the R-Matrix calculations

Table 2. Effective Quantum Numbers for the $(^1\Sigma^+)np\sigma$ series at R=1.127Å, from R-Matrix and RMPS Calculations

RMPS	R-Matrix
4.685	
5.682	
6.672	
7.658	7.727
	8.682
9.612	9.660
10.571	10.647
11.505	11.639

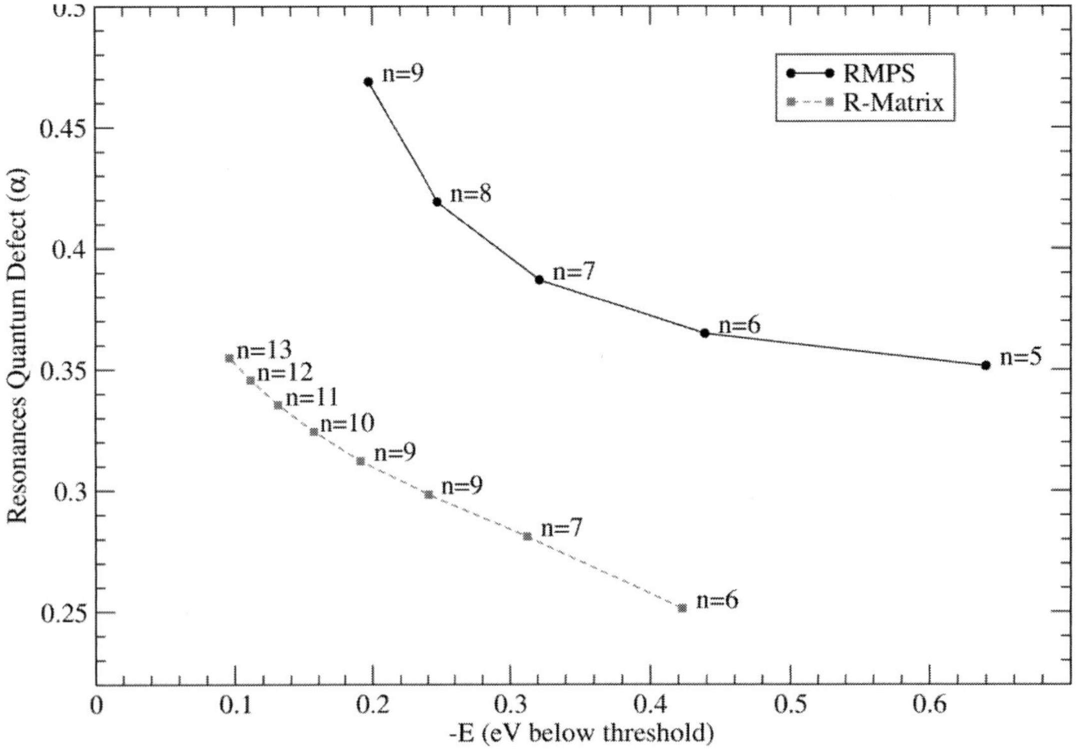

Figure 5. R-Matrix and RMPS Edlén plot at R=1.127Å, for the $(a^3\Pi)ns\sigma$ series. The Edlén plot shows the quantum defects in order of decending effective quantum number, by plotting the quantum defect, α, against the resonance position below the threshold.

5. Future Work

We have confirmed that the RMPS method is capable of giving significantly improved results for resonances in the CH system using a reduced model. Our next goal is to construct a final, large model which can be used to produce reliable data on these features. These resonances will then be analysised using a variety of methods, including the Timedel[34] module, with the aim of fully to resolving all the structures in the region of dense resonances just below the threshold. This should allow more accurate resolving of the crossing points of key resonances and the energy curves.

Acknowledgements

DM thanks EPSRC for PhD funding, and the US National Science Foundation and NASA for travel support.

References

[1] Tennyson J and Noble C J 1985 *J. Phys. B: At. Mol. Phys.* **18** 155–165

[2] Sarpal B K, Branchett S E, Tennyson J and Morgan L A 1991 *J. Phys. B: At. Mol. Opt. Phys.* **24** 3685–3699
[3] Tennyson J 1996 *At. Data Nucl. Data Tables* **64** 253–277
[4] Orel A E 1992 *Phys. Rev. A* **46** 1333
[5] Faure A and Tennyson J 2002 *J. Phys. B: At. Mol. Opt. Phys.* **35** 1865–1873
[6] Gorfinkiel J D and Tennyson J 2004 *J. Phys. B: At. Mol. Opt. Phys.* **37** L343–L350
[7] Sarpal B K, Tennyson J and Morgan L A 1994 *J. Phys. B: At. Mol. Opt. Phys.* **27** 5943–5953
[8] Schneider I F, Rabadán I, Carata L, Tennyson J, Andersen L H and Suzor-Weiner A 2000 *J. Phys. B: At. Mol. Opt. Phys.* **33** 4849–4861
[9] Motapon O, Fifirig M, Florescu A, Waffeu-Tamo F O, Crumeyrolle O, Varin-Breant G, Bultel A, Vervisch P, Tennyson J and Schneider I F 2006 *Plasma Phys. Science Technology* **15** 23–32
[10] Gorfinkiel J D and Tennyson J 2005 *J. Phys. B: At. Mol. Opt. Phys.* **38** 1607–1622
[11] Jones M and Tennyson J 2010 *J. Phys. B: At. Mol. Opt. Phys.* **43** 045101
[12] Halmová G and Tennyson J 2008 *Phys. Rev. Lett.* **100** 213202
[13] Halmová G, Gorfinkiel J D and Tennyson J 2008 *J. Phys. B: At. Mol. Opt. Phys.* **41** 155201
[14] Tarana M and Tennyson J 2008 *J. Phys. B: At. Mol. Opt. Phys.* **41** 205204
[15] Danby G and Tennyson J 1990 *J. Phys. B: At. Mol. Opt. Phys.* **23** 1005–1016 erratum 23, 2471 (1990)
[16] Franz J, Baluja K L, Zhang R and Tennyson J 2008 *Nucl. Instr. Meths. Phys. Res. B* **266** 419–424
[17] Zhang R, Baluja K L, Franz J and Tennyson J *J. Phys. B: At. Mol. Opt. Phys.*
[18] Hoffman K R, Dababneh M S, Hsieh Y, Kauppila W E, Pol V, Smart J H and Stein T S 1982 *Phys. Rev. A* **25** 1393–1403
[19] Charlton M, Griffith T C, Heyland G R and Wright G L 1983 *J. Phys. B: At. Mol. Phys* **16** 323–341.
[20] Armour E A G, Baker D J and Plummer M 1990 *J. Phys. B: At. Mol. Phys* **23** 3057–3074
[21] Pineau des Forets G, Flower D R, Hartquist T W and Dalgarno A 1986 *Mon. Not. R. astr. Soc* **220** 801–824
[22] Giusti-Suzor A and Lefebvre-Brion H 1977 *Astrophys. J.* **214** 101–103
[23] Takagi H, Kosugi N and Le Dourneuf M 1991 *J. Phys. B: At. Mol. Phys.* **24** 711–732
[24] Amitay Z, Zajfman D, Forck P, Hechtfischer U, Seidel B, Grieser M, Habs D, Repnow R, Schwalm D and Wolf A 1996 *Phys. Rev. A* **54** 4032
[25] Mul P M, Mitchell J B A, D'Angelo V S, Defrance P and McGowan J W 1981 *J. Phys. B: At. Mol. Phys.* **14** 1353–1361
[26] Forck P, Broude C, Grieser M, Habs D, Kenntner J, Liebmann J, Repnow R, Amitay Z and Zajfman D 1994 *Phys. Rev. Lett.* **72** 2002–2005
[27] Tennyson J 1988 *J. Phys. B: At. Mol. Opt. Phys.* **21** 805–816
[28] Morgan L A, Tennyson J and Gillan C J 1998 *Computer Phys. Comms.* **114** 120–128
[29] Faure A, Gorfinkiel J D, Morgan L A and Tennyson J 2002 *Computer Phys. Comms.* **144** 224–241
[30] Tennyson J 2010 *Phys. Rep.* **491** 29–76
[31] Barinovs G and van Hemert M C 2004 *Chem. Phys. Lett.* **399** 406–411
[32] Tennyson J and Noble C J 1984 *Computer Phys. Comms.* **33** 421–424
[33] Stibbe D T and Tennyson J 1996 *J. Phys. B: At. Mol. Opt. Phys.* **29** 4267–4283
[34] Stibbe D T and Tennyson J 1998 *Computer Phys. Comms.* **114** 236–242
[35] Noble C J, Dörr M and Burke P G 1993 *J. Phys. B: At. Mol. Opt. Phys.* **26** 2983–3000
[36] Quigley L and Berrington K 1996 *J. Phys. B: At. Mol. Opt. Phys.* **29** 4529–4542
[37] Carata L, Orel A E, Raoult M, Schneider I F and Suzor-Weiner A 2000 *Phys. Rev. A* **62** 052711

Eighth International Conference on Dissociative Recombination (DR2010) IOP Publishing
Journal of Physics: Conference Series **300** (2011) 012018 doi:10.1088/1742-6596/300/1/012018

Electron–molecular cation reactive collisions: from channel mixing to competitive processes

O. Motapon[1,2], D. Backodissa[2], F. O. Waffeu Tamo[1,2,3], D. Tudorache[4], K. Chakrabarti[2,5], J. Zs. Mezei[2,*], F. Lique[2,3], A. Bultel[7], L. Tchang-Brillet[3], O. Dulieu[6], J. Tennyson[8], A. Wolf[9], X. Urbain[10], and I. F. Schneider[2,*]

[1]Physics Department, University of Douala, P. O. Box 24157, Douala, Cameroon
[2]LOMC CNRS-FRE-3102, Université du Havre, 76058 Le Havre, France
[3]LERMA CNRS-UMR-8112, Observatoire de Paris, 92195 Meudon, France
[4]Laboratoire EM2C, CNRS UPR288, École Centrale Paris, 92295 Châtenay-Malabry, France
[5]Dept. of Mathematics, Scottish Church College, 1 & 3 Urquhart Sq., Kolkata 700 006, India
[6]Laboratoire Aimé-Cotton CNRS-UPR-3321, Université Paris-Sud, F-91405 Orsay, France
[7]CORIA CNRS-UMR-6614, Université de Rouen, F-76801 Saint-Etienne du Rouvray, France
[8]Dept. of Physics and Astronomy, University College London, London WC1E 6BT, UK
[9]Max-Planck-Institut für Kernphysik, Saupfercheckweg 1, D-69117 Heidelberg, Germany
[10]Institute of Condensed Matter and Nanosciences, PAMO - Louvain-la-Neuve, Belgium
*Part of this work has been performed while IFS and JZsM were visitors of Laboratoire Aimé–Cotton, Université Paris-Sud

E-mail: `ioan.schneider@univ-lehavre.fr`

Abstract.
The competition between dissociative recombination, vibrational excitation, and dissociative excitation of molecular cations in electron-impact collisions is discussed within the formalism of the Multichannel Quantum Defect Theory. Illustrative results are given for the HD^+/HD and CO^+/CO systems.

1. Introduction

Dissociative recombination (DR) of molecular cations with electrons [1, 2]:

$$AB^+(N_i^+, v_i^+) + e^-(\varepsilon) \longrightarrow [A + B]_{(\epsilon)} \tag{1}$$

is one of the simplest reactive collisions; it drives the charged particle densities in astrophysical ionized media, fusion plasmas in the divertor region, hypersonic entry plasmas and in many other cold media of technological interest [3, 4]. Eq. (1) is appropriate for diatomic systems, but can be generalized for polyatomic ones: N_i^+ and v_i^+ stand for the initial rotational and vibrational quantum number in the relevant modes of the target cation, considered in its ground electronic state, ε for the energy of the incident electron and ϵ for the relative kinetic energy release of the products. During the last 60 years, DR has been the subject of an increasing number of experimental and theoretical studies. However, this process is just one among several reactions triggered by the electron impact on molecular cations. Indeed, elastic, inelastic or superelastic electron scattering:

$$AB^+(N_i^+, v_i^+) + e^-(\varepsilon_i) \longrightarrow AB^+(N_f^+, v_f^+) + e^-(\varepsilon_f) \quad (2)$$

corresponding to ε_f equal, smaller or larger than ε_i respectively, and, if ε_i exceeds the dissociation energy of the target cation, electron impact ion dissociation, i.e. dissociative excitation:

$$AB^+(N_i^+, v_i^+) + e^-(\varepsilon_i) \longrightarrow [A + B^+]_{(\epsilon)} + e^-(\varepsilon_f) \quad (3)$$

strongly compete with DR. These are also very important processes for the chemistry and for the energetics of the molecular species in cold reactive media.

In this article, after highlighting the major elements of the theoretical background (section 2), we will address some of these processes starting from the simplest mechanisms and advancing towards the complex ones (section 3), illustrating our approach with concrete results on the HD^+/HD and CO^+/CO systems. Section 4 will provide our conclusions and perspectives.

2. Theoretical Background

During the processes (1, 2, 3), intermediate collision complexes in so-called super excited states are formed, and their account allows the writing of the preceding reactions in the global form:

$$AB^+(N_i^+, v_i^+) + e^-(\varepsilon) \longrightarrow AB^*, AB^{*(c)}, AB^{**}, AB^{**(c)} \longrightarrow \text{products, energy} \quad (4)$$

in which AB^* stands for states bound from both electronic and vibrational point of view, $AB^{*(c)}$ for states from the mono-electronic continuum (superscript (c)), AB^{**} for states bound from the electronic point of view but dissociative (i.e. from the vibrational continuum), and $AB^{**(c)}$ for states from the mono-electronic continuum *and* dissociative. The occurrence of AB^* states, often called 'Rydberg resonances', is the main reason for the building of an approach of the DR [5, 6, 7, 8, 9] based on the Multichannel Quantum Defect Theory (MQDT) [10, 11, 12, 13]. The processes (1-3) result from the coupling between ionization and dissociation *channels*, i.e. groups of states characterized by a common set of quantum numbers and by the same fragmentation threshold (either for ionization, or for dissociation), having the energy below or above this threshold. More specifically, within a *quasi-diabatic* representation [6, 7, 14], an *ionization* channel is built starting from the ground electronic state of the ion in one of its ro-vibrational levels N^+v^+, and is completed by gathering all the mono-electronic states of a given orbital quantum number l, describing an 'optical' electron. These mono-electronic states describe, with respect to the N^+v^+ threshold, either a 'free' electron - in which case the total state $AB^{*(c)}$ corresponds to (auto)ionization - or to a bound one - in which case the total state AB^* corresponds to a temporary capture into a Rydberg state. Meanwhile, a *dissociation* channel relies on an electronically bound state AB^{**} whose potential energy in the asymptotic limit is situated below the total energy of the system.

Accordingly, the ionization channels gather together AB^* and $AB^{*(c)}$ states, and the dissociation channels correspond to AB^{**} states. As for those of $AB^{**(c)}$ type, they can be organized either into dissociation, or into ionization channels, but the latter option has been prefered so far. Given the total energy of the molecular system, a channel is *open* if this energy is higher than the energy of its fragmentation threshold, and *closed* in the opposite case. In the modeling of the reactions (1-3), relying exclusively on the open channels means accounting for the *direct* mechanism. The inclusion of the closed channels - including AB^* states - allows for the *indirect* mechanism, which interferes with the direct one resulting in the *total* process.

Our MQDT approach starts with the building of the interaction matrix $\boldsymbol{\mathcal{V}}$, performed in the 'A-region' [15], where the Born-Oppenheimer context is appropriate for the description of the collision system. The good quantum numbers in this region are N, M, and Λ, associated respectively to the total angular momentum and its projections on the z-axis of the laboratory-fixed and of the molecule-fixed frames. In the A-region, the states belonging to an ionization channel may be modeled reasonably well with respect to hydrogenic states in terms of the quantum defect μ_l^Λ, which is dependent on the internuclear distance R, but assumed to be *independent of energy*. An ionization channel is coupled to a dissociation one, labeled d_j, on *electronic* level first, through an R-dependent *scaled* 'Rydberg-valence' interaction term, $V_{d_j,l}^{(e)\Lambda}$, which is assumed to be *independent of the energy* of the electronic states pertaining to the ionization channel. Integrating this coupling over the internuclear distance gives elements of the interaction matrix $\boldsymbol{\mathcal{V}}$:

$$\mathcal{V}_{d_j,lN^+v^+}^{NM\Lambda}(E,E) = \langle\chi_{Nd_j}^\Lambda|V_{d_j,l}^{(e)\Lambda}|\chi_{N^+,v^+}^\Lambda\rangle \tag{5}$$

where E is the total energy and $\chi_{d_j}^\Lambda$ and χ_{N^+,v^+}^Λ are the nuclear wave-functions corresponding to a dissociative state and to an ionization channel, respectively. This procedure applies in each Λ-subspace, and results in a block-diagonal global interaction matrix.

Starting from the interaction matrix $\boldsymbol{\mathcal{V}}$ and from the zero-order Hamiltonian $\boldsymbol{H_0}$, we build the reaction K-matrix, which satisfies the Lippmann-Schwinger equation [16]:

$$\boldsymbol{\mathcal{K}} = \boldsymbol{\mathcal{V}} + \boldsymbol{\mathcal{V}}\frac{\mathbf{1}}{\boldsymbol{E} - \boldsymbol{H_0}}\boldsymbol{\mathcal{K}} \tag{6}$$

In order to express the result of the short-range interaction in terms of phase-shifts, we perform a unitary transformation of our initial basis into a new one, corresponding to *eigenchannels*, via the diagonalization of the reaction matrix $\boldsymbol{\mathcal{K}}$:

$$\boldsymbol{\mathcal{K}}\mathbf{U} = -\frac{\mathbf{1}}{\boldsymbol{\pi}}\tan(\boldsymbol{\eta})\mathbf{U} \tag{7}$$

In the external 'B-region' [15] the Born-Oppenheimer model is no longer valid for the neutral molecule, and a frame transformation [17, 18, 19] is performed, via the projection coefficients:

$$\begin{aligned}\mathcal{C}_{lN^+v^+,\Lambda\alpha} &= \left(\tfrac{2N^++1}{2N+1}\right)^{1/2}\langle l\,(\Lambda-\Lambda^+)\,N^+\Lambda^+|lN^+N\Lambda\rangle\\ &\times\frac{1+\tau^+\tau(-1)^{N-l-N^+}}{\left[2(2-\delta_{\Lambda^+,0})(1+\delta_{\Lambda^+,0}\delta_{\Lambda,0})\right]^{1/2}}\times\textstyle\sum_v U_{lv,\alpha}^\Lambda\langle\chi_{N^+v^+}^{\Lambda^+}|\cos(\pi\mu_l^\Lambda(R)+\eta_\alpha^\Lambda)|\chi_{Nv}^\Lambda\rangle\end{aligned} \tag{8}$$

$$\mathcal{C}_{d_j,\Lambda\alpha} = U_{d_j\alpha}^\Lambda\cos\eta_\alpha^\Lambda \tag{9}$$

which can be represented by matrix $\boldsymbol{\mathcal{C}}$. The other solutions, represented by matrix $\boldsymbol{\mathcal{S}}$, are obtained as $\mathcal{S}_{lN^+v^+,\Lambda\alpha}$ and $\mathcal{S}_{d_j,\Lambda\alpha}$ by replacing cosine with sine in Eqs. (8) and (9). In the equations, χ_{Nv}^Λ is the vibrational wavefunction of the neutral system in the interaction (A) region. The quantities τ^+ and τ are related to the reflection symmetry of the ion and neutral wave function respectively, and take the values $+1/-1$ for symmetric/antisymmetric states respectively.

Matrices $\boldsymbol{\mathcal{C}}$ and $\boldsymbol{\mathcal{S}}$ are the building blocks of the generalized scattering matrix $\boldsymbol{X}$, involving all the channels, open ('o') and closed ('c'), and organized in 4 sub-matrices:

$$\boldsymbol{X} = \frac{\boldsymbol{\mathcal{C}} + i\boldsymbol{\mathcal{S}}}{\boldsymbol{\mathcal{C}} - i\boldsymbol{\mathcal{S}}} \qquad \boldsymbol{X} = \begin{pmatrix}\boldsymbol{X_{oo}} & \boldsymbol{X_{oc}}\\ \boldsymbol{X_{co}} & \boldsymbol{X_{cc}}\end{pmatrix} \tag{10}$$

Imposing boundary conditions leads to the physical scattering matrix [10]:

Journal of Physics: Conference Series **300** (2011) 012018 doi:10.1088/1742-6596/300/1/012018

$$\boldsymbol{S} = \boldsymbol{X}_{oo} - \boldsymbol{X}_{oc}\frac{1}{\boldsymbol{X}_{cc} - \exp(-\mathrm{i}2\pi\boldsymbol{\nu})}\boldsymbol{X}_{co} \tag{11}$$

where the diagonal matrix $\boldsymbol{\nu}$ is formed with the effective quantum numbers $\nu_{N^+v^+} = [2(E_{N^+v^+} - E)]^{-1/2}$ (in atomic units) associated with each vibrational threshold $E_{N^+v^+}$ of the ion situated *above* the current energy E (and consequently labelling a *closed* channel).

For a molecular ion initially on the level $N_i^+v_i^+$ recombining with an electron of energy ε, the cross section of capture into *all* the dissociative states d_j of the same symmetry can be written:

$$\sigma^{N,sym}_{diss\leftarrow N_i^+v_i^+} = \frac{\pi}{4\varepsilon}\frac{2N+1}{2N_i^+ + 1}\rho^{sym}\sum_{l,\Lambda,j}|S^{N\Lambda}_{d_j,lN_i^+v_i^+}|^2 \tag{12}$$

Here ρ^{sym} is the ratio between the multiplicities of the neutral and the target ion.

3. Illustrations with new results

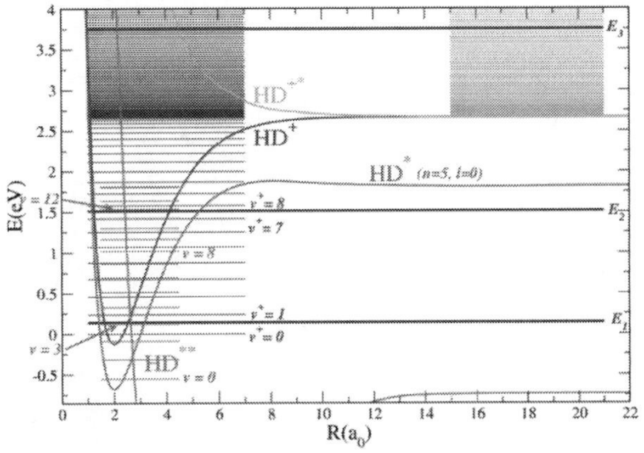

Figure 1. HD^+/HD states (potential energy curves and vibrational levels) relevant for the electron/HD^+ reactive collisions: HD^+ ${}^2\Sigma_g^+(1s\sigma_g)$ (blue), HD^+ ${}^2\Sigma_u^+(2p\sigma_u)$ (orange), HD^{**} ${}^1\Sigma_g^+(2p\sigma_u^2)$ dissociative state (red), and HD^* ${}^1\Sigma_g^+(1s\sigma_g, 5s\sigma_g)$ Rydberg state (green). The violet lines represent illustrative values of the *total* energy of the system, see the text.

In the present work, we will restrict ourselves to the non-rotational case, when rotational excitation and rotational couplings are neglected. In this case, one has to perform separate calculation within each symmetry block Λ and eventually sum over this quantum number the resulting cross sections. For a given Λ, the equations (5, 8, 9,12) become:

$$\mathcal{V}^{\Lambda}_{d_j,lv^+}(E,E) = \langle\chi^{\Lambda}_{d_j}|V^{(e)\Lambda}_{d_j,l}|\chi_{v^+}\rangle \tag{13}$$

$$\mathcal{C}_{lv^+,\Lambda\alpha} = \sum_v U^{\Lambda}_{lv,\alpha}\langle\chi_{v^+}(R)|\cos(\pi\mu_l^{\Lambda}(R) + \eta_{\alpha}^{\Lambda})|\chi_v(R)\rangle \qquad \mathcal{C}_{d,\Lambda\alpha} = U^{\Lambda}_{d\alpha}\cos\eta^{\Lambda}_{\alpha} \tag{14}$$

$$\sigma^{sym,\Lambda}_{diss\leftarrow v_i^+} = \frac{\pi}{4\varepsilon}\rho^{sym,\Lambda}\sum_{l,j}|S^{\Lambda}_{d_j,lv_i^+}|^2 \qquad \sigma^{sym}_{diss\leftarrow v_i^+} = \sum_{\Lambda}\sigma^{sym,\Lambda}_{diss\leftarrow v_i^+} \tag{15}$$

In order to illustrate the concept of open and closed channels, and to be more specific, we address the case of the HD^+/HD system, for which a relevant energy diagram - restricted to the ${}^1\Sigma_g^+$ symmetry of the neutral - is given in Figure 1. For simplicity, we will assume that a single partial wave of the electron is dominant in the ionization channels. The thick, violet, horizontal lines stand for possible values of the *total* energy of the system. E_1 corresponds to the case where the only *open* ionization channel is that associated to the ion ground vibrational state ($v^+=0$). This channel, gathering together states of the type $HD^{*(c)}$ (according to section 2), and the dissociative channel, consisting on the state HD^{**}, are always open; all the other channels, which are of ionization type, associated to $v^+=1$, 2, etc, are *closed*, and correspond to families of states of the type HD^*. A Born-Oppenheimer - and therefore, *approximate* - picture of some of these states, i.e. those labelled by ${}^1\Sigma_g^+(1s\sigma_g 5s\sigma_g)$, v=0, 1, 2, etc., is shown in green in Figure 1. E_2 represents a total energy for which ionization channels associated to $v^+=0-7$ are open, and those associated to $v^+ \geq 8$ are closed.

3.1. The direct process at low energy: recombination versus vibrational (de-)excitation

When the closed channels are either not available or neglected, the $\boldsymbol{X}$-matrix from eq. (10) reduces to its $\boldsymbol{X}_{oo}$ component and all the resonant part from the second term of the $\boldsymbol{\mathcal{S}}$-matrix (eq. (11)) is missing. In the case when (i) one single dissociation channel '*d*' is open only,

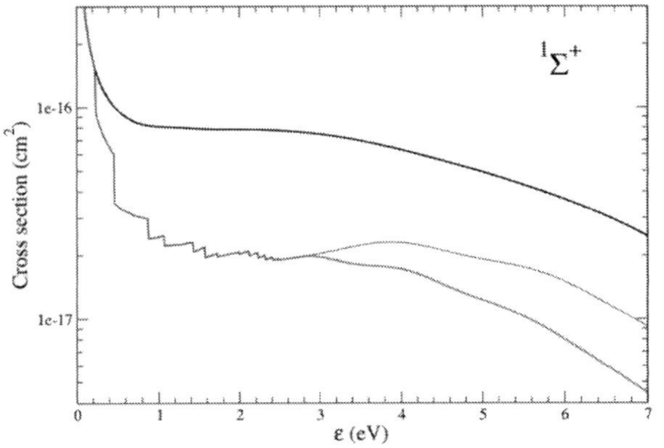

Figure 2. Direct dissociative recombination of ${}^2\Sigma_g^+(1s\sigma_g)$ $v_i^+=0$ HD^+ ion into HD^{**} ${}^1\Sigma_g^+(2p\sigma_u^2)$ state. Black curve: only $v^+=0$ included in the calculation. Blue curve: all bound vibrational levels ($v^+=0$-20) included. Red curve: bound vibrational levels *and* discretized vibrational levels from the continuum ($v^+=21$ - 355) included.

(ii) the first order solution of the Lippman-Schwinger equation (6) is adopted, and (iii) the R-dependence of the quantum defects is neglected, the direct DR cross section can be written as a product of a capture cross section $\sigma_{dv_i^+}^{(cap)}$ and a 'survival' factor $f_{dv_i^+}^{(surv)}$ [7]:

$$\sigma_{dv_i^+} = \sigma_{dv_i^+}^{(cap)} \cdot f_{dv_i^+}^{(surv)} \qquad \sigma_{dv_i^+}^{(cap)} = \frac{\pi}{\varepsilon} \rho^{sym,\Lambda} \widetilde{\xi}_{v_i^+}^2 \qquad f_{dv_i^+}^{(surv)} = \frac{1}{[1+\sum_{v^+} \widetilde{\xi}_{v^+}^2]^2} \tag{16}$$

These quantities rely all on the strength of the Rydberg-valence interaction:

$$\widetilde{\xi}_{v^+}^2 = \sum_l \xi_{v^+l}^2 \qquad \xi_{v^+l} = \pi \cdot \mathcal{V}_{d,lv^+}^{\Lambda} \tag{17}$$

Note that even if we do not use approximations (ii) and (iii) invoked above, the computed direct cross section is very close to that modeled by the simple formulas (16). Figure 2 displays the cross section for DR of a ground state HD^+ ion into HD^{**} $^1\Sigma_g^+(2p\sigma_u^2)$ dissociative state in three cases. The simplest one comes from the account of only a single ionization channel, the entrance one, i.e. $v^+=0$. DR here is in competition with elastic scattering only and, according to the preceding equations, one finds the cross section given by the smooth black curve in Fig. 2. In a second case, we involve further ionization channels, which means that we allow for more autoionization through *vibrational excitation*. At very low energy the cross section – the blue curve in Fig. 2 – evolves smoothly, and is identical to that from the first case invoked. However, it drops suddenly as each new vibrational threshold opens, in agreement with the simple predictive formulas (16). When the last vibrational (bound) level is reached ($v^+=20$), the cross section continues to decrease without displaying any further step-like structure. Note that autoionization through vibrational excitation results in the decrease of the DR cross section by a factor of up to 5 with respect to the first case discussed.

3.2. The direct process at high energy: recombination versus dissociative excitation

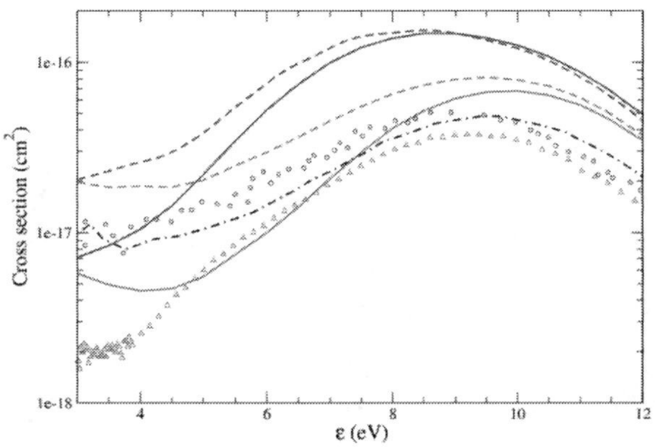

Figure 3. Dissociative recombination of $^2\Sigma_g^+(1s\sigma_g)$ $v_i^+=0$ HD^+ ion. Blue (upper)/red (lower) curves: dissociative excitation neglected/accounted, dashed/continuous corresponding to first-order/second-order calculations respectively. Black dot-dashed curve: computation of Takagi [20]. Circles: Tanabe *et al* [21]. Triangles: Al-Khalili *et al* [22].

At energies higher than the dissociation threshold of the ion - e.g. E_3 in Figure 1 - we have to allow for the autoionization with respect to ion states from the continuum part of the vibrational spectrum, i.e. *dissociative excitation* (DE). For the two available electronic states of the ionic core, $^2\Sigma_g^+(1s\sigma_g)$ and $^2\Sigma_u^+(2p\sigma_u)$, we have discretized their vibrational continua by adding a 15 eV high potential energy step at R=25 a_0. This corresponds, for every partial wave of the incident electron, to 334 further ionization channels for the ground core (responsible for what we call DE of the first kind, DE1, relying on states of $AB^{*(c)}$ type), and to 382 ionization channels for the excited core (responsible for what we call DE of the second kind, DE2, relying on states of $AB^{**(c)}$ type). The red curve in Figure 2 represents the cross section of DR assisted by DE1 only, and shows that this process decreases it with respect to that affected by vibrational excitation (blue curve) by a factor up to 2. We have performed extensive calculations of DE-assisted DR cross section for capture into all the dissociative states within all the relevant symmetries, $^1\Sigma_g^+$, $^{1,3}\Sigma_u^+$, $^{1,3}\Pi_{u,g}$, and for a broad range of energy, 0 – 12 eV. The computations have been carried out in the first and second order of the K-matrix, and relied on molecular data previously used [9, 23, 24]. Our final best result, the full red curve in Figure 3, is in reasonable agreement with previous theoretical estimations [20] and storage-ring measurements [21, 22].

3.3. The indirect process: quantum interference between direct dissociation and temporary capture into bound Rydberg states

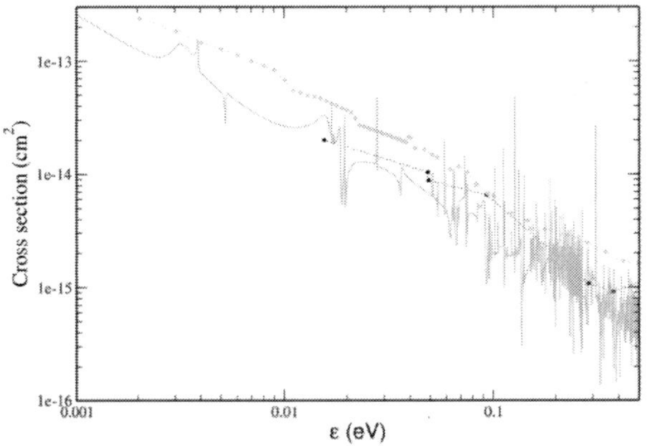

Figure 4. Dissociative Recombination of ground state CO^+. Blue full curve: MQDT total (direct and indirect) cross section [25]. Black/green dots: merged beam [33]/ storage ring [34] results respectively.

As shown in Section 2, allowing for closed channels means taking into account the temporary capture into bound electron-ion states, which induce local resonances in the shape of the cross section, according to Eq. (11). In a Born-Oppenheimer picture, this corresponds to the capture into a Rydberg bound state. One may notice that the levels v=3 and v=12 of the Rydberg state having as potential the green curve appearing in Figure 1 are situated at the energies E_1 and E_2 respectively, which suggests the occurrence of resonant features in the cross section close to these total energies. Since numerous computations of the total - i.e. direct and indirect - process are nowadays available [9], we prefer to illustrate this quantum interference mechanism on the recently studied CO^+/CO system. This is done in Figure 4, which is based on a number of previous molecular structure calculations [26, 27, 28, 29, 30, 31, 32]. The MQDT cross section displayed in this figure comes from the account of $^1\Sigma^+$, $^1\Pi$ and $^3\Pi$ symmetries of CO. For each symmetry, the four lowest dissociative states have been included. The very rich resonant structure covers the range of the measured cross sections obtained in merged-beam [33] and storage ring [34] experiments. According to the authors of ref. [34], the possible presence of ions in the electronically excited A $^2\Pi$ state in the merged-beam device of ref. [33] might explain the observed discrepancy between the two measurements. However, the average theoretical cross section is found lower than these observations.

4. Conclusion and perspectives

Using the Multichannel Quantum Defect Theory, we have illustrated the competition between different processes, subject to quantum interference: dissociative recombination, vibrational excitation, dissociative excitation and resonant capture into super-excited bound states. Dissociative excitation has been included in our approach after extending our collision formalism to the case of two active electronic states of the cation. The computed cross sections for HD^+/HD and CO^+/CO systems are in good agreement with the measurements. A huge amount of cross sections and rate coefficients [25, 35, 36] will complete those already displayed in the present article, corresponding to a broad range of energy and coming from careful state-to-state analysis.

5. Acknowledgements

We thank Ch. Jungen, J. Robert, D. Benredjem, H. Takagi, S. L. Guberman, H. Buhr, M. Lange and S. Niyonzima for numerous discussions. We acknowledge the scientific and financial support from the European Spatial Agency/ESTEC 21790/08/NL/HE, the International Atomic Energy Agency, EFDA, the French Research Federation for Fusion Studies, the ANR-contract 'SUMOSTAI', the CNRS/INSU/'Physique et Chimie du Milieu Interstellaire' programme, the Triangle de la Physique/PEPS/'Physique théorique et ses interfaces', the CPER Haute-Normandie/CNRT/'Energie, Electronique, Matériaux', the PPF/CORIA-LOMC/ 'Energie-Environnement', IEFE:Rouen-Le Havre, and Conseil Régional de la Haute Normandie. XU is senior research associate of the Belgian Fund for Scientific Research - FNRS.

References

[1] M Larsson and A E Orel, 'Dissociative recombination of molecular ions', Cambridge Univ. Press, 2008.
[2] A I Florescu-Mitchell and J B A Mitchell, *Phys. Rep.* **430**, 277 (2006).
[3] A Wolf, L Lammich and P Schmelcher (editors), *J. Phys. Conf. Series* **4**, Proceedings of the Sixth International Conference on Dissociative Recombination: Theory, Experiments and Applications, 12-16 July 2004, Mosbach, Germany (2005).
[4] W J van der Zande (editor), *J. Phys. Conf. Series* **192**, Proceedings of the Seventh International Conference on Dissociative Recombination: Theory, Experiments and Applications, 18-23 July 2007, Ameland, The Netherlands (2009).
[5] U Fano, *Phys. Rev.* **124**, 1866 (1961).
[6] N Bardsley, *J. Phys. B* **1**, 349, 365 (1968).
[7] A Giusti-Suzor, *J. Phys. B: At. Mol. Phys.* **13**, 3867 (1980).
[8] A Giusti-Suzor, J N Bardsley and C Derkits, *Phys. Rev. A* **28**, 682(1983).
[9] O Motapon, F O Waffeu Tamo, X Urbain and I F Schneider, *Phys. Rev. A* **77**, 052711 (2008).
[10] M J Seaton, *Rept. Prog. Phys.* **46**, 167 (1983).
[11] C H Greene and C Jungen, *Adv. At. Mol. Phys.* **21**, 51 (1985).
[12] C Jungen (editor), 'Molecular Applications of Quantum Defect Theory', IoP Publishing, Bristol (1996).
[13] C Jungen, 'Elements of Quantum Defect Theory', in M. Quack and F. Merkt (editors), Handbook of High Resolution Spectroscopy, Wiley, Chichester, New York (2010), *Phys. Rev. A* **28**, 682 (1983).
[14] V Sidis and H Lefèbvre-Brion, *J. Phys. B: At. Mol. Phys.* **4**, 1040 (1971).
[15] C Jungen and O Atabek, *J. Chem. Phys.* **66**, 5584 (1977).
[16] V Ngassam, A Florescu, L Pichl, I F Schneider, O Motapon and A Suzor-Weiner, *Euro. Phys. J. D* **26**, 165 (2003).
[17] U Fano, *Phys. Rev. A* **2**, 353 (1970).
[18] E S Chang and U Fano, *Phys. Rev. A* **6**, 173 (1972).
[19] B Vâlcu et al, *Euro. Phys. J. D* **1**, 71 (1998).
[20] H Takagi, Physica Scripta **T96**, 52-60 (2002).
[21] T Tanabe et al, *Phys. Rev. Lett.* **75**, 1066 (1995).
[22] A Al-Khalili et al., *Phys. Rev. A* **68**, 042702 (2003).
[23] J Tennyson, *At. Data Nucl. Data Tables* **64**, 253 (1996).
[24] V Ngassam, O Motapon, A Florescu, L Pichl, I F Schneider and A Suzor-Weiner, *Phys. Rev. A* **68**, 032704 (2003).
[25] D. Backodissa et al, in preparation.
[26] W -Ü L Tchang-Brillet et al, *J. Chem. Phys.* **96**, 6735 (1992).
[27] K Chakrabarti and J Tennyson, *J. Phys. B: At. Mol. Opt. Phys.* **39**, 1485 (2006).
[28] K Chakrabarti and J Tennyson, *J. Phys. B: At. Mol. Opt. Phys.* **40**, 2135 (2007).
[29] H Lefébvre-Brion and B R Lewis, *Mol. Phys.* **105**, 1625 (2007).
[30] S L Guberman, *Workshop on Planetary Atmospheres*, **communication** (2007).

[31] G J Vázquez, J M Amero, H P Liebermann and H Lefébvre-Brion, *J. Chem. Phys.* **113**, 13395 (2009).
[32] H Lefébvre-Brion, H P Liebermann and G J Vázquez, *J. Chem. Phys.* **132**, 024311 (2010).
[33] J B A Mitchell and H Hus, *J. Phys. B: At. Mol. Phys.* **18**, 547 (1985).
[34] S Rosen at al, *Phys. Rev. A* **57**, 4462 (1998).
[35] F O Waffeu Tamo et al, in preparation.
[36] K Chakrabarti et al, in preparation.

Eighth International Conference on Dissociative Recombination (DR2010) IOP Publishing
Journal of Physics: Conference Series **300** (2011) 012019 doi:10.1088/1742-6596/300/1/012019

Dissociative Recombination of Small Polyatomic Molecules

Stephen T. Pratt[‡] and Christian Jungen[†*]
[‡]Argonne National Laboratory, Argonne, IL 60439 USA
[†]Laboratoire Aimé Cotton du CNRS, Bâtiment 505, Université de Paris-Sud, F91405, Orsay, France
[*]Department of Physics and Astronomy, University College London

E-mail: stpratt@anl.gov and Christian.Jungen@lac.u-psud.fr

Abstract. Simplified analytical expressions for dissociative recombination driven by the Jahn-Teller and Renner-Teller interactions are discussed. These expressions allow straightforward predictions of isotope effects, and the results are discussed for both H_3^+ and HCO^+ and their isotopomers. Although the predicted effects are generally in good agreement with experiment, discrepancies indicate areas worthy of more extensive investigation. Some general features of isotope effects in the dissociative recombination of other systems are also discussed.

1. Introduction

The dissociative recombination of an electron and a molecular ion involves the coupling of ionization and dissociation continua. As it was originally envisioned [1], this process can occur by capture of the electron into a dissociative electronic state of the neutral, resulting in the production of neutral fragments. Subsequently, it was realized that indirect processes can also play an important role in this process [2]. In this scheme, the electron is captured into a quasi-bound Rydberg state of the neutral, which is subsequently predissociated by a dissociative state. The accurate theoretical description of dissociative recombination is thus exceedingly difficult, and a full treatment of the process is limited to diatomic ions and a small number of polyatomic ions like H_3^+. Because dissociative recombination is important in a wide range of applications such as interstellar chemistry, planetary atmospheres, low-temperature plasmas, and plasma-assisted combustion [3], there is a significant desire to treat larger polyatomic systems. This situation provides motivation for the development of approximate approaches to predict dissociative recombination rate constants. We have recently developed such an approach to describe resonance capture processes at low collision energy ($< \sim 1$eV) in systems like H_3^+ and HCO^+, where the dominant capture processes into vibrationally excited neutral Rydberg states are driven by the Jahn-Teller [4] and Renner-Teller interactions [5], respectively.

In this approximate approach, it is assumed that the indirect process for dissociative recombination dominates, that capture of slow electrons into vibrationally excited Rydberg states is the rate limiting step, and that dissociation is sufficiently fast that re-ionization can be ignored. Thus, the dissociative recombination cross section is equal to the capture cross section. The cross section for capture into vibrationally excited Rydberg states is related to the matrix elements for the inverse process of vibrational autoionization, that is, the decay of a vibrationally excited Rydberg state through the conversion of vibrational energy of the ion core into electronic energy of the Rydberg/escaping electron. The vibrational autoionization width is given by a variation of the formula originally developed by Herzberg and Jungen [6], but generalized to include processes driven by the Jahn-Teller or Renner-Teller effect. With this approach, the dissociative recombination cross section can be calculated with a minimal amount of input data. In our previous papers, these data were obtained from spectroscopic analyses of low-lying, bound Rydberg states of the species of interest. However, the input data can also be obtained from quantum chemistry calculations, as has been demonstrated recently by Douguet and coworkers [7].

Such simplified calculations are in no way aimed at replacing high-level benchmark calculations like those of Kokoouline, Greene, and coworkers [8-11]. Rather, such calculations are meant to provide a first-order picture of what is expected, and thus provide a means of highlighting unexpected experimental results. One of the useful aspects of these approximate expressions is that they allow the simple prediction of isotope effects, which can then be compared with experiment. In contrast, for example, a fully developed calculation of the dissociative recombination cross section for D_3^+ requires nearly as much effort as the original calculation for H_3^+ [8]. In what follows, we focus on this aspect of the work. We begin by reviewing the expressions for electron capture in H_3^+, and then compare the predicted isotope dependence with experiment. The isotope dependence for H_3^+ vs. D_3^+ is straightforward to predict, but the isotope dependence for the mixed isotopomers H_2D^+ and HD_2^+ involve more subtle approximations. Comparison of these predictions with experiment serves to illustrate surprising aspects for the latter, and suggests areas for future experimental and theoretical efforts. Following the discussion of the isotopomers of H_3^+, we present a brief discussion of expectations for isotope effects in other systems, and compare these with experiment.

2. Dissociative recombination H_3^+ and its isotopomers

Kokoouline, Greene, and coworkers [9] showed that the Jahn-Teller interaction provides the key mechanism for electron capture by H_3^+ at low energies, and they performed high-level calculations that produced dissociative recombination cross sections that were for the first time in good agreement with experiment [10]. In our simplified description of dissociative recombination, we included only the linear Jahn-Teller effect, and were thus limited to capture processes with $\Delta v_2 = 1$. However, the propensity rule for vibrational autoionization (the inverse of this capture process) suggests that $\Delta v = 1$ processes will dominate [6,12].

For efficient dissociative recombination, the electron partial wave must be penetrating, which requires $\ell = 0$ or 1 for H_3^+. Of these, only the $\ell = 1$ channel has a degenerate component, pe', that can have a Jahn-Teller interaction, and our analysis is limited to that channel. The Jahn-Teller interaction is described by the parameter D, which measures the lowering of the potential by the linear Jahn-Teller interaction in units of the vibrational frequency, ω. The quantity $D\omega$ is purely electronic and independent of isotope. For H_3 and D_3, analysis of bound state spectroscopic data by Herzberg et al. [13,14] allowed the characterization of the Jahn-Teller interaction in the 3pe' Rydberg states, and the isotopic independence of $D\omega$ was confirmed. This information was used to describe the Jahn-Teller interaction throughout the pe' channel by scaling the quantity $D^{1/2}$ by the spacing between levels $2\mathcal{R}/n^{*3}$, as required by quantum defect theory. Here $\mathcal{R}$ is the Rydberg constant and n^* the effective principal quantum number. With this scaled interaction, the golden rule expression for the vibrational autoionization rate of a single level is [4]:

$$\Gamma_a = 4\pi \frac{2R}{n^{*3}}(D\omega^2) \tag{1}$$

When used with the Breit-Wigner cross section formula, and averaged over all the resonances in a single series, this expression leads to an average cross section

$$\langle\sigma\rangle \approx \frac{2\pi^2}{k^2}\left(\frac{\Gamma_a}{\Delta}\right) \tag{2}$$

where $\Delta = 2\mathcal{R}/n^{*3}$, the spacing between levels in the series, and k is the wavenumber of the electron. The $1/n^{*3}$ dependence of Δ is cancelled by the $1/n^{*3}$ dependence of Γ_a, making the term in parentheses in equation 2 constant. The cross sections are converted to rate coefficients by multiplying by the relative velocity of the collision. The isotope dependence of $<\sigma>$ is contained in Γ_a, and goes as $D\omega^2$. Because $D\omega$ is isotope independent, the isotope dependence simply reduces to that of ω.

For H_2D^+ and D_2H^+, there is strictly speaking no true Jahn-Teller effect. However, the electronic surfaces have the same shape and similar $\Delta v = 1$ couplings between the surfaces. Although the degenerate ν_2 vibration of H_3^+ and D_3^+ is split into two components in H_2D^+ and D_2H^+, the splittings are quite small (~130 cm^{-1} and 110 cm^{-1}, respectively, out of ~2000 cm^{-1}). As a result, we use the $D\omega$ value from the spectroscopic analysis of H_3 and D_3 for all four isotopomers, and use the average ω values for ν_2 and ν_3 in H_2D^+ and D_2H^+.

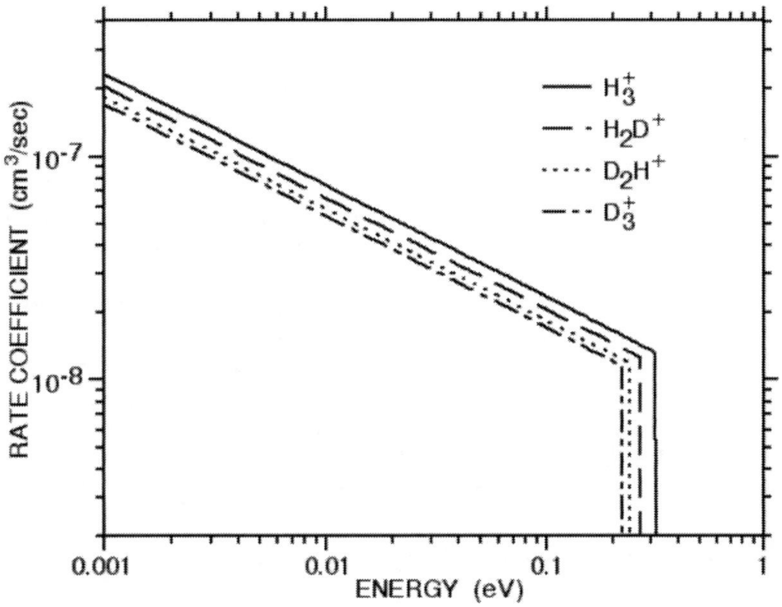

Figure 1. The rate coefficients for the four isotopomers of H_3^+ calculated by using equation 2 and spectroscopic data for the required input parameters. The drop in the cross section occurs at the energy of the fundamental frequency of the corresponding ν_2^+ vibration of the ion.

Figure 1 shows the rate coefficients for the four isotopomers as predicted by using equation 2, the value of $D\omega$ from Herzberg et al. [13,14], and the experimental vibrational frequencies for the ions. Two features of these curves deserve mention. First, the rate increases with decreasing mass of the ion. This result is consistent with the expectation that vibrational excitation in electron-ion collisions is more difficult for heavier molecules. Second, the rate coefficients drop to zero at the $\nu_2^+ = 1$ vibrational threshold. Because we consider only $\Delta v = 1$ processes, there is no mechanism to capture higher energy electrons. Although other mechanisms exist for the capture of higher energy electron, the $\Delta v = 1$ propensity rule suggests that the cross sections will fall considerably at the $v^+ = 1$ thresholds if capture into vibrationally excited Rydberg states is a dominant mechanism.

In the experimental determination of dissociative recombination cross sections and rate coefficients, there is typically a finite spread in the velocity distribution, and in ion-beam and storage ring experiments, the longitudinal and transverse velocity spreads are typically different. Thus, the theoretical cross sections must be convoluted with this velocity distribution to make a useful comparison with experiment. When this convolution is made, the agreement of the H_3^+ curve in Figure 1 with the experimental data is quite good, as seen in figure 2 of reference [4]. Figures 2 - 4 show the comparison of the convoluted theoretical curves and the experimental data for H_2D^+, D_2H^+, an D_3^+, respectively [16,17]. Overall, the agreement is quite good, but several features are evident that deserve further consideration.

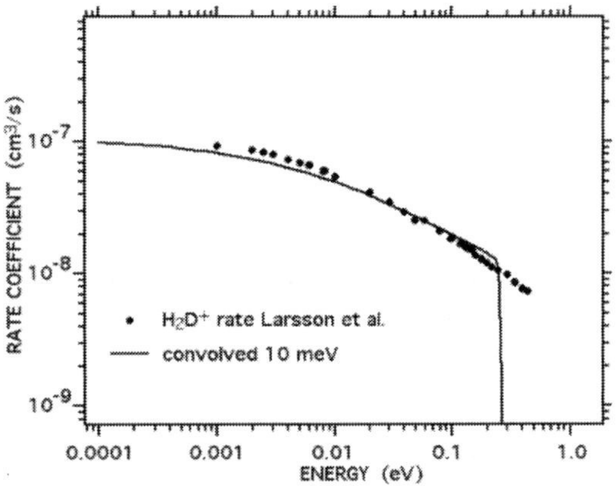

Figure 2. Comparison of the experimental [16] and theoretical rate coefficients for H_2D^+. The theoretical curve from equation 2 is convoluted with a 10 meV enegy spread.

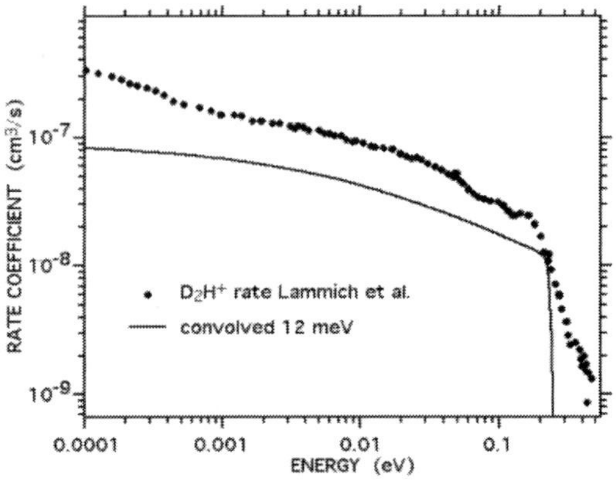

Figure 3. Comparison of the experimental [17] and theoretical rate coefficients for D_2H^+. The theoretical curve from equation 2 is convoluted with a 12 meV energy spread.

First, while the magnitude of the rate coefficient is reproduced extremely well for H_2D^+, the experimental curve [16] shows no drop near the $v_2^+ = 1$ threshold, a feature that is evident in all of the other isotopomers. The otherwise excellent agreement between experiment and theory suggests that this might be an experimental issue. Indeed, the H_2D^+ cross section is the most difficult to measure due to interferences from D_2^+.

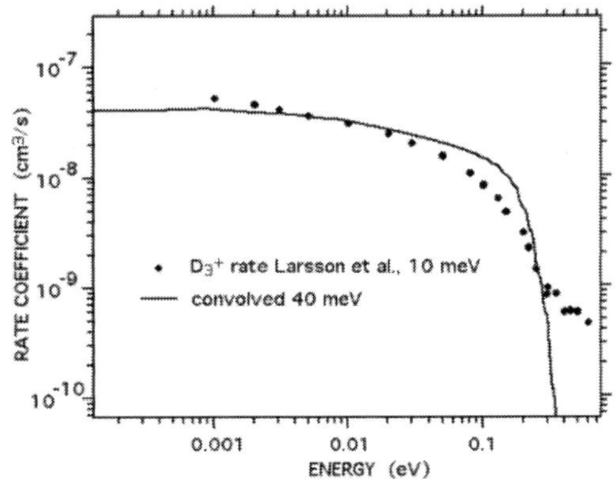

Figure 4. Comparison of the experimental [16] and theoretical rate coefficients for D_3^+. The reported experimental energy spread was 10 meV, but we have used the value chosen by Kokoouline and Greene [8] of 40 meV, which provides a better match to the experiment.

The second issue is that the D_3^+ data [16] appear to have a much larger velocity spread than originally specified. For figure 4 we have used the velocity spread chosen by Kokoouline and Greene [8] in their high-level calculation of D_3^+. Again, the overall agreement between the simple calculation and the experiment is quite good.

Finally, the last issue is that the experimental data for D_2H^+ [17] show a much higher rate coefficient than the corresponding data for the other isotopomers. Indeed, the H_3^+, H_2D^+, and D_3^+ data follow the predicted trend of decreasing rate coefficients. In contrast, the D_2H^+ rate coefficients actually lie higher than both H_2D^+ and H_3^+. While this observation suggests that the experimental results might be flawed, a new storage ring experiment [18] is in excellent agreement with (actually slightly higher than) the earlier measurements. This observation thus suggests that the dissociative recombination mechanism for D_2H^+ may involve additional processes.

The observations on the isotopomers of H_3^+ demonstrate the utility of simple formulae like equation 2. While such equations cannot be expected to reproduce the detailed experimental results, they provide a baseline from which to consider the experimental results, and thus highlight features or issues of real importance where the expectations break down, and where there is still more work to be done.

3. Dissociative recombination of HCO^+ and DCO^+

In a series of papers, Mikhailov et al. [15] and Douguet et al. [19,20] have performed an ever improving series of calculations on the low-energy dissociative recombination of HCO^+ and DCO^+. Key to these calculations was the realization that the dissociative recombination process was dominated by the indirect mechanism involving capture into vibrationally excited Rydberg states, and that the Renner-Teller interaction in the bending vibration is the driving force for the electron capture process. This observation led us to consider a simplified description for dissociative recombination in HCO^+ that is similar in spirit to the one developed for H_3^+. In particular, it was assumed that electron capture was the rate-limiting step in the dissociative recombination process, that capture driven by the Renner-Teller interaction in the pπ channel was the dominant factor, and that the parameters required to describe the interactions could be obtained by properly scaling parameters extracted from the spectroscopic analysis of low-lying, bound Rydberg states of HCO. The dissociative recombination cross section can be written in the same form as equation 2, but with the autoionization width given by a somewhat more complicated expression describing the Renner-Teller interaction.

The Renner-Teller interaction is described by using the three-state model of Gauyacq and Jungen [21] adapted to the quantum defect framework. The expression for the vibrational autoionization width in a single channel is given by:

$$\Gamma_a = 2\pi \frac{2R}{n^{*3}} \left[\frac{d\mu_{\ell\lambda\ell'\lambda'}(Q_2)}{dQ_2} \right]^2_{Q_2=0} \left| \langle \ell v_2^+, K | Q_2 | \ell' v_2^{+\prime}, K \rangle \right|^2 . \tag{3}$$

Here, $\mu_{\ell\lambda,\ell'\lambda'}(Q_2)$ is an off-diagonal element of the quantum defect matrix describing the Renner-Teller interaction, Q_2 is the mass-weighted normal coordinate for bending, ℓ is the orbital angular momentum of the Rydberg electron, and λ is the projection of ℓ on the molecular axis. The matrix elements of Q_2 can be evaluated in a harmonic oscillator basis [22], and the quantum defect matrix elements can be extracted from the spectroscopic data as discussed in [5]. As in the case of H_3^+, only $\Delta v = 1$ processes were considered initially, although expressions for larger Δv were also developed.

Expressions 2 and 3 provide dissociative recombination rates that are in excellent agreement with the theoretical rate coefficients of Mikhailov et al. [15]. The theoretical curves reproduce the shape of the experimental dissociative recombination data for HCO^+ for energies below approximately 0.1 eV (the fundamental bending frequency of HCO^+), but the magnitude of the theoretical cross section is too low by a factor of approximately 3 - 5. Above ~0.1 eV, the simple formula for the cross section goes to zero, while the experimental cross section does not fall nearly as fast. The improved calculations of Douguet et al. [19,20], which include the other vibrational modes of HCO^+ as well as the effect of the large dipole moment of HCO^+, lie only marginally closer to the experimental curve in the energy region below 0.1eV, where our simplified approach gives a non-zero cross section. However, their new calculations do reproduce the slower falloff of the cross section above the $v^+ = 1$ bending threshold where the simplified approach, which includes only $\Delta v^+ = 1$ processes associated with the bending mode, predicts the cross section to go to zero. Interestingly, because the input parameters (and in particular the quantum defect matrices) in our earlier work were derived from experiment, they should have effectively already included the effects of the ion dipole moment.

Indeed, the primary effect of the inclusion of the ion core dipole moment in the calculations of Ref. 20 is that it yields improved quantum defect matrices. Thus, the improved agreement observed by Douguet et al. may have more to do with their consideration of a greater number of partial waves and interactions, and with their inclusion of the two additional vibrational modes.

As in the case of H_3^+, one of the advantages of simple expressions for the dissociative recombination rate is that the isotope effect can be readily evaluated. The mass dependence of the dissociative recombination cross section in equation 2 is contained in the autoionization width, Γ_a, given in equation 3. In this expression, both the quantum defect derivative and the matrix element of the displacement, Q_2, have a mass dependence. In a harmonic oscillator basis, the mass dependence of the matrix element of Q_2 is contained in a factor of $1/\omega_2$, which is squared in the expression for the autoionization width. The mass dependence of the quantum defect derivative in equation 3 can be shown explicitly by changing from the mass weighted normal coordinate to the bending angle as follows:

$$\frac{d\mu_{\ell\lambda,\ell'\lambda'}}{dQ_2} = \frac{d\rho}{dQ_2}\frac{d\mu_{\ell\lambda,\ell'\lambda'}}{d\rho} = \left[\frac{m_H m_C m_O r_{HC}^2 r_{CO}^2}{m_H m_C r_{HC}^2 + m_H m_O (r_{HC}+r_{CO})^2 + m_C m_O r_{CO}^2}\right]^{-\frac{1}{2}} \frac{d\mu_{\ell\lambda,\ell'\lambda'}}{d\rho} . \quad (4)$$

In this expression, ρ is the bending angle, the m_i are the masses of the different atoms, and r_{ab} is the bondlength between a and b. The quantum defect derivative with respect to ρ is an electronic quantity that is independent of the atomic masses. If it is assumed that $r_{HC} = r_{DC}$ in HCO and DCO, respectively, and that r_{CO} is the same in both species, the isotope effect on the quantum defect derivative can be readily evaluated. The quantum defect derivative is also squared in equation 3.

Using the known vibrational frequencies of HCO^+ and DCO^+, Equation 2 predicts the dissociative recombination cross section for HCO^+ will be 1.26 times faster than that for DCO^+. This value can be compared with both previous theory and experiment. In particular, Mikhailov et al. predicted the dissociative recombination rate for HCO^+ is ~30% higher than that for DCO^+, in good agreement with the present result. Later, the improved calculations of Douguet et al. [19] gave a rate for HCO^+ that was only ~10% higher than that for DCO^+. To our knowledge, there are only two measurements of the thermal rates for the dissociative recombination of HCO^+ and DCO^+. Adams et al. [23] reported thermal rates α(95 K) that are ~10% higher for HCO^+ than for DCO^+, while Korolov et al. [24] reported thermal rates α(300 K) that are ~18% higher for HCO^+ than for DCO^+. In any event, the simple model discussed here appears to provide a reasonable first order estimate of the isotope effect in this system. Storage ring measurements for DCO^+, and perhaps new measurements for HCO^+, are expected to provide more insight into this behavior.

4. Isotope effects in the dissociative recombination of other systems

Capture into vibrationally excited Rydberg states is generally expected to show isotope effects, particularly in hydride ions. However, the relatively simple isotope effects observed in H_3^+ and HCO^+ will not necessarily be observed in larger polyatomic ions. In particular, because the mass dependence can depend on the vibrational mode, the isotope effect may become more complicated when capture occurs through multiple vibrational modes. Nevertheless, if one mode dominates the capture process, the observed isotope effect in the dissociative recombination cross section may still have a relatively simple form. Indeed, experimental data on several small polyatomic ions do show relatively simple isotope effects.

As an example, between approximately 0.1 and 0.3 eV, the dissociative recombination rate for H_3O^+ is approximately 40% higher than that for D_3O^+ [25]. Above 0.3 eV, the D_3O^+ cross section falls off at slightly lower energy than the H_3O^+ cross section, owing to the smaller vibrational frequency of the former and the decreased efficiency for capture with $\Delta v > 1$. However, below about 0.1 eV, the two cross sections are nearly equal, which suggests some other effect may be important at very low energies.

In contrast, up to the $v^+ = 1$ vibrational thresholds, the dissociative recombination rate for NH_4^+ is consistently nearly a factor of two greater than that for ND_4^+ [26]. Similarly, in the same low-energy region, the dissociative recombination rate for CH_5^+ is also approximately a factor of two greater than that for CD_5^+ [27]. The simplicity of the isotope effect suggests that a small number of vibrational modes dominate the capture process. The direction of this isotope effect is also consistent with expectations, but it remains to be seen if it can be reproduced either with simple models or high-level calculations. As a final example, the dissociative recombination of the protonated ammonia dimer shows a somewhat unexpected isotope effect, with the cross section for $D^+(ND_3)_2$ actually somewhat greater than that for $H^+(NH_3)_2$ for all energies below ~0.3 eV [28]. In contrast, the protonated ammonia trimer shows the expected behavior, with the $H^+(NH_3)_3$ cross section above the $D^+(ND_3)_3$ cross section. These observations suggest that additional work is necessary to understand the low-energy dissociative recombination of these species.

5. Conclusions

We have discussed how simple formula for indirect dissociative recombination via capture into vibrationally excited Rydberg states can provide insight into isotope effects in dissociative recombination cross sections. As discussed above, such formulae are not a substitute for high-level, first principles calculations of these cross sections. However, they can provide a valuable first order prediction that can be readily compared to experiment. In addition, for many larger hydrocarbon systems that may be important in planetary atmospheres, plasma-assisted combustion, and other applications, large-scale calculations are currently not feasible. In such cases, we believe these simple expressions can play an important role.

Finally, the simple formulae discussed here only include indirect capture into vibrationally excited Rydberg states, and are not expected to work well in all cases. In particular, if the direct process for dissociative recombination is important, the true cross section will deviate significantly from the prediction of the present model. Indeed, the direct capture process is known to be important for the dissociative recombination of NO^+ [29,30], which is isoelectronic with HCO^+. When the direct mechanism is important, the comparison of the predictions of the present model with experiment will provide an estimate of the relative importance of the two mechanisms.

Acknowledgments

S. T. P. was supported at Argonne by the U.S. Department of Energy, Office of Science, Office of Basic Energy Sciences, Division of Chemical Sciences, Geosciences, and Biosciences under contract No. DE-AC02-06CH11357. Travel by S. T. P. was partially supported by the Université Paris Sud. Ch. J. was supported in part by the ANR (France) under Contract No. 09-BLAN-020901. Ch. J. has also benefitted from support by the E. Miescher Foundation (Basel, Switzerland), and has received travel support from the U.S. National Science Foundation and NASA.

References

[1] D. R. Bates, *Phys. Rev.* **78**, 492 (1950).
[2] J. N. Bardsley, *J. Phys. B* **1**, 365 (1968).
[3] M. Larsson and A. E. Orel, *Dissociative Recombination of Molecular Ions* (Cambridge University, New York, 2008).
[4] Ch. Jungen and S. T. Pratt, *Phys. Rev. Lett.* **102**, 023201 (2009).
[5] Ch. Jungen and S. T. Pratt, *J. Chem. Phys.* **129**, 164311 (2008)
[6] G. Herzberg and Ch. Jungen, *J. Mol. Spectrosc.* **41**, 425 (1972).
[7] N. Douguet, Ph. D. Thesis, University of Central Florida, 2010.
[8] V. Kokoouline and C. H. Greene, *Phys. Rev. A* **68**, 012703 (2003).
[9] V. Kokoouline, C. H. Greene, and B. D. Esry, *Nature* (London) **412**, 891 (2001).
[10] S. Fonseca dos Santos, V. Kokoouline, and C. H. Greene, *J. Chem. Phys.* **127**, 124309 (2007).
[11] V. Kokoouline and C. H. Greene, *Phys. Rev. A* **72**, 022712 (2005).

[12] R. S. Berry, *J. Chem. Phys.* **45**, 1228 (1966).
[13] G. Herzberg and J. K. G. Watson, *Can. J. Phys.* **58**, 1250 (1980).
[14] G. Herzberg H. Lew, J. J. Sloan, and J. K. G. Watson, *Can. J. Phys.* **59**, 428 (1981).
[15] I. A. Mikhailov, V. Kokoouline, A. Larson, S. Tonzani, and C. H. Greene, *Phys. Rev. A* **74**, 032707 (2006).
[16] M. Larsson, H. Danared, Å. Larson, A. Le Padellec, J. R. Peterson, S. Rosén, J. Semaniak, and C. Strömholm, *Phys. Rev. Lett.* **79**, 395 (1997).
[17] L. Lammich, D. Strasser, H. Kreckel, M. Lange, H. B. Pedersen, S. Altevogt, V. Andrianarijaona, H. Buhr, O. Heber, D. Schwalm, A. Wolf, and D. Zajfman, *Phys. Rev. Lett.* **91**, 143201 (2003).
[18] V. Zhaunerchyk, R. D. Thomas, W. D. Geppert, M. Hamberg, M. Kaminska, E. Vigren, and M. Larsson, *Phys. Rev. A* **77**, 034701 (2008).
[19] N. Douguet, V. Kokoouline, and C. H. Greene, *Phys. Rev. A* **77**, 064703 (2008).
[20] N. Douguet, V. Kokoouline, and C. H. Greene, *Phys. Rev. A* **80**, 062712 (2009).
[21] D. Gauyacq and Ch. Jungen, *Mol. Phys.* **41**, 383 (1980).
[22] Ch. Jungen and A. J. Merer, in *Molecular Spectroscopy; Modern Research*, edited by K. N. Rao (Academic, New York, 1976), Vol. 1, p.127.
[23] N. G. Adams, D. Smith, and E. Alge, *J. Chem. Phys.* **81**, 1778 (1984).
[24] I. Korolov, R. Plasil, T. Kotrik, P. Dohnal, and J. Glosik, *Int. J. Mass Spectrom.* **280**, 144 (2009).
[25] M. J. Jensen, R. C. Bilodeau, C. P. Safvan, K. Seiersen, L. H. Andersen, H. B. Pedersen, and O. Heber, *Astrophys. J.* **543**, 764 (2000).
[26] J. Öjekull, P. U. Andersson, M. B. Någård, J. B. C. Pettersson, A. M. Derkatch, A. Neau, S. Rosén, R. Thomas, M. Larsson, F. Österdahl, J. Semaniak, H. Danared, A. Källberg, M. af. Ugglas, and N. Markovic, *J. Chem. Phys.* **120**, 7391 (2004).
[27] M. Kaminska, V. Zhanunerchyk, E. Vigren, M. Danielsson, M. Hamberg, W. D. Geppert, M. Larsson, S. Rosén, R. D. Thomas, and J. Semaniak, *Phys. Rev. A* **81**, 062701 (2010).
[28] J. Öjekull, P. U. Andersson, M. B. Någård, J. B. C. Pettersson, A. Neau, S. Rosén, R. D. Thomas, M. Larsson, J. Semaniak, F. Österdahl, H. Danared, A. Källberg, and M. af. Ugglas, *J. Chem. Phys.* **125**, 194306 (2006).
[29] L. Vejby-Christensen, D. Kella, H. B. Pedersen, and L. H. Andersen, *Phys. Rev. A* **57**, 3267 (1998).
[30] I. F. Schneider, I. Rabadán, L. Carata, L. H. Andersen, A. Suzor-Weiner, and J. Tennyson, *J. Phys. B* **33**, 4849 (2000).

Eighth International Conference on Dissociative Recombination (DR2010) IOP Publishing
Journal of Physics: Conference Series **300** (2011) 012020 doi:10.1088/1742-6596/300/1/012020

Multi-channel quantum defect theory to dissociative recombination with and without electronic resonance scattering

H Takagi[1], **M Tashiro**[2]

[1] Center for Natural Science, Kitasato University, 1-15-1 Kitasato, Minami-ku, Sagamihara, Kanagawa, 252-0373 Japan
[2] Institute for Molecular Science, Nishigo-Naka 38, Myodaiji, Okazaki, 444-8585, Japan

E-mail: [1]takagi@kitasato-u.ac.jp

Abstract. The multichannel quantum defect theory (MQDT) of dissociative recombination (DR) is summarized by classifying into two cases whether an electronic resonance state as the dissociative channel exists or not. The DR and dissociative excitation of the isotopes of hydrogen molecular ion are calculated using recent sophisticated MQDT method where the electronic resonance state exists. On the other hand, an adiabatic quantum defect and a mixing matrix of the partial waves in $HeH^+ + e$ collisions are investigated. These parameters are crucial in MQDT calculation without electronic resonance state. A new approach is proposed by unifying the two cases in a above MQDT treatment.

1. Introduction

The mechanism of dissociative recombination (DR) has been considered as a capture process of an incident electron into an electronic resonance state and sequential stabilization by molecular dissociation [1]. This mechanism is possible when there is a potential-curve crossing between the resonance state and the target molecular ionic state at the energy lower than the total energy of collision system. The DR cross section had been considered to be quite small unless the molecular system has such a potential-curve crossing. However, large cross section has been confirmed for the DR of HeH^+ [2], where there is no potential-curve crossing. Hereafter, we abbreviate 'potential-curve crossing' as 'crossing'.

There are two kinds of theoretical approaches depending on whether the crossing exists or not. In the crossing case, the resonance state is the main dissociative channel and the Rydberg states contribute to DR through the indirect process [3], where the incident electron is temporally captured into a rotationally and vibrationally excited Rydberg state. In the non-crossing case, the dissociative channels are the Rydberg states. It is well known that the multichannel quantum defect theory (MQDT) can describe the dynamics by the strong non-adiabatic interactions (NAI) among the Rydberg states [4]. We here summarize and investigate MQDT method for DR with and without the crossing. As an extension of these theories, we shall propose a new approach based on MQDT. Examples will be shown for cases both with and without crossings.

2. MQDT formulation with and without the crossing

2.1. With the crossing

In the crossing case, DR is caused by two different types of interactions. One is the configuration interaction (CI) between the initial continuum electronic state and the resonance state, which is dissociative. The other is the non-adiabatic interaction (NAI) between the initial state and the Rydberg state of which nuclear motion is bound. The so called "two-step method" enables to solve the dynamic process induced by these two interactions stepwise [5]. This two-step method is nothing but MQDT method with crossing.

To make clear the theories, the basis functions have to be compared. In the crossing case, we distinguish the nuclear motion depending on the electronic configuration; The dissociative nuclear motion $F_\varepsilon(R)$ is associated with a doubly excited electronic state $\varphi_d(r;R)$, and the vibrational nuclear motion $\chi_v(R)$ is connected with a singly excited (ether Rydberg or continuun) state $\phi_\epsilon(r;R)$, where r and R represent electronic and nuclear coordinates respectively. The total wave function is represented in the basis set of $(\{\varphi_d(r;R)\}\otimes\{F_\varepsilon(R)\})\cup(\{\phi_\epsilon(r;R)\}\otimes\{\chi_v(R)\})$, where $\{X\}$ indicates a set composed of elements of type X, $\otimes$ means direct product, and $\cup$ represents union.

The configuration interaction (CI) between $\phi_\epsilon(r;R)$ and $\varphi_d(r;R)$ induces transitions between $\{\phi_\epsilon(r;R)\}\otimes\{\chi_v(R)\}$ and $\{\varphi_d(r;R)\}\otimes\{F_\varepsilon(R)\}$, which are driving DR in the crossing case. Assuming an adiabatic representation for the CI, the coupling element is given by

$$V^{NJ}_{v\epsilon,d\varepsilon} = \langle\chi^N_v(R)|V_{\epsilon,d}(R)|F^J_\varepsilon(R)\rangle_R, \tag{1}$$

where

$$V_{\epsilon,d}(R) = \langle\phi_\epsilon(r;R)|H^{ele}(r;R)|\varphi_d(r;R)\rangle_r. \tag{2}$$

In the above equation, $H^{ele}(r;R)$ is the Hamiltonian at fixed nuclei coordinate, and N and J are the angular momentum of the molecular ion and the total system respectively [6]. It is required to solve the scattering problem by the CI with the accuracy beyond the perturbation theory. This could be achieved by solving the Lippmann-Schwinger equation fully taken account the off-the-energy-shell contribution beside the on-the-energy shell [7, 8]. The off-the-energy-shell contribution induces the transition among $\{\phi_\epsilon(r;R)\}\otimes\{\chi_v(R)\}$ other than between $\{\phi_\epsilon(r;R)\}\otimes\{\chi_v(R)\}$ and $\{\varphi_d(r;R)\}\otimes\{F_\varepsilon(R)\}$. The solution of the scattering problem gives new basis functions of which the CI coupling elements constitute diagonal matrix.

The residual interaction is NAI mainly among $\{\phi_\epsilon(r;R)\}\otimes\{\chi_v(R)\}$. The scattering problem by the NAI could be effectively solved by MQDT adopting the new basis functions diagonalizing the CI coupling. The NAI affects the DR at the low energies when an incident electron is temporally captured into a Rydberg state by exciting rotational motion of molecular ion [3, 8].

2.2. Without the crossing: discretization

In the non-crossing case, where the doubly excited state $\{\varphi_d(r;R)\}$ is missing, the DR is driven only by the NAI, which is described by MQDT with dissociative motion [9, 10]. Large cross section of the DR process is obtained when the adiabatic quantum defect (AQD) and the mixing parameter of the partial waves of incident electrons vary strongly with the internuclear distance [9]. The precise information on those electronic parameter is crucial to the DR. To include the dissociative continuum states besides the vibrational states, a discretizing method of continuum states is required not only to the basis expansion but also to set in MQDT as we shall see in subsection 2.3. The dissociative state $\chi^{\mathsf{S}}_j(R)$ is an extension of the vibrational state $\chi_v(R)$ to the positive energy.

$$\chi^{\mathsf{S}}_j(R) = \frac{1}{\sqrt{\Delta}}\int_{\varepsilon_j-\frac{\Delta}{2}}^{\varepsilon_j+\frac{\Delta}{2}}\chi^{\mathsf{E}}_\varepsilon(R)d\varepsilon \tag{3}$$

$$\simeq \sqrt{\Delta}\chi^{\mathsf{E}}_{\varepsilon_j}(R), \tag{4}$$

where the superscripts S and E indiacte normalization by state and energy, respectively. Furthermore ε is energy of relative nuclear motion, and ε_j represents the energy of state j , which is discretized with the energy interval of Δ. The state normalization means $\langle\chi^{\mathsf{S}}_i|\chi^{\mathsf{S}}_j\rangle = \delta_{i,j}$ with the Kronecker's delta δ, and the energy normalization does $\langle\chi^{\mathsf{E}}_{\varepsilon'}|\chi^{\mathsf{E}}_{\varepsilon}\rangle = \delta^D(\varepsilon'-\varepsilon)$, with the Dirac's δ function δ^D. Using the basis set $\{\phi_\epsilon(r;R)\}\otimes(\{\chi_v(R)\}\cup\{\chi^{\mathsf{S}}_j(R)\})$ in MQDT, the DR process without crossing can be described. The function $\phi_\epsilon(r;R)$ represents both the Rydberg states and continuum electronic states. MQDT analytically imposes the normalization condition upon the basis functions according to the energy ϵ. MQDT including the dissociative channel $\chi^{\mathsf{S}}_j(R)$ also impose the proper normalization condition upon $\chi^{\mathsf{S}}_j(R)$ depending on the energy ε_j [9].

2.3. MQDT with dissociative channels

In MQDT, the energy normalized Coulombic wave function is adopted in the inner region, regardless of whether the energy ϵ is negative or positive. The asymptotic function of the inner region is expressed as

$$\Psi(r_e,\bar{r},R) \overset{r\to\infty}{\sim} \sum_{i,i',j}\left\{\delta_{i'i}\phi_i^{\mathsf{E}(s)}(r_e) + K_{i'i}(R)\phi_i^{\mathsf{E}(c)}(r_e)\right\}\psi_i(\bar{r})\chi_j^{i\mathsf{S}}(R), \tag{5}$$

where the radial coordinate of a outer electron in Rydberg/continuum state is r_e, and the other coordinates of all electrons are represented by $\bar{r}$, i, i' represent electronic channels, $K_{i'i}(R)$ is electronic K matrix element at R, and $\chi_j^{i\mathsf{S}}$ represents nuclear state j, which indicates both vibrational and dissociative states. The function $\phi_i^{\mathsf{E}(s)}(r_e)$ is the energy normalized Coulombic function with an asymptotic form of a sine function. The function $\phi_i^{\mathsf{E}(c)}(r_e)$ is another independent Coulombic function with a advanced phase difference of $\pi/2$ relative $\phi_i^{\mathsf{E}(s)}(r_e)$. The nuclear state $\chi_j^{i\mathsf{S}}(R)$ is determined by the electronic state i in the Born-Oppenheimer representation. The wave function in the inner region contains all information of the interactions in the scattering and is almost independent of the normalization condition in the outer region [4].

The wave function of the outer region is represented as

$$\Psi(r_e,\bar{r},R) \overset{r\to\infty}{\sim} \sum_{j,i,j',i'}\{\ \delta_{ij,i'j'}\phi_{i'}^{\mathsf{E}(s)}(r_e)\psi_{i'}(\bar{r})\chi_{j'}^{i'\mathsf{S}}(R) + K_{ij,i'j'}\phi_{i'}^{\mathsf{E}(c)}(r_e)\psi_{i'}(\bar{r})\chi_{j'}^{i'\mathsf{S}}(R)\ \}, \tag{6}$$

if the channel of nuclear motion is closed as vibrational state or 'closed dissociative state' [9]. The indices $(i'j'), (ij)$ represent physical channels and $K_{ij,i'j'}$ is the K matrix, which is determined by the frame-transformation between the inner and outer regions [4, 11]. When the channels are composed of the partial waves of incident electron ℓ^+, vibrational and dissociative Rydberg states j, and rotational states of molecular ion N, we obtain the K matrix K [9].

$$\mathsf{K} = \mathcal{R}_{oo} + \mathcal{R}_{oc}(\mathcal{R}_{cc} - \tan\pi\nu_c)^{-1}\mathcal{R}_{co}, \tag{7}$$

where $\tan\pi\nu_c$ is the diagonal matrix composed of effective principal quantum numbers ν, sufix $o(c)$ indicates open (closed) channel, and

$$\mathcal{R}^J_{j'N'\ell^{+\prime},jN\ell^+} = \sum_\Lambda \langle N'|\Lambda\rangle^{J\ell^{+\prime}\Lambda^+}\sum_{\tilde{\ell}}\langle\chi_{j'}^{\mathsf{S}N'\Lambda^+}|\ K^{\Lambda\tilde{\ell}}_{\ell^{+\prime},\ell^+}(R)\ |\chi_j^{\mathsf{S}N\Lambda^+}\rangle\langle\Lambda|N\rangle^{J\ell^+\Lambda^+} \tag{8}$$

for J total angular momentum. Good quantum numbers in the inner region are the angular momentum around the molecular axis Λ (Λ^+ for molecular ion) and the eigen state of partial waves $\tilde{\ell}$. The factor $\langle\Lambda|N\rangle^{J\ell^+\Lambda^+}$ is represented by

$$\langle\Lambda|N\rangle^{J\ell^+\Lambda^+} = \sqrt{\frac{2N+1}{2J+1}}C(\ell^+NJ;\Lambda-\Lambda^+,\Lambda^+), \tag{9}$$

using the Clebsch Gordan coefficient C [12, 4]. The cross section of dissociative recombination is defined by the ratio of the flux of dissociated fragments to the incident electron's flux. Thus, we should take energy normalization for the dissociative states. The eigen state of the system does not change from equation (6) except for a constant multiplier C. The asymptotic form using the alternative normalization for electric closed channel and nuclear open channel is

$$\Psi(r_e,\bar{r},R) \overset{r\to\infty}{\sim} \sum_{j,i,j',i'} C_{jij'i'}\{\ \delta_{ij,i'j'}\phi_{i'}^{\mathsf{S}}(r_e)\psi_{i'}(\bar{r})\chi_{j'}^{i'\mathsf{E}(s)}(R) + K_{ij,i'j'}^{(ecno)}\phi_{i'}^{\mathsf{S}}(r_e)\psi_{i'}(\bar{r})\chi_{j'}^{i'\mathsf{E}(c)}(R)\ \}, \tag{10}$$

where the function $\chi_{j'}^{i'\mathsf{E}(s)}(R)$ is the energy normalized nuclear wave function of which asymptotic form is a sine function, and the function with superscript (c) has cosine-type asymptotic form. The relation between the two K matrix elements is given by

$$K_{ij,i'j'}^{(ecno)} = \Delta^{-1/2}\nu_{i'}^{-3/2}K_{ij,i'j'}. \tag{11}$$

2.4. Energy higher than several eV

At the collision energies higher than several eV, the electronic excitation to Rydberg states having various excited ionic core can occur. The potential energy curves of these Rydberg states are usually dissociative. Thus, the doubly excited states should be expanded by the basis set of $(\{\varphi_d(r;R)\}\otimes\{F_\varepsilon(R)\})\cup\left(\{\varphi_\epsilon(r;R)\}\otimes\{\chi_j^{ex,\mathsf{S}}(R)\}\right)$, where $\varphi_\epsilon(r;R)$ represents both core excited Rydberg state and electronically inelastic scattering state whereas $\varphi_d(r;R)$ is valence type electronic state, and $\chi_j^{ex,\mathsf{S}}(R)$ represents discretized nuclear dissociative motion under the core excited electronic state. In such high energy region, the Rydberg states having ground state ion core also could dissociate. The basis function in the non-crossing case could be adopted for the single electron excited states: $\{\phi_\epsilon(r;R)\}\otimes(\{\chi_v(R)\}\cup\{\chi_j^{g,\mathsf{S}}(R)\})$, where the superscript g indicates the nuclear motion under the ground electronic state of molecular ion. Using this basis set in the two-step method [5], we can calculate DR cross section [13]. However, we cannot help adopting the first order perturbation theory to solve the scattering problem by the CI except for the lowest doubly excited state.

2.5. New approach

A new method alternative to the 'two-step method' can be possible by using a basis set of $(\{\varphi_\epsilon(r;R)\}\cup\{\phi_\epsilon(r;R)\})\otimes\left(\{\chi_v(R)\}\cup\{\chi_j^{ex,\mathsf{S}}(R)\}\cup\{\chi_j^{g,\mathsf{S}}(R)\}\right)$. First, by solving the electron scattering problem in the space of $(\{\varphi_\epsilon(r;R)\}\cup\{\phi_\epsilon(r;R)\})$, we obtain the adiabatic K matrix $K_{i'\ell^{+\prime},i\ell^+}^{\Lambda\tilde{\ell}}(R,\epsilon)$, where i' and i indicate electronic scattering states including inelastic scattering. The smoothed K matrix in MQDT is given by

$$\mathcal{R}_{i'j'N'\ell^{+\prime},ijN\ell^+}^{\epsilon J} = \sum_{\Lambda}\langle N'|\Lambda\rangle^{J\ell^{+\prime}\Lambda^+}\sum_{\tilde{\ell}}\langle\chi_{j'}^{i'N'\Lambda^+\mathsf{S}}|\ K_{i'\ell^{+\prime},i\ell^+}^{\Lambda\tilde{\ell}}(R,\epsilon)\ |\chi_j^{iN\Lambda^+\mathsf{S}}\rangle\langle\Lambda|N\rangle^{J\ell^+\Lambda^+} \tag{12}$$

instead of equation (8) in the MQDT without the crossing. Owing to taking into account the energy ϵ dependence, the basis set $\{\varphi_\epsilon(r;R)\}\otimes\{\chi_j^{ex,\mathsf{S}}(R)\}$ is expected to represent even the valence type electronic state $\varphi_d(r;R)F_\varepsilon(R)$. The physical K matrix is obtained by imposing the physical boundary condition as for the non-crossing case (subsection 2.2, 2.3) [9]. We show

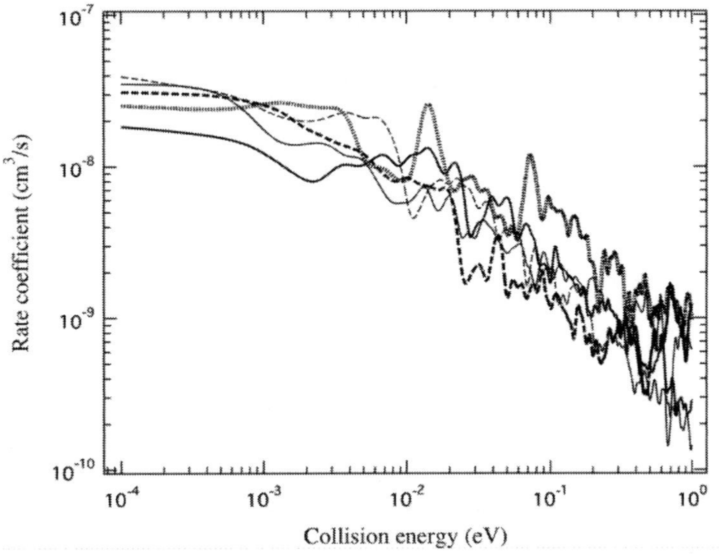

Figure 1. DR rate coefficients of five isotopes of the hydrogen molecular ion of the ground vibrational state. Bold dotted line: H_2^+, bold solid line: D_2^+, bold dashed line: T_2^+, thin solid line: HD^+, thin dashed line: DT^+. An thermal distribution of 300 K is assumed for the initial rotational state. For the collision energy distribution, the transversal and longitudinal energy fluctuations are assumed to be 1 meV and 0.05 meV, respectively.

the smoothed K matrix of the conventional two-step method below by fitting its form with equation (12) for comparison.

$$\mathcal{R}^{\epsilon J}_{k'N'\ell^{+\prime},kN\ell^+} = \sum_{\Lambda} \langle N'|\Lambda\rangle^{J\ell^{+\prime}\Lambda^+} \sum_{\alpha\tilde{\ell}} U^{J\Lambda}_{k',\alpha} \langle\chi^{N'\Lambda^+}_{v'}|\, K^{\alpha\Lambda\tilde{\ell}}_{k'\ell^{+\prime},k\ell^+}(R)\, |\chi^{N\Lambda^+}_{v}\rangle U^{J\Lambda}_{\alpha,k} \langle\Lambda|N\rangle^{J\ell^+\Lambda^+}, \quad (13)$$

where the matrix U diagonalizes the K matrix of the electron scattering by the CI, that eigen state is α, and k and k' indicate either the vibrational state $\chi_v(R)$ or the dissociative state $F_\varepsilon(R)$ in section 2.1.

If this new method works well, it has great advantage because this method requires only the adiabatic electronic information on the-energy-shell. This means that the cross sections of DR and other dynamical processes can be accurately calculated only if the electron scattering problem at fixed nuclei is solved.

3. DR of isotopes of H_2^+

3.1. Energy lower than 1 eV

Using the method briefly introduced in section 2.1, the cross sections for DR of H_2^+ and HD^+ at low energies (< 1eV) are calculated [8, 7]. This method seems to work well although it is difficult to extend to a wider collision energy region or to other complex molecules. Using the method of MQDT, the cross sections of DR of D_2^+, T_2^+, DT^+ are calculated. The complete set of the DR cross sections for various vibrational and rotational initial states will be provided as a data base from the Japan Atomic Energy Agency and International Atomic Energy Agency. In Figure 1, we show the calculated cross sections under the condition of a typical storage ring experiment: the initial rotational state is in the thermal distribution of 300K, the ground vibrational state is assumed, and the transversal and longitudinal energy widths are assumed to be 1 meV and 0.05 meV respectively. There is a tendency that the heavier molecular ion has the smaller DR cross section. The energy dependent structure are very different for the different isotopes.

3.2. Higher energy

When the energy of an incident electron becomes high or the target molecular ion is highly excited, the electronic excited states of molecular ions contribute to the DR process and the

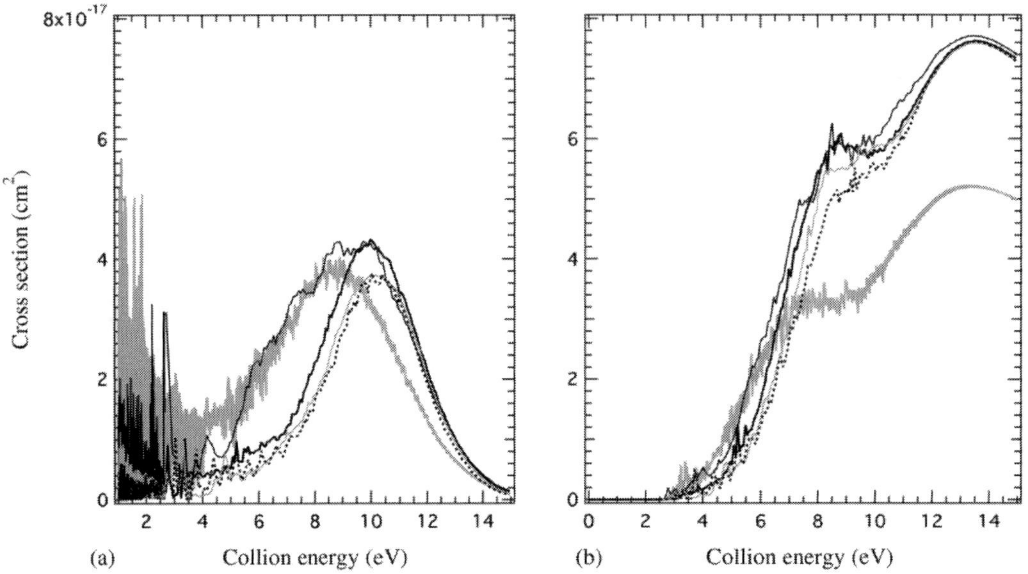

Figure 2. Cross sections of dissociative recombination (a) and dissociative excitation (b) for the initial vibrational state $v = 0$. Bold grey line: H_2^+, bold black line: D_2^+, bold dotted line: T_2^+, thin black line: HD^+, thin grey line: DT^+.

dissociative excitation (DE) process also occurs. In section 2.4, the method to calculate those processes is briefly introduced. We here take into account the effect of AQD of $p\sigma, p\pi, d\sigma$ partial waves [4, 14], which are neglected in the previous paper [13]. This means that the NAI is taken into account in the present calculation although it is less than perfect. For the CI of the lowest doubly excited state, we used the newest solution of scattering state mentioned section 2.1. The calculated cross section for the isotopes of H_2^+ is shown in Fig. 2. The DR cross section of H_2^+ and HD^+ ($v^+ = 0$) is distributed in a wider energy region than for the other isotopes (Figure 2(a)). In the DE cross section of $v^+ = 0$ (Figure 2(b)), the cross section of H_2^+ become smaller than the cross section calculated when NAI is neglected [13]. Other isotopes have larger cross section than H_2^+.

4. DR of HeH^+

We need to investigate the DR process of a non-crossing system in detail to develop the new method proposed in section 2.5. The important factor for the DR process is the integral of the adiabatic K matrix in equation (8). The adiabatic K matrix is represented by using the AQD $\mu_{\tilde{\ell}\Lambda}(R)$.

$$K^{\Lambda\tilde{\ell}}_{\ell^{+\prime},\ell^+}(R) = M_{\ell^{+\prime},\tilde{\ell}}(R)\ \tan\pi\mu_{\tilde{\ell}\Lambda}(R)\ M_{\tilde{\ell},\ell^+}(R), \qquad (14)$$

where the $M_{\tilde{\ell},\ell^+}(R)$ is the mixing matrix element for the partial waves. As a typical non-crossing system, we investigate these quantities of the HeH^+ + e system.

According to the previous study [9], large internuclear dependence of the AQD and mixing matrix brings large DR cross section. Especially the most important factor is that the AQD of $p\sigma$ partial wave changes one unit when the internuclear distance varies from small to large distance. This change is expected by the correlation diagram between the united atom limit (Li*) and the separated atom limit (He + H*). However,we could not find clearly these characteristic

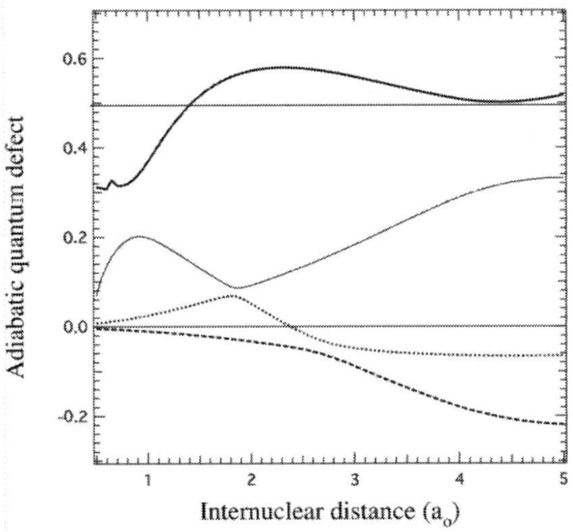

Figure 3. Adiabatic quantum defect of the Σ ($\Lambda = 0$) state. Bold solid line: $\tilde{\ell} = 0$, thin solid line: $\tilde{\ell} = 1$, dotted line: $\tilde{\ell} = 2$, dushed line: $\tilde{\ell} = 3$.

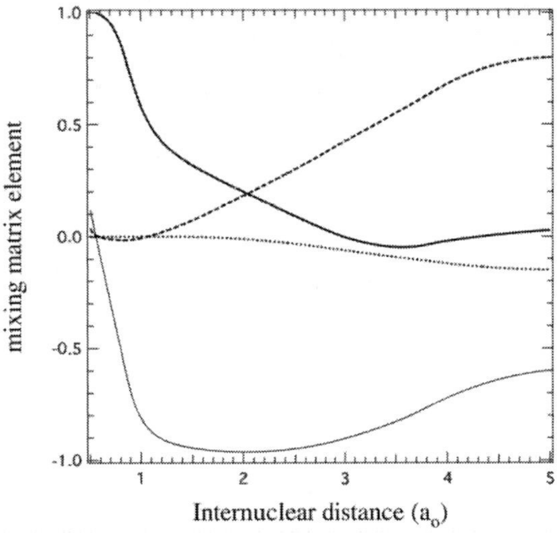

Figure 4. Mixing matrix elements of the ($\tilde{\ell} = 0, \Lambda = 0$) eigen state. Bold solid line: $\ell^+ = 0$, thin solid line: $\ell^+ = 1$, dushed line: $\ell^+ = 2$, dotted line: $\ell^+ = 3$.

in the AQD calculated by Haxton and Greens, which are obtained using an accurate R-matrix calculation [10]. We thus calculate the AQD and the mixing matrix to investigate the character of the Rydberg states.

The UK R-matrix code was used for the electron scattering calculation, where the basis set of cc-pVTZ was employed in the full CI calculation [15]. The adopted partial waves are $\ell^+ = 0 \sim 3$ and $\Lambda \leq 2$ of which origins are centered at the gravity center of the molecule, and the R-matrix radius is 15 bohr. The AQD is shown in Figure 3 for $\Lambda = 0$, which contribute to the DR most dominantly. The values at the nearest internuclear distances are smoothly connected by line in the figure. The AQD connected by the same line is labeld by $\tilde{\ell}$. The value of $\tilde{\ell}$ itself does not have physical meaning but only simbol. The mixing matrix elements are shown in Figure 4 for $\tilde{\ell} = 0, \Lambda = 0$.

The $p\sigma$ AQD is about 0.07 at 0.5 a_o, and it exceeds 0.5 at 1.4 a_0. It decreases at the larger than 2.5 a_o after reaching the maximum value 0.6 at 2.5 a_o. At the large internuclear distance, the $d\sigma$ component increases. It should be $p\sigma$ partial wave centered on the H atom physically. We should take care of the asymptotic behavior of the electronic states for the dissociative processes.

References

[1] Bates D R 1950 *Phys. Rev.* **77** 718
[2] Strömholm C *et al.* 1996 *Phys. Rev.* A **54** 3086
[3] Bardsley J N 1968 *J. Phys.* B **1** 349
[4] Jungen C and Atabek O 1977 *J. Chem. Phys.* **66** 5584
[5] Giusti A 1980 *J. Phys.* B **12** 3867
[6] Takagi H 1993 *J. Phys.* B **26** 4815
[7] Takagi H, Hara S and Sato H 2009 *J. Phys.* Conf. ser. **192** 012003
[8] Takagi H, Hara S and Sato H 2009 *Phys. Rev.* A **79** 012715
[9] Takagi H 2004 *Phys. Rev.* A **70** 022709
[10] Haxton D J and Greene C H 2009 *Phys. Rev. A* **79** 022701
[11] Seaton M J 1969 *J. Phys.* B **2** 5

Journal of Physics: Conference Series **300** (2011) 012020 doi:10.1088/1742-6596/300/1/012020

[12] Fano U 1970 *Phys. Rev. A* **2** 353
[13] Takagi H 2002 *phys. scr.* **T96** 52
[14] Nakashima K, Takagi H and Nakamura H 1987 *J. Chem. Phys.* **86** 726
[15] Morgan L A, Tennyson J and Gillan C J 1998 *Comput. Phys. Commun.* **114** 120

Journal of Physics: Conference Series **300** (2011) 012021 doi:10.1088/1742-6596/300/1/012021

Recombination in low temperature Ar^+ dominated plasma

P Dohnal, T Kotrík, R Plašil, I Korolov and J Glosík
Department of Surface and Plasma Science, Faculty of Mathematics and Physics, Charles University, Prague, Czech Republic

E-mail: juraj.glosik@mff.cuni.cz

Abstract. The results of measurements in low temperature Ar^+ dominated afterglow plasma are presented. The study was realized using Cryo-FALP apparatus operable in temperature range 77 – 300 K and at pressures 200 – 2000 Pa. We have observed decay of electron number density due to collisional radiative recombination (CRR). The measured ternary recombination rate coefficient of CRR of Ar^+ ions with electrons and its temperature dependence are in good agreement with theoretical predictions of $T^{-4.5}$ dependence.

1. Introduction

Collisional Radiative Recombination (CRR) was studied many times theoretically and experimentally. Experiments were carried out mostly at electron temperatures 1000-4000 K and in only few cases also at 300 K. To our knowledge there are up to now no experimental measurements of rate coefficient of CRR at temperatures below 300 K. Present study is devoted to measurement of rate coefficients of collisional radiative recombination at 77-200 K. Bates [1], Mansbach and Keck [2] and later Stevefelt [3] derived analytical formula for apparent binary rate coefficient of CRR:

$$\alpha_{\mathrm{CRR}} = 3.8\times10^{-9}T_{\mathrm{e}}^{-4.5}n_{\mathrm{e}} + 1.55\times10^{-10}T_{\mathrm{e}}^{-0.63} + 6\times10^{-9}T_{\mathrm{e}}^{-2.18}n_{\mathrm{e}}^{0.37}\mathrm{cm}^3\mathrm{s}^{-1}, \quad (1)$$

where T_e is electron temperature given in K and n_e is electron number density given in cm^{-3}. This formula was affirmed to certain extend also by later quantum mechanical calculations [4,5]. In our experimental conditions ($n_e \sim 10^9$ cm^{-3}, $T_e < 200$ K) the second and the third term can be neglected and only the first term representing recombination through collisions with electrons is significant. This term describes ternary recombination where in the first step the ion collides with an electron and highly excited neutral is formed. In the following second step this highly excited neutral collides with second electron and recombination process is terminated. The actual recombination process can be described by equation: $Ar^+ + e^- + e^- \rightarrow Ar + e^-$. For this ternary electron assisted process we can introduce the ternary recombination rate coefficient $K_{\mathrm{CRR}} = \alpha_{\mathrm{CRR}}/n_e = 3.8\times10^{-9}T_e^{-9/2}$ cm^6s^{-1}. Here we present the results of measurements of K_{CRR} using Cryo-FALP apparatus.

2. Experiment

The experiments were carried out using Flowing Afterglow with Langmuir Probe apparatus designed to study slow recombination processes at temperatures down to 77 K [6,7,8]. Scheme of the modified Cryo-FALP is shown in figure 1. The flow tube can be formally divided to three sections with different temperature regimes. Section A: discharge region and Ar port at 300K. Section B: transition section with large temperature gradient 300-100 K. Section C: tube cooled down to 77 K. Plasma is formed by microwave discharge in pure helium in the upstream part of the section A, then plasma is carried along the flow tube by a fast flow of helium. Ar is added to the flow tube shortly after discharge and in sequence of ion-molecule reactions Ar^+ dominated plasma is formed [9]. Plasma is carried through the section B where it is precooled to 100-150 K. In the section C plasma is already termalized with the temperature equal to the temperature of the flow tube wall. Langmuir electrostatic probe is used to monitor plasma parameters along the section C (details can be found e.g. in refs. [10]).

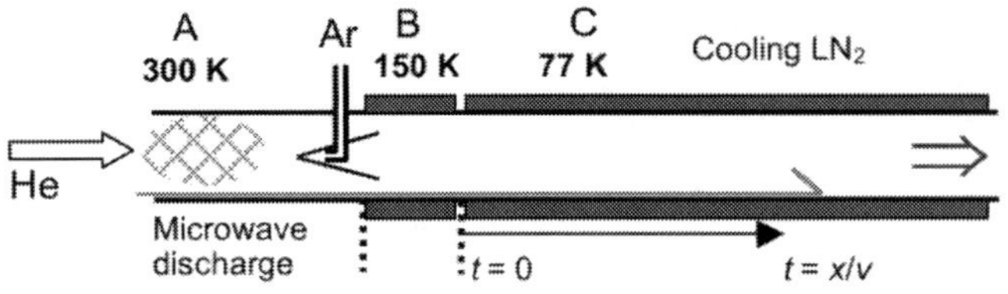

Figure 1. Cryo-FALP. Sections: A: discharge and Ar port, B: transition area with temperature gradient, C: cold region with Langmuir probe. Position of the probe determines decay time, $t = x/v_{PLASMA}$

Because of a very pronounced temperature dependence of CRR, it is critical to ensure constant plasma temperature over the section C. In the experiment it is necessary to find experimental conditions with constant temperature of helium flowing through the section C. In figure 2 three decay curves measured at fixed He pressure but with different flow rates (F) of He are plotted. At constant pressure (p) the flow rate is proportional to He velocity, $v_{He} \sim F/p$. We found that the decay curves measured with low He velocities can be fitted over the whole section C. For details concerning data analysis see below. In data analysis the effect of He velocity was considered and majority of experiments were carried out in a low velocity regime. In figure 3 we plotted apparent binary recombination rate coefficient α_{eff} obtained from measured decay curves. The dependence of α_{eff} on electron density clearly indicates three-body character of Ar^+ recombination - CRR.

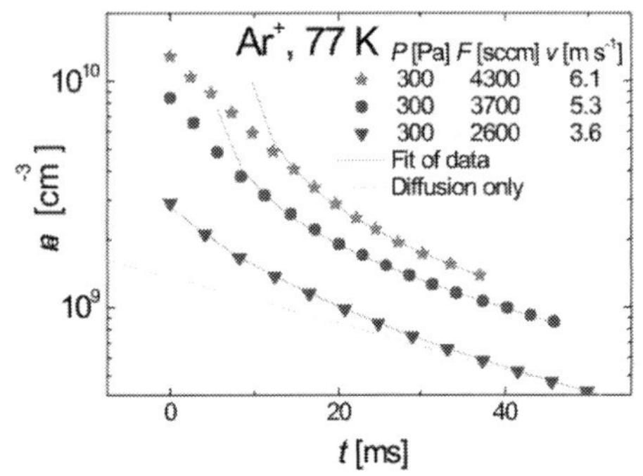

Figure 2. The measured decay of electron density at 77 K and 300 Pa using three different flows of He. The straight dotted line corresponds to the diffusion losses. The data are fitted with function obtained by analytical solution of balance equation (2) under the assumption that two constants, K_{CRR} and τ_D, characterize the decay.

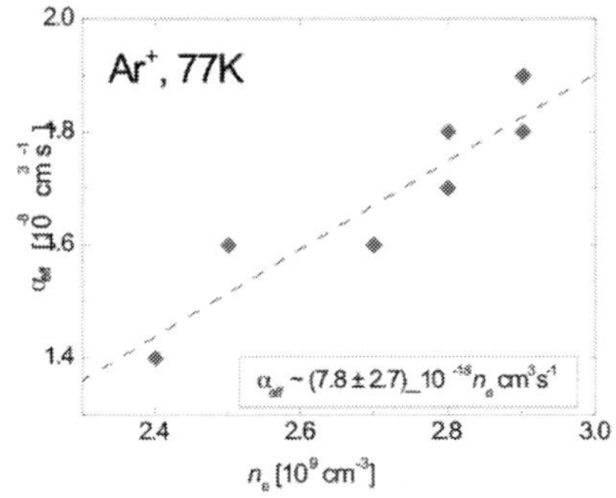

Figure 3. Apparent binary recombination rate coefficient α_{eff} obtained from measured decay curves in low velocity regime.

The evolution of the electron density in diffusion and CRR dominated quasineutral plasma ($[Ar^+] = n_e$) can be described by differential equation:

$$\frac{dn_e}{dt} = -K_{CRR}\,[Ar^+]n_e^2 - \frac{n_e}{\tau_D} = -K_{CRR}\;n_e^3 - \frac{n_e}{\tau_D} \qquad (2)$$

where τ_D is characteristic diffusion time. If K_{CRR} and τ_D are constant in decaying plasma, then these constants can be obtained from measured electron density decay. By integration of balance equation (2) we can obtain dependence:

$$Y = \ln\left(\frac{n(t_b)}{n(t_a)}\right) + (t_b - t_a)\frac{1}{\tau_D} = -K_{CRR}\int_{t_a}^{t_b} n^2(t)dt = -K_{CRR}X\text{, where } X = \int_{t_a}^{t_b} n^2(t)dt\ .$$

(3)

From this follows that plot of Y versus X should be linear with a slope given by K_{CRR}. We have used this "integral analysis" to calculate K_{CRR} and τ_D from measured decay curves (for details see ref. [11] where similar analysis was introduced for binary recombination). The plots used in "integral analysis" are shown in figure 4. To obtain temperature dependence of K_{CRR} and τ_D we cooled section C to 77K by liquid nitrogen and section B to ~ 150 K and then we turned off the cooling. During the subsequent slow warming up of the tube we continuously monitored the decay of plasma along the flow tube. Such set of "decay curves" plotted using "integral analysis" is shown in figure 4.

3. Results and discussion

Because the recombination of Ar^+ is very slow, the results can be influenced by formation of Ar_2^+ and their subsequent recombination. To exclude this possibility we measured K_{CRR} at different [Ar] and we found that for [Ar] = $(1\text{-}5)\times10^{-13}cm^{-3}$ the obtained K_{CRR} are not influenced by Ar_2^+ formation. The data obtained at 77 K are shown in figure 5.

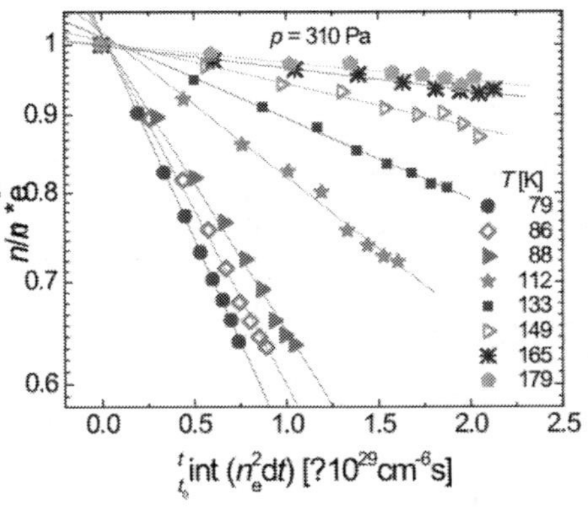

Figure 4. Curves used in "integral analysis" to obtain K_{CRR} and τ_D. Slopes of the straight lines give K_{CRR}.

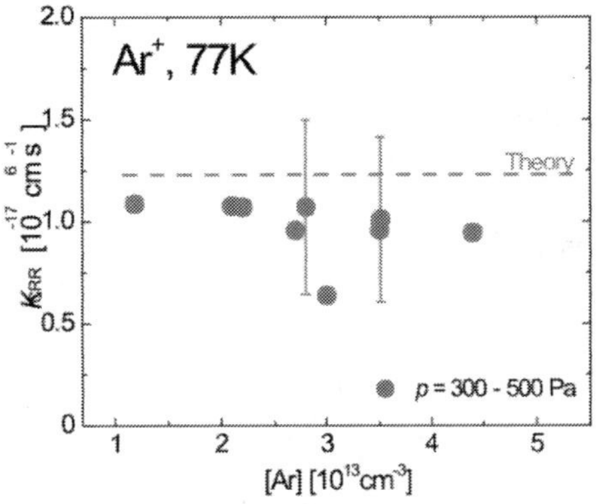

Figure 5. The dependence of K_{CRR} on [Ar].

We measured the decay curves at many pressures and helium velocities over the temperature range 77-200 K. The examples of measured data are plotted in figure 6. The data measured at low velocities overlap and are independent on actual velocity. The theoretical dependence (1) is plotted by dashed line (collisional term only) and by dotted lines (all terms in equation (1)) for three different electron number densities. The actual temperatures were verified by determining τ_D from the measured decay curves. The time constant $\tau_D \sim\lambda^2/D_+$, where λ is characteristic diffusion length and D_+ is coefficient of ambipollar diffusion for Ar^+ in He. When using dependence of D_+ on [He] and temperature we obtain $\tau_D \sim \lambda^2[He]/\mu_0 T$, where μ_0 is reduced zero field mobility (for details see ref. [12]). This indicates that plot of $[He]/\tau_D$ versus T should be linear and the slope is given by μ_0. In the inset of figure 6 we plotted this dependence for low He velocity. We obtained linear dependence which is in excellent agreement with calculated dependence using μ_0 measured in independent experiments [13,14].

Presented preliminary experimental results show that the agreement of measured K_{CRR} with theory [3] is excellent and that the theory is valid also for temperatures down to 77 K. Further studies of CRR in Ar^+ dominated plasma are in progress.

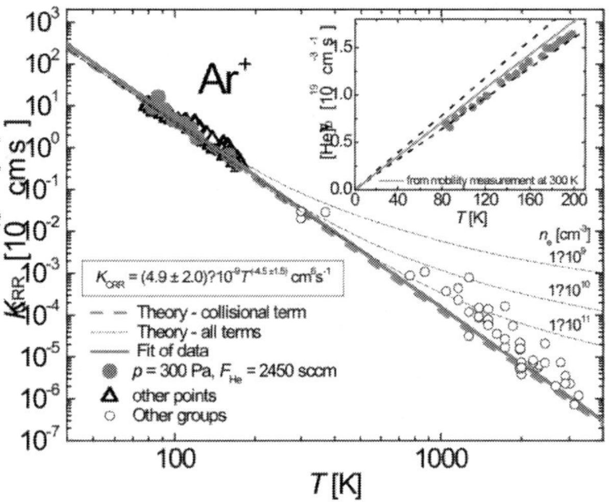

Figure 6. The measured temperature dependence of K_{CRR} and the data obtained in other high temperature experiments (see compilation in ref. [2,15,16,17]). In the inset $[He]/\tau_D$ versus T are plotted. Indicated is linear dependence calculated by using measured reduced mobility μ_0 [13,14] with corresponding error (dashed line).

Acknowledgments
This work is a part of the research plan MSM 0021620834 financed by the Ministry of Education of the Czech Republic and was partly supported by GACR (202/07/0495, 205/09/1183, 202/09/0642, 202/08/H057), by GAUK 92410, GAUK 54010, GAUK 25709 and GAUK 86908.

[1] Bates D R, Kingston A E and McWhirter R W P 1962 *Proc. Roy. Soc. A* **267** 297
[2] Mansbach P and Keck J 1969 *Phys. Rev.* **181** 275
[3] Stevefelt J, Boulmer J and Delpech J-F 1975 *Phys. Rev. A* **12**, 1246
[4] Pert G J 1990 *J. Phys. B: At. Mol. Opt. Phys.* **23** 619
[5] Hu S X 2007 *Phys. Rev. Lett.* **98** 133201-1
[6] Glosik J, Zakouril P, Hanzal V and Skalsky V 1995 *Int. J. Mass Spectrom. Ion Processes* **149/150** 187
[7] Glosik J, Bano G, Plasil R, Luca A and Zakouril P 1999 *Int. J. Mass Spectrom.* **189** 103
[8] Glosik J, Novotny O, Pysanenko A, Zakouril P, Plasil R, Kudrna P and Poterya V 2003 *Plasma Sources Sci. Technol.* **12** 117
[9] Plasil R, Glosik J, Poterya V, Kudrna P, Rusz J, Tichy M and Pysanenko A 2002 *Int. J.Mass Spectrom.* **218** 105
[10] Chudacek O, Kudrna P, Glosik J, Sicha M, and Tichy M 1995 *Contrib. Plasma Phys.* **35** 503
[11] Korolov I, Kotrık T, Plasil R, Varju J, Hejduk M and Glosık J 2008 *Contrib. Plasma Phys.* **48(5-7)** 521
[12] McDaniel E W 1964 *Collision phenomena in ionized gases* (New York: John Wiley & Sons)
[13] Lindinger W and Albritton D L 1975 *J. Chem. Phys.* **62(9)** 3517
[14] Johnsen R and Biondi M A 1979 *Phys. Rev. A* **20** 221
[15] Berlande J, Cheret M, Deloche R, Gonfalone A and Manus C 1970 *Phys. Rev. A* **1** 887
[16] Skrzypkowski M P, Johnsen R, Rosati R E, and Golde M F 2004 *Chem. Phys.* **296** 23
[17] Curry B P 1970 *Phys. Rev. A* **1(1)** 166

Eighth International Conference on Dissociative Recombination (DR2010) IOP Publishing
Journal of Physics: Conference Series **300** (2011) 012022 doi:10.1088/1742-6596/300/1/012022

Small Molecule Ion Chemistry in ITER

J. Brian A. Mitchell, S. Carles, J.L. LeGarrec

Institut de Physique de Rennes
UMR du CNRS No. 6251
Université de Rennes I
35042, Rennes, France
mitchell@univ-rennes1.fr h

Abstract. The new ITER reactor will have special problems with heat loads on exposed plasma surfaces. A good knowledge of the plasma chemistry is essential to be able to predict these heat loads and to find efficient ways of cooling the plasma near these surfaces. This paper addresses the data needs for light element reactions with particular emphasis on electron impact reactions.

1. Introduction

The International Thermonuclear Experimental Reactor ITER [1] is beginning construction at Cadarache in the south of France. This is an enormous project that is scheduled to begin operation after 2019 and the first use of tritium fuel to achieve thermonuclear ignition, is scheduled for after 2027. The technological challenges facing this experimental machine are daunting and successful operation will require amongst other things, a good knowledge of the atomic and molecular processes near the walls of the reactor and in the divertor region. A special issue of Nuclear Fusion, published in 2007 [1] gives an in-depth analysis of the technological and scientific issues involved in the design of ITER. One of the key engineering problems is that the heat load to the walls in normal operation will be of the order of 10 MW/m^2 with excursions to 20 MW/m^2 during unstable plasma events such as ELMs (Edge Localized Modes). To put this in perspective, an arc welder delivers a heat load of about 40 MW/m^2 so clearly, such high heat loads expected during operation, can be very destructive to the plasma facing components of the reactor. These problems are most apparent in the divertor region into which plasma is scraped off from the main toroidal region (the pedestal). This allows helium "ash" to be removed from the burning plasma by pumping and the divertor is also one of the means of extracting energy from the reactor.

2. High Temperature Energy Loss Processes

In the divertor region, the plasma is cooler than in the central core and one can expect to find electrons with energy in the range (0.1-200 eV) at a density of ~ 10^{14} cm^{-3}. In the core, where the thermonuclear reaction will take place, once the machine is fuelled with tritium and deuterium, the electron energies are in the range 0.2-20 keV. In earlier tokamaks such as the Princeton Large Torus (PLT), it was found that serious energy loss in the plasma could be attributed to the emission of photons due to dielectronic recombination:

$$e + A^{z+} \rightarrow A^{(z-1)+**} \rightarrow A^{(z-1)+*} + h\upsilon$$

and electron impact excitation followed by radiative relaxation:

$$e(E) + A^{z+} \rightarrow e(E') + A^{Z+*} \rightarrow\rightarrow e(E') + A^{z+} + h\upsilon$$

of multiply charged ions of heavy metal impurities (A^{z+}) , generated by the interaction of the plasma with the walls of the reactor. In these reactions in the tokamak, the ions are typically highly charged and thus the emitted photons are in the x-ray range and thus carry away a large amount of energy. The key point however, is that both of these collision processes require the ion target to contain electrons that can be excited. Following the realization of the significance of these reactions, carbon tiles were implemented as wall coverings and for plasma facing components and this allowed the energy of the plasmas in future machines (TFTR, JET, JT60) to be increased up to ignition conditions, for any carbon atom impurities entering the plasma, are rapidly fully ionized so that both dielectronic recombination and electron impact excitation loss processes are eliminated. The only recombination process that occurs is the radiative recombination reaction;

$$e + C^{6+} \rightarrow C^{5+} + h\upsilon$$

that is at least two orders of magnitude slower that dielectronic recombination.

3. Wall Materials

While carbon has been the material of choice for the walls of most tokamaks up to the present, it does however, present different problems that have become apparent through usage. Firstly it is a reactive material and is rapidly eroded by interaction with the hydrogen plasma leading to the generation of hydrocarbon ions that serve as nuclei for the formation of carbon dust. This dust is electrostatically charged and can be carried along the magnetic field lines and be dumped elsewhere in the machine, in the divertor regions for example. This is a nuisance rather that a fatal flaw but the fact that when running in a thermonuclear mode, the hydrogen is replaced with radioactive tritium and this becomes bound up in the carbon tiles and the dust, this becomes a very serious problem both from fuel depletion and from safety points of view. Although carbon will be used for initial experiments, it will be eventually be replaced by beryllium as the plasma facing material on the walls of ITER.

4. Plasma Cooling

While the goal of tokamaks is to heat up, sustain and confine a plasma at very high temperature so as to allow isotopes of hydrogen to fuse together and produce energy, in the divertor and near the walls, one needs to cool down the plasma to avoid material damage to the components. In order to do this, impurities such a neon or argon are puffed into the divertor where the energy of the electrons is taken up in ionizing these gases which, becoming fully ionized, do not cause problems if they drift back into the central hot core region. Other elements that can perhaps be found in the divertor region are nitrogen and oxygen which may be due to issues connected with vacuum technology (leaks, outgassing of water, etc.). With the electron temperature being lower in the divertor region, the plasma is very different from that in the main core. Indeed molecules are formed due to recombination on surfaces and these ro-vibrationally excited species return to the plasma causing a very rich chemistry to be generated. Knowledge of this chemistry and in particular of the ion chemistry, is very important, for it is desirable that ions are neutralized prior to reaching the surfaces of the components so that the recombination energy is not released in the surface. Thus dissociative recombination plays an important role in this region and it is necessary to have good state-to-state knowledge of this process. Molecular ions can also be formed in the volume through three-body processes and by ion-molecule reactions. In this paper we shall discuss just the elements with atomic number 1 to 4 i.e. hydrogen, helium, lithium and beryllium. Lithium is included for it may be used for diagnostic processes. Since this paper covers a very wide range of subjects, with a few exceptions, we shall not give specific references to excited state distributions, reaction cross sections etc. The reader is directed towards the standard literature and previous proceedings of this conference for further information on these processes.

5. Light Element Chemistry

In this paper, for the light elements H, He, Li and Be, the most important species that are likely to be found in the divertor and at the plasma edge are given in Table I.:

Table I: Primary species

- H^+, He^+, He^{++}, Li^+, Li^{++}, Be^{z+} dominant boundary species
- H^*, H^-, He, He*, $H_2(v)$
- $H_2^+(v)$, $H_3^+(v)$, $He_2^+(v)$, $HeH^+(v)$, $HeH^{+*}(v)$, $LiH^+(v)$, $BeH^+(v)$
- $NeH^+(v)$, $ArH^+(v)$, $HeNe^+(v)$

An important thing to note is that many of these species will be present in excited states. This is particularly true of H, He, He^+ and HeH^+, all of which have long lived electronically excited states and the molecular ions will initially be in ro-vibrationally excited states and this will greatly affect the reaction cross sections. As mentioned above, neon and argon will be present if deliberately introduced into the plasma for cooling purposes.

Information on the formation, reaction and destruction mechanisms for these species is crucial for accurate and complete models of the plasma chemistry in ITER. The International Atomic Energy Authority (IAEA) has begun a Collaborative Research Program (CRP) specifically dealing with small light elements and their reactions [2]. For hydrogen molecular ions, the following reactions listed in Table II must be taken into account. Of particular note is the need to understand the initial distribution of vibrational states of the hydrogen molecule formed on the surface, that for the hydrogen molecular ion created by electron impact and the subsequent life history of these distributions. State-to-state reaction cross sections therefore need to be known and evaluated as to their relative importance. Although the table deals mainly with electron impact processes, there will also be a strong photon flux in the divertor region and so photoionisation and photodissociation processes must also be taken account of. Possibly even photon stimulated processes may play a role although this has to be evaluated. For example almost nothing is known concerning the dissociative recombination of molecular ions in a strong radiation field. Given the uncertainties in this area, we shall consider that these are outside the scope of the present paper but nevertheless they should not be forgotten.

Table II: Hydrogen molecular ion reactions

- H_2, H_2^+,H_3^+
  - $H + H \rightarrow H_2(v)$.........................Surface recombination
  - $H^+ + H + X \rightarrow H_2^+(v) + X$.......3-body-process
  - $H^* + H \rightarrow H_2^+(v) + e$ ....Associative ionization
  - $H_2\,(v) + e \rightarrow H_2^+(v) + 2e$.........Ionization
  - $H_2(v) + e \rightarrow H + H + e$ ....Dissociation
  - $H_2^+(v) + e \rightarrow H + H$............ DR
  - $H_2^+(v) + e \rightarrow H + H^+ + e$...... DE
  - $H_2^+(v) + e \rightarrow H^+ + H^+ + 2e$......DI
  - $H_2^+(v) + H_2 \rightarrow H_3^+(v) + H$...... Ion-Molecule
  - $H_2^+(v{>}2) + He \rightarrow HeH^+ + H$... Ion-Molecule
  - $H_2^+(v{>}1) + Ne \rightarrow NeH^+ + H$... Ion-Molecule
  - $H_3^+(v) + e \rightarrow H + H + H$......... DR
  - $H_3^+(v) + e \rightarrow H_2 + H$...............DR
  - $H_3^+(v) + e \rightarrow H_2^+ + H^-$......... Ion-pair
  - $H_3^+(v) + e \rightarrow H_2^+ + H + e$.........DE
  - $H_3^+(v) + e \rightarrow H_2 + H^+ + e$......... DE

- $H_3^+(v) + e \rightarrow H^+ + H + H + e$ ... DE
- $H_3^+(v) + e \rightarrow H^+ + H^+ + H + 2e$... .DI
- $H_3^+(v) + e \rightarrow H_2^+ + H^+ + 2e$........DI
- $H_2^+ + H^- \rightarrow$ products................Ion-Ion processes

Molecular hydrogen will primarily be formed by the recombination of hydrogen atoms on surfaces and this leads to the production of vibrationally excited hydrogen molecules. Thus when ionized by electron impact, the resulting hydrogen ions will have a distribution of vibrational states populated, that is different from that formed from ground state ions. These vibrational states for homonuclear hydrogen and hydrogen isotopomers (D_2^+, T_2^+) have extremely long lifetimes (~10^6 s) against radiative decay. Vibrationally excited H_2^+ molecules can however, react with helium and neon through endothermic reactions (see below) and this could lead to a low vibrational state population for the surviving molecular ions. The reaction with argon is energy neutral and so this is not the case for the case of argon addition to the plasma. Heteronuclear isotopomers (HD^+, DT^+) will of course rapidly radiatively decay to their ground states.

While dissociative recombination is a reaction that is not only important but much desired in the divertor region, the high electron energies encountered will mean that in fact dissociative excitation may actually be dominant and dissociative ionisation must be included. This has been discussed in more detail in refs. [3,4].

Those of us who work with afterglow plasmas know well that He_2^+ (Table III) is an important ion that is efficiently formed by three-body processes even at quite low pressures. This ion has played a special role in the history of recombination research as discussed by Rainer Johnsen in these proceedings. Ion storage ring measurements have shown that while the ground state ion has an extremely low recombination cross section, that for excited He_2^+ ions is substantial. Here is a very good example of the need for information on excited state chemistry. There is also a peak in the DR cross section at higher electron energies, even for ground state ions, and the dissociative excitation cross section for ground state ions has been measured in a storage ring experiment [5]. Resonant dissociative excitation [6] plays a role here yielding measurable cross sections below the expected threshold for this reaction and this is a phenomenon that must be taken into account in theoretical studies of this process.

Table III: He_2^+ reactions

- He_2^+
  - $He^+ + He + X \rightarrow He_2^+ + X$.......3-body-process
  - $He_2^+(v) + e \rightarrow He + He$...............DR
  - $He_2^+(v) + e \rightarrow He + He^+ + e$..........DE
  - $He_2^+(v) + e \rightarrow He^+ + He^+ + 2e$........DI

HeH^+ is an ion that is expected to be present, being formed as shown in table II although three-body processes may also play a role, given that atomic hydrogen is likely to be more abundant than H_2 in the divertor. The reactions listed in table IV will then lead to its demise. As constantly re-stated in this review, the need for information on the reactivity of excited states is paramount and this means that experiments should be designed that allow excited states of heteronuclear ions to be examined. Much of the interest in dissociative recombination research has been concerned with ground state ions, especially for interstellar chemistry applications. Indeed one of the main features of the storage ring technique is its ability to allow ions to radiatively cool before measurements are performed. In fact, this is somewhat of a disadvantage for the data needed here and it shows the need for single-pass type experiments where the excited state can be adjusted by ion source conditions. An

Journal of Physics: Conference Series **300** (2011) 012022 doi:10.1088/1742-6596/300/1/012022

example of this is shown in figure 1, recalling merged beam measurements [7] of the recombination of HeH^+ ions where the excited state of the ions was tuned by varying the extraction voltage of the ion trap source. As seen, the ions extracted at an elevated extraction voltage (100V) display a cross section about a factor of 4-5 times greater than those extracted at a low voltage where the ions were ro-vibrationally much cooler. The actual excited state energy (though probably not the state distribution) can be determined by means of dissociative excitation threshold measurements coupled with theoretical studies (though this is complicated by the resonance DE mechanism). Though inexact, this type of measurement at least gives an indication of the overall effect of internal excitation on the reaction cross sections.

Table IV: HeH^+ reactions

- HeH+
  - $He^+ + H + X \rightarrow HeH^+ + X$.........3-body-process
  - $He + H^+ + X \rightarrow HeH^+ + X$ ....3-body process
  - $HeH^+(v) + e \rightarrow He^+ + H^+ + 2e$......Ionization
  - $HeH^+(v) + e \rightarrow HeH^* + e$...............Excitation
  - $HeH^+(v) + e \rightarrow He + H$...............DR
  - $HeH^+(v) + e \rightarrow He + H^+ + e$.........DE
  - $HeH^+(v) + e \rightarrow He^+ + H + e$.........DE
  - $HeH^+(v) + e \rightarrow He^+ + H^+ + 2e$.......DI
  - $HeH^+(v) + H_2 \rightarrow H_3^+(v) + He$......Ion-Molecule

Table V lists some reactions that may play a role in beryllium and lithium chemistry. Here there is much less information available though there have been experimental and theoretical studies of LiH^+ DR [8,9] , LiH_2^+ [10] and a theoretical calculation of the DR of BeH^+ [11]. Given the difficulty of preparing LiH^+ and the toxicity of BeH^+, it is unlikely that experimental studies will be forthcoming.

Table V: Lithium and beryllium reactions

- LiH^+ Formation ?
- $LiH^+ + e \rightarrow Li + H$...........DR
- $LiH^+ + H \nrightarrow$....................No reaction
- $LiH_2^+ + e \rightarrow Li + H_2$.........DR

- BeH^+ formation ?
- $BeH^+ + e \rightarrow Be + H$.........DR
- $BeH^+ + X \rightarrow$ products ?

Reactions with neon and argon (Table VI) are important as they will efficiently remove H_2^+ ions from the plasma and the effect of neon on the vibrational state distribution has been noted above. Experiment [12] have shown that ArH^+ does not recombine at low energies though excited ions might. This is because the channel leading to Ar + H(n=2) is endothermic and the likelihood of the ion recombining to ground state products is very low. The cross section for DR and DE does rise however at higher electron energies. Experiment has also shown that NeH^+ recombines at low energy [13] though theoretical calculations appear to show that this is only possible for vibrationally excited ions. The storage ring measurements of NeH^+ DR and DE were complemented by crossed beam studies of DE and DI [15].

Journal of Physics: Conference Series **300** (2011) 012022 doi:10.1088/1742-6596/300/1/012022

Given that ArH^+ formation will not create as much of a dis-equilibrium of the vibrational distribution of H_2^+ in the plasma, perhaps this element is a more desirable choice for plasma cooling operation than neon but this must await real in-situ experiments in the ITER divertor.

Table VI: Neon and argon reactions

- NeH^+
- $NeH^+ + e \rightarrow Ne + H$
- $NeH^+ + H_2 \rightarrow$ products

- ArH^+
- $ArH^+ + e \neq$
- $ArH^+ + H_2 \rightarrow H_3^+ + Ar$

The final set of reactions that are highlighted here are those concerning ro-vibrational excitation-de-excitation as listed in Table VII. These have been discussed in previous proceedings of this conference [16]. What is very striking about these reactions is their very large cross sections and so these must of necessity be included in plasma chemistry models.

Table VII:Electron impact rovibrational excitation

- $H_2^+(v,J) + e \rightarrow H_2^+(v',J') + e$
- $HeH^+(v,J) + e \rightarrow HeH^+(v',J') + e$
- $H_3^+(v,J) + e \rightarrow H_3^+(v',J') + e$
- $He_2^+(v,J) + e \rightarrow He_2^+(v',J') + e$
- $LiH^+(v,J) + e \rightarrow LiH^+(v',J') + e$
- $BeH^+(v,J) + e \rightarrow BeH^+(v',J') + e$
- $NeH^+(v,J) + e \rightarrow NeH^+(v',J') + e$
- $ArH^+ (v,J) + e \rightarrow ArH^+ (v',J') + e$

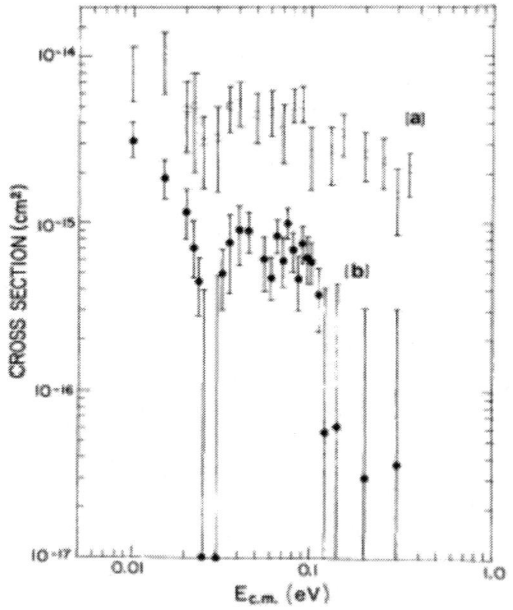

Fig. 1: Merged beam cross sections for the dissociative recombination of HeH^+ taken with high and low ion extraction voltages. The upper curve corresponds to metastable excited ions that displayed a dissociative excitation threshold of about 2 eV.

6. Conclusions

Much of the data needed for light element chemistry modelling in ITER already exists though predominantly for ground state species. There is a need to understand what role excited states will play and how the excited state distributions will evolve with time in the machine. Unlike other tokamaks, pulse lengths in ITER can be up to 100 s long so this is an important point to take into account in order to understand the temporal evolution of the plasma in the divertor and wall regions. Experimental access to this type of information must surely be limited so there is a great need for theoretical studies to compliment the existing data and to show how reaction cross sections will be affected by different internal ion states. Although not discussed in detail above, the role of negative ions must also be considered. H^- ions can be formed on surfaces [17] and can re-enter the plasma where they will react with positive species via ion-ion processes. This is an area that also needs further study. ITER is a very exciting project and its success will depend on the resolution of all the technical difficulties that it will encounter. Certainly atomic and molecular processes will play an important role and there is a great challenge for the community to continue to study these processes so that we might understand their relative importance in regulating the temperature of the plasma in the near wall and divertor regions.

Other light elements that should be considered in the modelling of the cooler plasma regions in ITER include carbon (the initial wall material), nitrogen (another gas used as a coolant) and oxygen (absorbed and adsorbed on surfaces during periods when the machine is let up to air). The chemistry of these elements will be discussed elsewhere.

References

[1] Nuclear Fusion 47 (2007) *Progress in the ITER physics basis*
[2] B.J. Braams, *Light element atom, molecule and radical behaviour in the divertor and edge plasma regions*, International Nuclear Data Committee, IAEA, Summary Report of First Research Coordination Meeting, 2009.
[3] J.B.A. Mitchell in Atomic and Molecular Processes in Fusion Edge Plasmas, (ed. R.K. Janev), Plenum Press, New York, 1995, p. 225
[4] J.B.A. Mitchell in Atomic and Plasma-Material Interaction Data for Fusion, Vol. 9, IAEA, Vienna 2001, p. 97.
[5] H.B. Pedersen et al. Phys. Rev. A 72 (2005) 012712
[6] A.E. Orel and K.C. Kulander, Phys. Rev. A 54 (1996) 4992
[7] F.B. Yousif and J.B.A. Mitchell, Phys. Rev. A 40 (1989) 4318
[8] S. Krohn et al., Phys. Rev. Lett. 86 (2001) 4005
[9] R. Curik and C.H. Greene, Phys. Rev. Lett. 98 (2007) 173201
[10] D.J. Haxton and C.H. Greene, Phys. Rev. A 78 (2008) 052704
[11] J.B. Roos, A.E. Orel and A. Larson, Phys. Rev. A 80 (2009) 012501
[12] J.B.A. Mitchell et al., J. Phys. B 38 (2005) L175
[13] J.B. A. Mitchell et al.,J. Phys. B 38 (2005) 693
[14] H. Takagi (private communication)
[15] J. Lecointre et al. J. Phys. B 41 (2008) 045201
[16] A. Faure et al. J. Phys: Conference Series192 (2009) 012016
[17] L. Schiesko et al. Plasma Sources. Sci. Tech. 19 (210) 045016

Eighth International Conference on Dissociative Recombination (DR2010) IOP Publishing
Journal of Physics: Conference Series **300** (2011) 012023 doi:10.1088/1742-6596/300/1/012023

Experimental study of para- and ortho-H_3^+ recombination

R Plašil, J Varju, M Hejduk, P Dohnal, T Kotrík and J Glosík

Charles University, Faculty of Mathematics and Physics, Department of Surface and Plasma Science, V Holešovičkách 2, Prague 8, Czech Republic

E-mail: radek.plasil@mff.cuni.cz

Abstract. Recombination of H_3^+ with electrons is a key process for many plasmatic environments. Recent experiments on storage ring devices used ion sources producing H_3^+ with enhanced populations of H_3^+ ions in the para nuclear spin configuration to shed light on the theoretically predicted faster recombination of para states. Although increased recombination rates were observed, no *in situ* characterization of recombining ions was performed. We present a state selective recombination study of para- and ortho-H_3^+ ions with electrons at 77 K in afterglow plasma in a He/Ar/H_2 gas-mixture. Both spin configurations of H_3^+ have been observed *in situ* with a near infrared cavity ring down spectrometer (NIR-CRDS) using the two lowest energy levels of H_3^+. Using hydrogen with an enhanced population of H_2 molecules in para states allowed us to influence the [para-H_3^+]/[ortho-H_3^+] ratio in the discharge and in the afterglow. We observed an increase in the measured effective recombination rate coefficients with the increase of the fraction of para-H_3^+. Measurements with different fractions of para-H_3^+ at otherwise identical conditions allowed us to determine the binary recombination rate coefficients for pure para-H_3^+ ${}^{p}\alpha_{bin}(77\text{ K}) = (2.0\pm0.4)\times10^{-7}\text{ cm}^3\text{s}^{-1}$ and pure ortho-H_3^+ ${}^{o}\alpha_{bin}(77\text{ K}) = (4\pm3)\times10^{-8}\text{ cm}^3\text{s}^{-1}$.

1. Introduction

The recombination of H_3^+ with electron has been a long-standing issue in plasma physics for over 50 years, as both a theoretical and an experimental subject. Considering that this is one of the most important processes in the interstellar medium, planetary atmospheres [1,2] and in many technological types of plasma, it comes as no surprise that the enormous effort was put into its understanding. But despite best efforts the dissociative recombination of H_3^+ has kept its secrets for a long time. Recently theoretical predictions made significant progress when the non-Born-Oppenheimer Jahn-Teller coupling was added to the calculations. Predictions for binary dissociative recombination rate coefficients of H_3^+ ions with electrons at low energies (< 1 eV) [3,4,5,6] agree well with measurements from storage rings [7,8]. Large difference in the recombination rate for different nuclear spin (para and ortho) states of H_3^+ at temperatures below 300 K was predicted too [6]. Although measurements on storage rings are showing an increase of the recombination rate coefficient when using H_3^+ ions with enhanced population of para states [9,10,11], the beam experiments were made without *in situ* determination of the spin state and rotational excitation of recombining ions and so the theoretical predictions could not have been fully tested [11,12]. The ration of the thermal rate coefficients (para-H_3^+/ortho- H_3^+) observed in recent CRYRING experiment [9] is at 77 K equal ~1.5 while theory [5] predicts ratio ~3.

Journal of Physics: Conference Series **300** (2011) 012023 doi:10.1088/1742-6596/300/1/012023

In afterglow experiments, using a He/Ar/H_2 gas-mixture, H_3^+ ions can have several collisions with H_2 and thousands of collisions with He prior to recombination. Observed recombination rate coefficients are referred to as effective, because they are determined from plasma decay and several processes can influence the decaying plasma. By studying dependences of the effective recombination rate coefficient on experimental parameters it is possible to determine the rate coefficients for particular processes, the rate coefficient of binary recombination in particular.

Collision with H_2 can change the nuclear spin state of H_3^+ via formation of the $(H_5^+)^{\#}$ complex [13,14,15,16]. Number density of H_2 and its spin state composition are therefore important parameters in H_3^+ low temperature plasmatic experiments devoted to measurements of recombination rate coefficients. In such experiments the para-H_3^+/ortho-H_3^+ ratio has to be measured and considered in data evaluation and interpretation. For an extensive review of experiments and previous results see e.g. reviews [8,11,12,17] and book by *Larsson and Orel* [18].

We will use the following notation further: nuclear spin states with para and ortho symmetry of H_2 are designated as para-H_2 (or shortly $^{p}H_2$), ortho-H_2 ($^{o}H_2$) and equivalently for H_3^+. As normal hydrogen ($^{n}H_2$) we will designate hydrogen with populations of $^{p}H_2$ and $^{o}H_2$ corresponding to thermal equilibrium at 300 K (25% of $^{p}H_2$). Para enriched hydrogen ($^{e}H_2$) stands for hydrogen with populations of $^{p}H_2$ higher than 25%.

If the ortho- and para-H_3^+ populations are not in thermodynamic equilibrium, then their actual densities in an experiment have to be measured. In H_3^+ dominated low temperature plasma the sum of concentrations of both spin states of H_3^+ corresponds to the electron density, $[^{p}H_3^+] + [^{o}H_3^+] = n_e$. From the measured evolution of the electron density the effective recombination rate coefficient can be obtained (see [19] for the discussion). The balance equation for electrons is:

$$\frac{dn_e}{dt} = -({}^{p}\alpha_{\rm eff}\,{}^{p}f + {}^{o}\alpha_{\rm eff}\,{}^{o}f)n_e^2 - \frac{n_e}{\tau_D} = -\alpha_{\rm eff}n_e^2 - \frac{n_e}{\tau_D}, \qquad (1)$$

where ^{p}f and ^{o}f denotes the fraction of H_3^+ ions in para and ortho states respectively, τ_D is the time constant of ambipolar diffusion. $^{p}\alpha_{\rm eff}$ and $^{o}\alpha_{\rm eff}$ are effective recombination rate coefficients for pure para-H_3^+ and ortho-H_3^+, respectively. $\alpha_{\rm eff} = {}^{p}\alpha_{\rm eff}\,{}^{p}f + {}^{o}\alpha_{\rm eff}\,{}^{o}f$ is the overall effective recombination rate coefficient which is determined from the electron density decay during the afterglow.

Measuring the effective recombination rate coefficient $\alpha_{\rm eff}$ at least at two different para-H_3^+/ortho-H_3^+ ratios, at otherwise identical conditions, one can simply derive the separate effective recombination rate coefficients for para- and ortho-H_3^+. Thanks to spin selection rules at low temperatures < 300 K [15,16,20,21] one can change the para-H_3^+/ortho-H_3^+ ratio when using para enriched hydrogen as a reactant (see discussions and experiments in refs. [8,9,11,13,14,22]). Measurements at different helium concentrations offer the possibility to extract values of the binary and ternary recombination rate coefficients for the individual spin states (for a detailed discussion of the ternary process see previous papers of our group [23,24,25,26,27]).

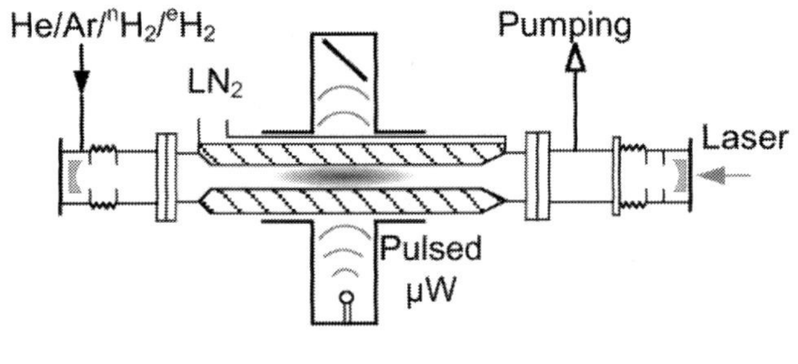

Figure 1. Schematic drawing of the used optical cavity, discharge tube and microwave cavity. 10 mW DFB diode laser with $\Delta\lambda \sim 2$ MHz was used in the experiment. With reflection of the mirrors R > 99.95% we obtained typical baseline ring-down time ~30 μs.

Journal of Physics: Conference Series **300** (2011) 012023 doi:10.1088/1742-6596/300/1/012023

2. Experimental apparatus

Pulsed microwave (μW) discharges were periodically ignited in a He/Ar/H_2 gas mixture. Temperatures of ~77 K were achieved by immersing the discharge tube into liquid nitrogen (see figure 1). Fast high voltage switch allowed us to turn off the incident μW power within 50 μs. Very low incident μW power (15 W) has been used to prevent heating of ions. Physical dimensions of the μW resonator limited the discharge column to ~5 cm of length.

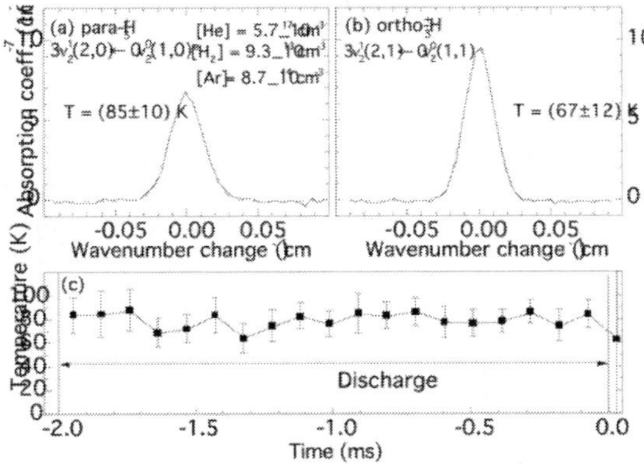

Figure 2. Panel (a) and (b): Observed absorption line profiles measured with para enriched H_2 used as precursor and 15 W of incident μW power. Other parameters are given in the legend. Panel (c): An example of evolution of ion temperature during the μW discharge.

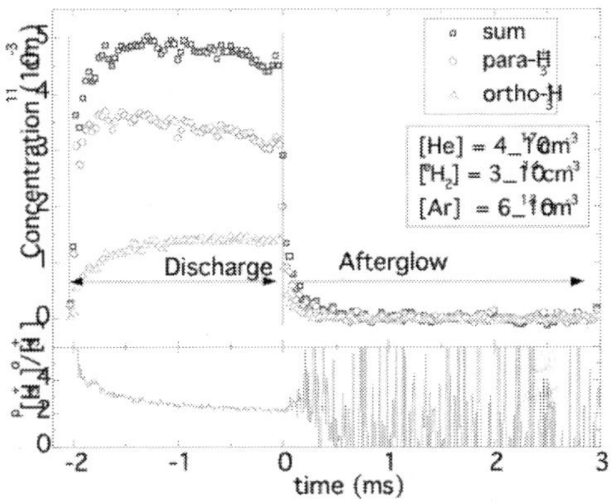

Figure 3. Upper panel: An example of the measured concentration evolutions during the whole discharge cycle using para enriched H_2 as precursor. Lower panel: Corresponding $[^pH_3^+]/[^oH_3^+]$ ratio.

To obtain decay curves of the two lowest rotational energy levels of H_3^+ we used a near infrared cavity ring-down spectrometer (NIR-CRDS) with synchronous detection capabilities. For each ring-down event the time of data acquisition start was recorded, relative to the discharge cycle. More detailed description of the used apparatus can be found in [28,29] and in references therein. Pressures from 200 Pa to 1000 Pa were used to make the Doppler broadening the main broadening mechanism, in order to extract ion temperature information from absorption line profiles. Para enriched hydrogen was created in line during the experiments, by passing the hydrogen through an Fe_2O_3 catalyst cooled to temperatures below 20 K.

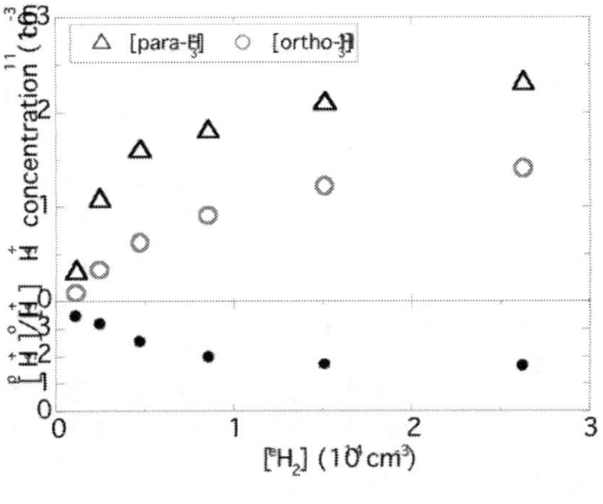

Figure 4. Upper panel: Example of $[^pH_3^+]$ and $[^oH_3^+]$ measured at the end of the discharge as a function of $^eH_2$ number density, with [He] ~2.7×10^{17} cm^{-3} and [Ar] ~ $[H_2]$. Lower panel: The corresponding dependence of the $[^pH_3^+]/[^oH_3^+]$ ratio on $[^eH_2]$.

3. Results

As mentioned in the introduction, all measurements were done on transitions from the two lowest ro-vibrational levels of H_3^+. That is for para-H_3^+ the transition $3\nu_2^1(2,0)\leftarrow0\nu_2^0(1,0)$ and $3\nu_2^1(2,1)\leftarrow0\nu_2^0(1,1)$ for ortho-H_3^+. Observed Doppler broadened absorption line profiles of H_3^+ indicate the ions temperature of (80±15) K during the discharge (see examples plotted in figures 2a and 2b). We have measured the line-widths for different time intervals during the discharge and the resulting ion temperature evolution (figure 2c) indicates that ions are not heated in the discharge. During the afterglow the concentration, and the corresponding absorption line intensity, falls down rapidly and obtaining a reliable temperature is difficult. Considering that already during the discharge the ion temperature is close to the buffer gas and wall temperature of 77 K, we can safely assume that it is not higher during the afterglow.

Since every ion undergoes several thousand collisions with He atoms prior to its recombination, we assumed that populations of higher rotational energy levels of both H_3^+ spin configurations are in thermal equilibrium in their own respective spin symmetry manifold. From observed absorption evolutions at line center, together with the information about ion temperature, the evolutions of [$^pH_3^+$] and [$^oH_3^+$] were calculated (see examples in figures 3,5,6). The overall [$^pH_3^+$]/[$^oH_3^+$] ratio is then calculated from the measured data for each set of experimental conditions.

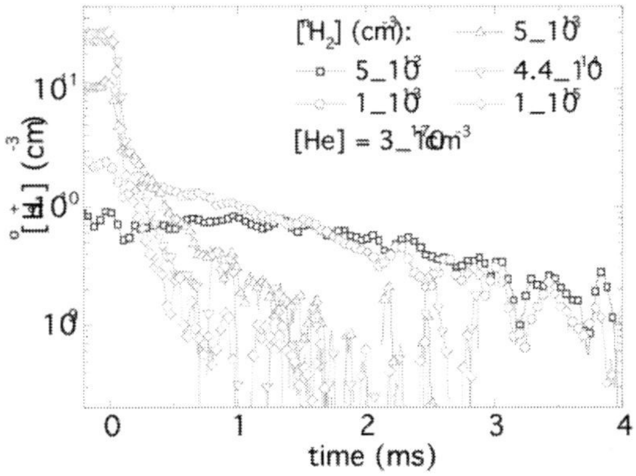

Figure 5. Examples of [$^oH_3^+$] decay curves measured at different [$^nH_2$] and at fixed [He] = 2.7×10^{17} cm^{-3}.

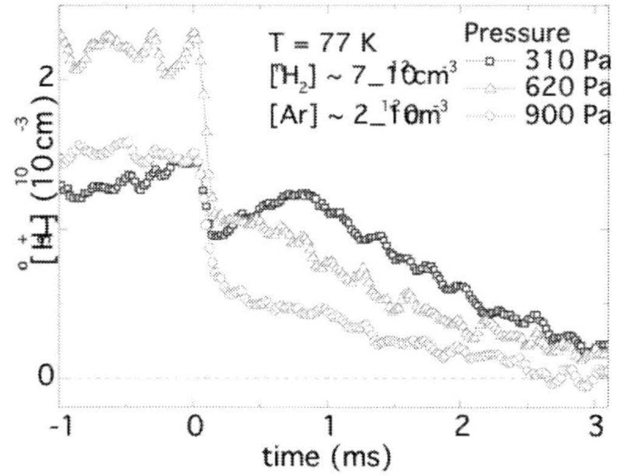

Figure 6. An example of [$^oH_3^+$] decay curves measured at different [He], and very low hydrogen density [$^nH_2$] = 7×10^{12} cm^{-3} and [Ar] = 2×10^{12} cm^{-3}.

Formation of H_3^+ during a μW discharge in gas mixture is a very complex process. The presence of all reactant gases in the discharge and afterglow calls for a very cautious setting of all experimental parameters in order to obtain reliable measurements from which the effective recombination rate coefficient can be determined. For reliable measurements we need at the beginning of the afterglow high plasma density $n_e \sim 10^{11}$ cm^{-3}. This leads in early afterglow to a plasma decay with time constant in the order of 0.1 ms. To exclude overlapping of H_3^+ formation with H_3^+ recombination, H_3^+ dominated afterglow has to be formed within 0.1 ms after discharge. To find the most suitable conditions for obtaining effective recombination rate coefficients, we have measured decay curves at a wide range of experimental conditions. The overall number density of H_3^+ ions formed in the discharge increases with increasing number density of H_2 ($^nH_2$ or $^eH_2$) (see figure 4). The kinetics of H_3^+ formation was discussed in many publications (see e.g. [30] and references therein) so we will not go to any details here. In present study we have used calculations to obtain optimal conditions for the experiment. In addition we measured [H_3^+] decay at a broad range of conditions to obtain dependences of [H_3^+] on helium and hydrogen densities. In figure 5, decay curves measured with several [H_2] are plotted. From the decay curves at H_2 concentrations lower than 5×10^{13} cm^{-3} (figure 5) one can clearly see that the H_3^+ formation process still continues in the afterglow and influences the shape of the decay curves thus making effective recombination rate coefficient extraction virtually impossible. Only at $[H_2] > 10^{14}$ cm^{-3} is the decay dominated by recombination and can be described by constant α_{eff} during the afterglow. The kinetic model is supporting the finding that the para/ortho H_3^+ conversion by reaction with H_2 is fast enough to keep the para/ortho H_3^+ ratio constant, even at high electron densities.

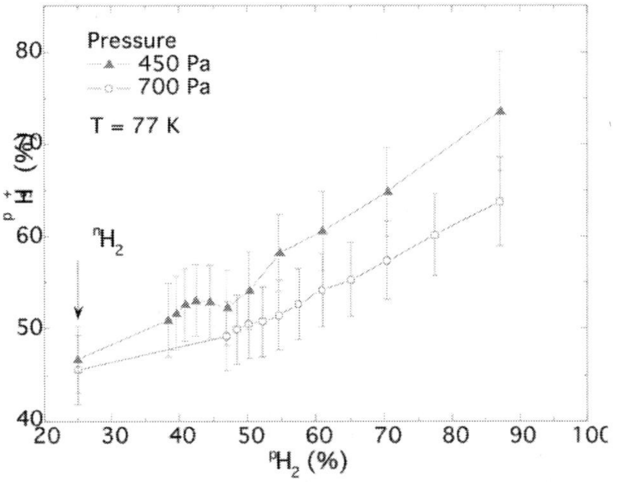

Figure 7. Dependence of the relative population of para-H_3^+ on the para enrichment of the precursor H_2. Measured at the end of the discharge where the ratio is constant (see figure 3). The arrow indicates data measured with normal hydrogen containing 25% of para-H_2 and 75% of ortho-H_2.

We stress here that, because of high electron number density, the plasma decay in present experiments is many times faster than decays observed in AISA and FALP experiments, where typically the first 10 ms of decays are excluded from evaluation of recombination rate coefficients [30,31].

We also studied dependence of decay in early afterglow on [He]. An example of data obtained at low $[H_2]$ and with several [He] are given in figure 6. Measurements at low $[H_2]$ ($\sim7\times10^{12}$ cm^{-3}), where the slow formation is most significant, show that formation of H_3^+ is faster at higher pressures. We stress again that these findings are not in contradiction with observations at the AISA apparatus where observation times were an order of magnitude longer.

Considering the observed influence of H_2 and He number densities on the formation of H_3^+, we adjusted both densities complementarily in order to obtained reliable decay curves for the determination of the effective recombination rate coefficients at various experimental conditions.

To change the ratio of $[^pH_3^+]/[^oH_3^+]$ in the discharge and afterglow $^eH_2$ has been used. The fraction of $^pH_2$ molecules after conversion by Fe_2O_3 catalyst has been determined by nuclear magnetic resonance to be 87±5%. Dependence of the $[^pH_3^+]/[^oH_3^+]$ ratio on the para enrichment of the precursor H_2 was measured at different conditions (see examples in figure 7). Final experiments with para enriched H_2 were performed with $[^pH_3^+]$ fraction of (68±5)%. In all measurements with $^nH_2$ the overall fraction was ~ 50%, independent of H_2, Ar and He number density (see figures 7,8).

For each set of experimental conditions densities of $[^pH_3^+]$ and $[^oH_3^+]$ were measured. Decay curves measured with $^eH_2$ show that the $[^pH_3^+]/[^oH_3^+]$ ratio is constant during the afterglow (see examples in figure 8). As was already mentioned, electron density decays were obtained by summing measured $[^pH_3^+]$ and $[^oH_3^+]$. The electron density decay curves were fitted to obtain the values of effective recombination rate coefficients α_{eff}. The α_{eff} obtained in experiments with $^nH_2$ and $^eH_2$, are plotted in figure 9. It is evident that the values of α_{eff} are higher for experiments with para enriched hydrogen. The measurements indicate very low values for the rate of the ternary helium assisted recombination of both para- and ortho-H_3^+ (figure 9). This is consistent with our previous experiments with normal H_2 [24]. Nevertheless, further measurements are needed to make quantitative conclusions about the particular values of both ternary rate coefficients. Extrapolating effective recombination rate coefficients linearly to zero He density, for both ${}^{e}H_2$ and ${}^{n}H_2$ measurements, and using the average fraction $[{}^{p}H_3^+]$ of 68% and 50%, for ${}^{e}H_2$ and ${}^{n}H_2$, respectively, we obtained the following binary recombination rate coefficients: for para-H_3^+, ${}^{p}\alpha_{bin} = (2.0\pm0.4)\times10^{-7}$ cm^3s^{-1}, and ortho-H_3^+, ${}^{o}\alpha_{bin} = (4\pm3)\times10^{-8}$ cm^3s^{-1}.

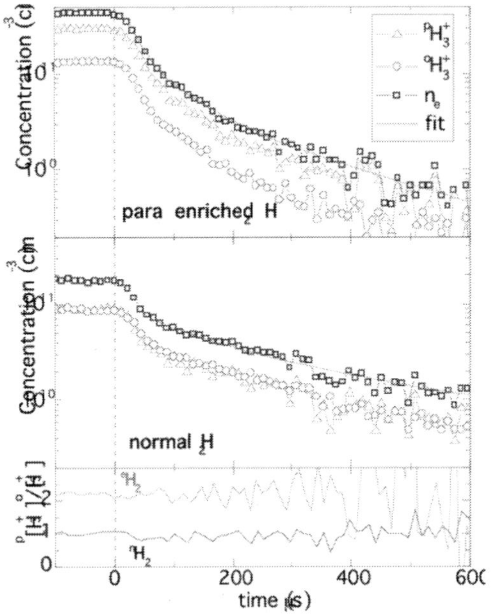

Figure 8. The measured decay curves of ${}^{p}H_3^+$ and ${}^{o}H_3^+$, using para enriched hydrogen (${}^{e}H_2$, top panel) and normal hydrogen (${}^{n}H_2$, middle panel). Obtained n_e in both experiments are also plotted. The variations of the ratio $[{}^{p}H_3^+]/[{}^{o}H_3^+]$ in experiments with ${}^{n}H_2$ and ${}^{e}H_2$ are plotted in lower panel.

4. Summary

Using NIR-CRDS we have measured the concentration evolution of para- and ortho-H_3^+ during the microwave discharge and afterglow in He/Ar/H_2 gas-mixture at various experimental conditions in order to determine optimal conditions for measurements of recombination rate coefficient of H_3^+ ions. The measurements with normal hydrogen and hydrogen with higher relative population of para-H_2 (${}^{n}H_2$ and ${}^{e}H_2$) showed that the $[{}^{p}H_3^+]/[{}^{o}H_3^+]$ ratio is higher when using para enriched hydrogen. By measuring α_{eff} at two different relative populations of para- and ortho-H_3^+ over a broad range of [He] we determined binary recombination rate coefficients for pure para-H_3^+ and pure ortho-H_3^+ at 77 K. The obtained values are in excellent agreement with theoretical predictions [6]. Our measurements show that at 77 K the binary recombination rate coefficient for para-H_3^+ is approximately 5 times larger than for ortho-H_3^+.

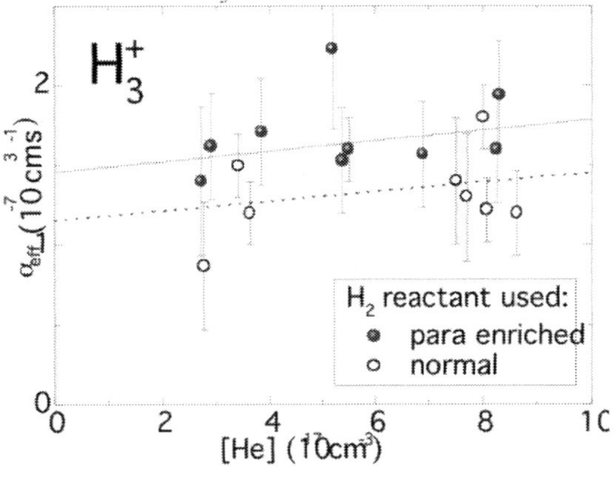

Figure 9. The effective recombination rate coefficients measured with normal and para enriched H_2 in He/Ar/H_2 afterglow experiment at 77 K. NIR-CRDS was used for *in-situ* measurements of para-H_3^+ and ortho-H_3^+ number densities.

This is the first time that the recombination rate coefficients for para-H_3^+ and ortho-H_3^+ were determined in experiment with direct *in situ* determination of densities of para-H_3^+ and ortho-H_3^+ ions. Further studies are in progress.

Acknowledgements
This work is a part of the research plan MSM 0021620834 and grant OC10046 financed by the Ministry of Education of the Czech Republic and was partly supported by GACR (202/07/0495,202/08/H057, 205/09/1183, 202/09/0642), by GAUK 92410, GAUK 86908, GAUK 54010 and by COST Action CM0805 (The Chemical Cosmos).

References

[1] Herbst E 2000 *Phil. Trans. R. Soc. Lond. A* **358** 2523
[2] Miller S *et al.* 2000 *Phil. Trans. R. Soc. Lond. A* **358** 2485
[3] Kokoouline V, Greene C H and Esry B 2001 *Nature* **412** 891
[4] Kokoouline V and Greene C H 2003 *Phys. Rev.* A **68** 012703
[5] dos Santos S, Kokoouline V and Greene C H 2007 *J. Chem. Phys.* **127** 124309
[6] Pagani L, Vastel C, Hugo E, Kokoouline V, Greene C H, Bacmann A, Bayet E, Ceccarelli C, Peng R and Schlemmer S 2009 *Astron. Astrophys.* **494** 623
[7] McCall B J *et al.* 2003 *Nature* **422** 500
[8] Kreckel H *et al.* 2005 *Phys. Rev. Lett.* **95** 263201
[9] Tom B A *et al.* 2009 J. *Chem. Phys.* **130** 031101
[10] Petrignani A *et al.* 2009 *J. Phys.: Conf. Ser.* **192** 012022
[11] Kreckel H *et al. Phys. Rev. A,* in print 2010
[12] Kreckel H *et al. present volume*
[13] Cordonnier M, Uy D, Dickson R M, Kerr K E, Zhang Y and T. Oka 2000 *J. Chem. Phys.* **113** 3181
[14] Gerlich D, Windisch F, Hlavenka P, Plasil R and Glosik J 2006 *Phil. Trans. R. Soc. Lond. A* **364** 3007
[15] Park K and Light J C 2007 *J. Chem. Phys.* **126** 044305
[16] Oka T 2004 *J. Mol. Spectrosc.* **228** 635
[17] Johnsen R 2005 *J. Phys.: Conf. Ser.* **4** 83
[18] Larsson M and Orel A E 2008 *Dissociative R ecombination of Molecular Ions* (Cambridge: Cambridge University Press)
[19] Varju J, Roučka Š, Kotrík T, Plašil R and Glosík J 2010 *J. Phys.: Conf. Ser.* **227** 012026
[20] Quack M 1977 *Mol. Phys.* **34** 477
[21] Hugo E, Asvany O and Schlemmer S 2009 *J. Chem. Phys.* **130** 164302
[22] Tom B A, Mills A A, Wiczer M B, Crabtree K N and McCall B J 2010 *J. Chem. Phys.* **132** 081103
[23] Glosik J, Korolov I, Plasil R, Novotny O, Kotrik T, Hlavenka P, Varju J, Mikhaylov I A, Kokoouline V and Greene C H 2008 *J. Phys. B: At. Mol. Opt. Phys.* **41** 191001
[24] Glosík J, Plašil R, Korolov I, Kotrík T, Novotný O, Hlavenka P, Dohnal P, Kokoouline V and Greene C H 2009 *Phys. Rev.* A **79** 052707
[25] Glosík J, Plasil R, Korolov I, Novotny O, Kotrik T 2009 *J. Phys.: Conf. Ser.* **192** 012005
[26] Kotrík T, Dohnal P, Korolov I, Plašil R, Roučka Š, Glosík J, Greene C H and Kokoouline V 2010 *J. Chem. Phys.* **133** 034305
[27] Glosik J, Plasil R, Kotrik T, Dohnal P, Varju J, Hejduk M, Korolov I, Roucka S and Kokoouline V 2010 *Molec. Phys.* **108** 2253
[28] Macko P, Báno G, Hlavenka P, Plašil R, Poterya V, Pysanenko A, Votava O, Johnsen R and Glosík J 2004 *Int. J. Mass Spectrom.* **233** 299
[29] Hlavenka P, Plašil R, Báno G, Korolov I, Gerlich D, Ramanlal J, Tennyson J and Glosík J 2006 *Int. J. Mass Spectrom.* **255** 170

[30] Plašil R, Glosík J, Poterya V, Kudrna P, Rusz J, Tichy M and Pysanenko A 2002 *Int. J. Mass Spectrom.* **218** 105
[31] Glosík J, Novotný O, Pysanenko A, Zakouřil P, Plašil R, Kudrna P, Poterya V 2003 *Plasma Source Science & Technology* **12(4)** S117

Discrepancies between H_3^+ recombination rate coefficients measured in ion storage rings, afterglows, and theory. Are they "real" or "apparent"?

Rainer Johnsen

University of Pittsburgh

E-mail: rj@pitt.edu

Abstract. While theoretical Jahn-Teller calculations of the binary dissociative recombination of H_3^+ ions with electrons agree quite well with the results obtained in ion storage rings, many plasma afterglow measurements have yielded either much lower (by factors of 10 or more) or higher values (by factors of 3 to 4). This paper examines possible causes of these apparent discrepancies. Numerical afterglow simulations indicate that the very low recombination coefficients obtained in afterglows at very low concentrations of neutral hydrogen were affected by the presence of slowly recombining ion species (Ar^+ and ArH^+) and do not really pose a challenge to the storage ring data and theoretical results. On the other hand, the faster recombination observed in afterglow plasmas seems to be due to third-body assisted mechanisms that occur in addition to binary recombination.

1. Introduction

Experimental and theoretical studies of the dissociative recombination of "simple" triatomic H_3^+ ions have a long and controversial history. The results obtained by the storage-ring method [1,2] now agree rather well (except for some finer details) with the theoretical results based on the Jahn-Teller mechanism [3] and earlier doubts about the process have largely disappeared. The binary rate coefficients inferred from ion storage-ring data can be expressed by the power law

$$\alpha(T_e)[cm^3/s] = -1.3\times10^{-8} + 1.27\times10^{-6}T_e^{-.48} \qquad (1)$$

as a function of the electron temperature T_e. If one accepts the 300 K storage-ring value ($\sim7\times10^{-8}$ cm^3/s) as a "benchmark", then the coefficients observed in many plasma afterglow measurements are either "too large" by factors of 2 to 3, or "too small" by factors of 10 and more. The question then arises if plasma recombination involves different or additional recombination mechanisms or if the afterglow measurements were simply "flawed". The resolution of these apparent discrepancies is the subject of this paper. I will first discuss the unusually small values.

2. Slow recombination at low hydrogen concentrations

An extensive series of afterglow measurements by the Prague experimenters [4] seems to show that the H_3^+ recombination rate drops far below the binary value (by a factor of 10 and more) when the hydrogen concentration in their experiments was reduced to below about 1×10^{12} cm^{-3} and similar results were obtained for D_3^+ ions [5]. Several suggestions have been made to rationalize these observations: Perhaps, the equilibrium between the ortho and para spin modifications of H_3^+ was not maintained at low H_2 concentrations and the low values are due to slowly recombining ortho H_3^+ ions. However, this explanation is incompatible with results of both theory and other experiments: The effect of the nuclear spin states is actually rather small, at least at 300 K. Likewise, tentative suggestions that the ions in those experiments were in some excited vibrational state that recombines slowly have been ruled out by recent theoretical calculations. Considerable latitude in devising H_3^+ recombination mechanisms existed before the storage ring data and theory became well established, but at this time it is difficult to accept recombination mechanisms in which the presence of hydrogen is necessary for recombination. The principal function of hydrogen in afterglow measurements is simply to provide the molecules needed to produce H_3^+ ions. But is it obvious that the H_3^+ ions in the cited measurements at low [H_2] were formed sufficiently fast to become the dominant recombining species? The H_3^+ ions were produced by the reaction scheme: $Ar^+ + H_2 \rightarrow ArH^+ + H$ followed by $ArH^+ + H_2 \rightarrow Ar + H_3^+$. The reactions are quite fast (rate coefficients near 10^{-9} cm^3/s) but it takes roughly 10 msec at [H_2] = 10^{11} cm^{-3} to produce H_3^+ ions, while recombination of an ion with $\alpha \sim 10^{-7}$ cm^3/s (at $n_e = 10^{10}$ cm^{-3}) proceeds at a time scale of $1/\alpha n_e$ =1 msec. In most other experiments [H_2] is made sufficiently large to form H_3^+ ions in a time short compared to the recombination time scale!

To examine the question in some more detail, I constructed a simple numerical model that simulates the afterglow processes and the methods of analysis, called the "advanced analysis" by the Prague group [6]. They determine the recombination coefficients from graphs of the quantity

$$-\left(\frac{1}{n_e^2}\frac{dn_e}{dt}+\frac{\nu_D}{n_e}\right) \qquad (2)$$

vs. $1/n_e$, using experimentally measured values of the electron densities n_e and the ambipolar diffusion loss rate ν_d. My numerical model shows that an input value of $\alpha = 10^{-7}$ cm^3/s leads to same inferred value only if [H_2] is somewhat larger than 10^{12} cm^{-3}. For [H_2] =10^{11}, the simulated data (see Figure 1) approach the input recombination rate asymptotically in the limit $1/n_e \rightarrow \infty$. If one extrapolates the data to $1/n_e = 0$, as the authors did (see e.g. [6]) for reasons that I do not fully understand, one obtains a much smaller, incorrect value. The model also returns the relative abundance of the ions as a function of time and one sees that non-recombining Ar^+ and ArH^+ decay fairly slowly and that H_3^+ does not become dominant until very late in the afterglow. If one carries out simulations for much smaller assumed H_3^+ recombination coefficients ($\alpha = 2\times10^{-8}$ cm^3/s) one finds that H_3^+ dominates much earlier. In that case the experiment would have worked correctly. However, an experiment that contradicts the larger ("default") values should be capable, in principle, to measure them! I should emphasize that the extensive data collected at higher [H_2] are not affected by such problems.

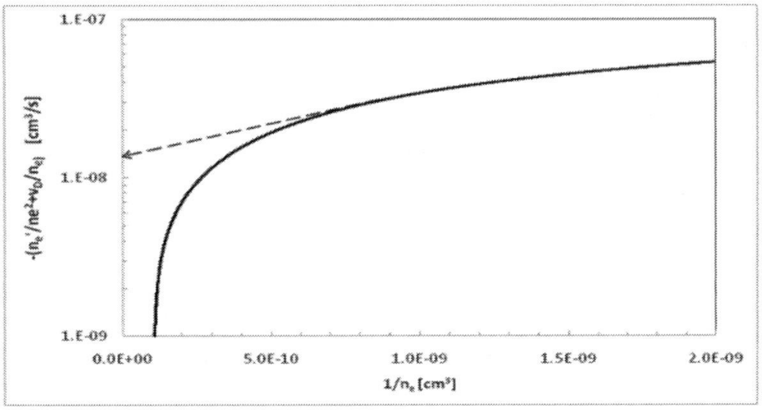

Figure 1 Simulation of an afterglow in an He/Ar/H_2 mixture for [H_2]= 1×10^{11} cm^{-3} for an H_3^+ recombination coefficient of 1×10^{-7} cm^3/s. Extrapolation to $1/n_e=0$ returns a far smaller recombination coefficient.

3. Third-body assisted recombination

Many afterglow experiments have yielded H_3^+ recombination coefficients that are considerably larger than the binary value obtained in storage ring experiments. A nearly complete compilation for data obtained in helium/argon/hydrogen mixtures can be found in the paper by Glosik et al [4] and these observations strongly suggest some form of third-body assisted recombination. The well-known collisional-radiative mechanisms for atomic ions in which either electrons or neutral particles stabilize the recombination are definitely too slow (by about two orders of magnitude) to explain observations. Several ideas for more efficient three-body recombination have been proposed:

One might assume that the scattering resonances in electron/molecular-ion collisions have sufficiently long lifetimes to permit interactions with third particles (atoms, molecules, electrons) which reduce the probability of releasing the electron by autoionization. Gougousi et al [7] proposed that vibrational resonances might live long enough to be stabilized by angular-momentum changing (*l*-mixing) collisions and reactions with ambient H_2 molecules. In hindsight, the lifetimes used in their estimates were unrealistically long. Glosik et al [4] devised a mechanism in which rotational resonances undergo l-mixing in collisions with ambient He atoms and form a population of long-lived Rydberg molecules with n>40. This is an improvement over the earlier model, in as much as it uses more realistic, theoretically calculated lifetimes, but I think the model has several shortcomings: The *l*-mixing efficiency was taken as independent of principal quantum number n instead of declining as $1/n^{2.7}$, as indicated by theory [8]. The final stabilizing process of the Rydberg molecules remains (a) unspecified and (b) it would have to be competitive with collisional reionization by electrons which is extremely fast for high Rydbergs.

I have explored a third possible three-body mechanism that is loosely based on the collisional dissociative mechanism of Collins [9]. I assume, as is done in the theory of collisional radiative recombination, that high Rydberg states with energies down to about -4kT below the ionization limit, formed by three-body collisions and destroyed by its inverse (collisional ionization),

$$e^- + H_3^+ + M \leftrightarrow H_3^*(n) + M \tag{3}$$

are in thermal Saha equilibrium. i.e.

$$\frac{[H_3^*(n)]}{[H_3^+]n_e} = K(n) = n^2\lambda_{th}^3 e^{E_n/kT} \tag{4}$$

where λ_{th} denotes thermal deBroglie wavelength $\lambda_{th} = (h^2/(2\pi m_e kT)^{1/2}$ (5)

Most of the these states will have high electronic angular momenta l which makes predissociation unlikely, but l-mixing due to electrons and/or atoms can reduce l to small values (I took $l=0$ in my estimate) and induce predissociation. Note that the l-mixing here goes from high l to low l. If states in a range of n-values are irreversibly destroyed with frequency ν_s, the recombination rate will exceed the binary value by

$$\Delta\alpha = \sum_{n_{min}}^{n_{max}} K(n)\nu_s(n) \tag{6}$$

The *l*-mixing cross section due to the electrons [10] and the corresponding rate coefficient rise sharply with increasing *n*

$$\sigma_{e,mix} = 4.4\times\pi a_0^2 n^5 \qquad k_{e,mix} = v_e\sigma_{e,mix} \tag{7}$$

while the l-mixing rate coefficient due to helium [8] falls off $n^{-2.7}$

$$k_{mix,He} = 3.1\times10^{-5}\frac{1}{n^{2.7}}[cm^3/s] \tag{8}$$

Both rates have to be divided by (n^2-1) to take into account that one needs the rates from a given *l* state to *l*=0. A further consideration shows that *l*-mixing will cease to be the rate limiting process for *n* greater than about 40 because predissociation decrease with *n* as approximately $1/n^3$.
Estimates based on this model indicate that He atoms (the carrier gas in the experiments) can have a significant effect. Fairly low Rydberg states with *n* from 10 to 40 are more important than higher states. The estimated three-body rate coefficient due to helium at T= 300 K comes out to be 2.6×10^{-25} [cm^6/s] which agrees with the experimental value [4] of $(2.5\pm1.2)\times10^{-25}$ [cm^6/s]). The close agreement is almost certainly fortuitous, but it is encouraging. More detailed work is clearly needed.

4. Conclusion

I have serious doubts about the validity of the very low H_3^+ recombination coefficients that were inferred from some afterglow measurements at low [H_2]. Perhaps I misunderstand some aspects of the experiments, but results that contradict many other findings require stronger substantiation than has been provided. There is good experimental evidence that three-body assisted recombination of H_3^+ occurs in plasma afterglows in addition to binary recombination. The theoretical models that have been proposed to rationalize these observations are still rather tentative, but they may serve as points of departure for more accurate treatments.

Acknowledgement:
The author gratefully acknowledges travel support from the US National Science Foundation and NASA.

References

[1] McCall B, Huneycutt J, Saykally R J, Djuric, N, Dunn, G, Semaniak J, Novotny O, Al-Khalili A, Ehlerding A, Hellberg F, Kalhori, S, Neau, A, Thomas R D, Paal A, Österdahl F, and Larsson M 2004, *Phys. Rev. A* **70** 052716

[2] Kreckel H, Motsch M, Mikosch J, Glosík J, Plašil R, Altevogt S., Andrianarijaona V, Buhr H, Hoffmann J, Lammich L, Lestinsky M, Nevo I, Novotny S, Orlov D A, Pedersen HB, Sprenger F, Terekhov A S, Toker J, Wester R, Gerlich D, Schwalm D, Wolf A and Zajfman D 2005 *Phys. Rev. Lett.* **95** 1

[3] Kokoouline V and Greene C H 2003, *Phys. Rev. A* **68** 012703.

[4] Glosík J, Plašil R, Korolov I, Kotrík T, Novotný O, Hlavenka, P, Dohnal P, Varju J , Kokoouline V, and Greene C H 2009 , *Phys. Rev. A* **79** 052707

[5] Glosík J, Korolov I, Plašil R, Kotrík T, Dohnal P, T, Novotný O, Hlavenka, P, Varju, J, Roučka Š, Greene C H and Kokoouline, V 2009, *Phys. Rev* A **80** 042706

[6] Plašil R, Glosík J, Poterya V, Kudrna P, Rusz J, Tichý M and Pysanenko A. 2002 *Int. J. Mass Spectr.* **218** 105

[7] Gougousi T, Johnsen R and Golde M F 1995, *Int. J. Mass Spectr. Ion Proc.*, **149-150** 131

[8] Hickman A P 1978 *Phys. Rev. A* **18,** 1339

[9] Collins C B 1965, "Collisional-dissociative recombination of electrons with molecular ions", *Phys. Rev.* **140** A1850

[10] Dutta S K, Feldbaum D, Walz-Flannigan A, Guest J R and Raithel G 2001, *Phys. Rev. Lett.* **86** 3993

Journal of Physics: Conference Series **300** (2011) 012025 doi:10.1088/1742-6596/300/1/012025

Dissociative Recombination Data Needs for the Aeronomy Community

Jane L Fox

Department of Physics, Wright State University, Dayton, OH 45435

E-mail: jane.fox@wright.edu

Abstract.
We discuss here the dissociative recombination (DR) reactions that are of importance to various processes in the thermospheres/ionosphere of the earth and planets, the data required to model these processes, and the desired format for these data. DR may be important for the chemistry of thermospheres/ionosphere, both as a loss process for ions and a production mechanism for minor neutrals. These neutrals may be metastable species that can participate in reactions not available to ground state species, or they may radiate, producing airglow or cooling. Because DR reactions are usually highly exothermic, they can also contribute to neutral heating. If the DR reactions are produced high enough in the atmosphere of a body with a small gravity well, the energetic neutrals formed can contribute to escape of species from the atmosphere, to the isotope fractionation of the escaping atoms, and thus to atmospheric evolution.

1. Introduction

The thermosphere and exosphere are the outermost regions of a planetary atmosphere. In the thermosphere, the neutral temperature increases with altitude above the mesopause, usually the coldest region of a planetary atmosphere, and eventually reaches a constant value, the exospheric temperature T_∞. The lower E and the upper F region above it are usually superimposed on the thermosphere and exosphere, and are characterized by the presence of simple (non-cluster) positive ions and free electrons. The F and E regions are formed by ionization of the constituents by photons in the main part of the solar EUV (100-1000Å), and by soft x-rays (10-100 Å), respectively. The D-region is below the E region, and its origin is absorption of hard x-rays, those with wavelengths less than 10 Å, and, on the Earth, by photoionization of NO by the solar Lyman-α line, which penetrates to low altitudes because of an accidental "window" in the Schumann-Runge continuum of O_2. The terrestrial D region is formed also by the absorption of cosmic rays. Where three-body collisions are important, cluster ions are formed, and free electrons are replaced by negative ions. At low altitudes, in addition to forming cluster ions, photoionization or charge transfer to meteoric atoms (e.g., Si, Fe, Mg, Ca, Na), when present, can take place (e.g., Plane [1]; Grebowsky et al. [2])

In the E and F_1 regions, the major ions are molecular, and are destroyed by mainly by DR. For the minor molecular ions, ion-molecule reactions compete with DR as a loss process. In both regions photochemical equilibrium prevails, which means the direct and chemical production rates of the ions are equal to the local chemical loss rates. Above the F_1 region is the F_2 region, where the major ions are atomic. Above the F_2 peak, the major atomic ion loss process is downward transport, and below the peak the atomic ions are destroyed in ion-molecule

Journal of Physics: Conference Series **300** (2011) 012025 doi:10.1088/1742-6596/300/1/012025

reactions. Molecular ions in this region, however, may be destroyed mainly by DR. Figure 1 from Bauer [3] shows the regions of the terrestrial ionosphere, and the major sources of ions in the various regions. These regions, while present on the dayside of all planets or satellites that have sufficiently dense atmospheres, may be smaller or larger, more or less important, or even "invisible" to some extent.

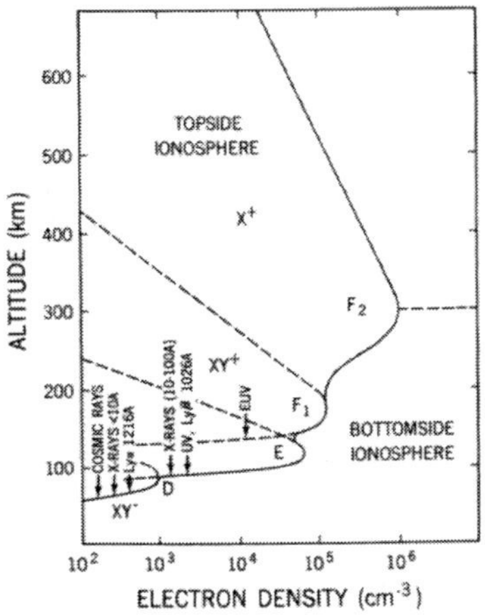

Figure 1. Electron density as a function of altitude that illustrates the regions of a planetary ionosphere by reference to the terrestrial ionosphere. The major sources of ions are also shown. Figure 48 from Bauer [3], reprinted with kind permission from Springer Science+Business Media.

In addition to its role as a loss process for molecular ions, DR is important in the E and F-regions of planetary ionospheres for several reasons, including as a production mechanism for minor neutral species. Because of the large exothermicity of simple DR reactions, the neutral atoms and fragments produced may be electronically excited; these species may be metastable, and go on to participate in reactions that are not available to ground state species; they may also radiate, producing airglow emissions and cooling. Because the products of DR reactions are neutral fragments, DR may also contribute to neutral heating to a greater extent than ion-molecule reactions, for which the exothermicity heats both the ions and the neutral products. We have discussed the role of DR in thermospheric heating in detail previously (Fox [4]), so here we confine our comments to the data needed to determine it. When DR occurs in the upper thermospheres or exospheres of planets or satellites with small gravity wells, such as Mars or Titan (but not Earth or Venus), the energetic neutrals produced in DR may escape from the atmosphere, thus contributing to atmospheric evolution. In addition, there is usually an isotope fractionation factor inherent in the DR mechanism, which operates in addition to that produced by the different altitude distributions of the molecular ion isotopomers (e.g., Fox and Hać [5]). Escape of atoms and isotope fractionation operating over the last 3.8 Gyr (the non-catastrophic period of planetary evolution) may be expected to cause the isotope ratios in these atmospheres to evolve so that they are currently different from the terrestrial values. We describe here the data needs to describe these processes and the format that is suitable for modeling.

2. Identifying ions for which DR is important for studies of ion chemistry

In oxidizing ionospheres, such as those of Venus, Earth and Mars, DR is important as a loss process mainly for the "terminal ions", which are, in general, those whose parent neutrals have low ionization potentials (IP's). Table 1 shows some IP's for atmospheric species. In most ion-molecule reactions ionization flows from species whose parent neutrals have high ionization potentials to those whose parent neutrals have low ionization potentials. This is just a simple restatement of the requirement that reactions must be exothermic, and is rigorously true only for charge transfer reactions. Figure 2 illustrates a portion of the ion chemistry of oxidizing atmospheres, where the neutral species are at the top and the ions, labeled and ordered by the IP of the parent neutral, are below. The ion-neutral reactions that interchange the ions are illustrated by the arrows. Most of the arrows in this diagram point downward, although a few reactions do not follow this "rule", such as the reaction of C^+ (IP_C=11.26 eV) with CO_2 to produce CO^+ (IP_{CO}=14.01 eV):

$$C^+ + CO_2 \rightarrow CO^+ + CO + 2.90\,\text{eV}. \tag{1}$$

This reaction is exothermic because of the very large binding energy of CO, 11.11 eV.

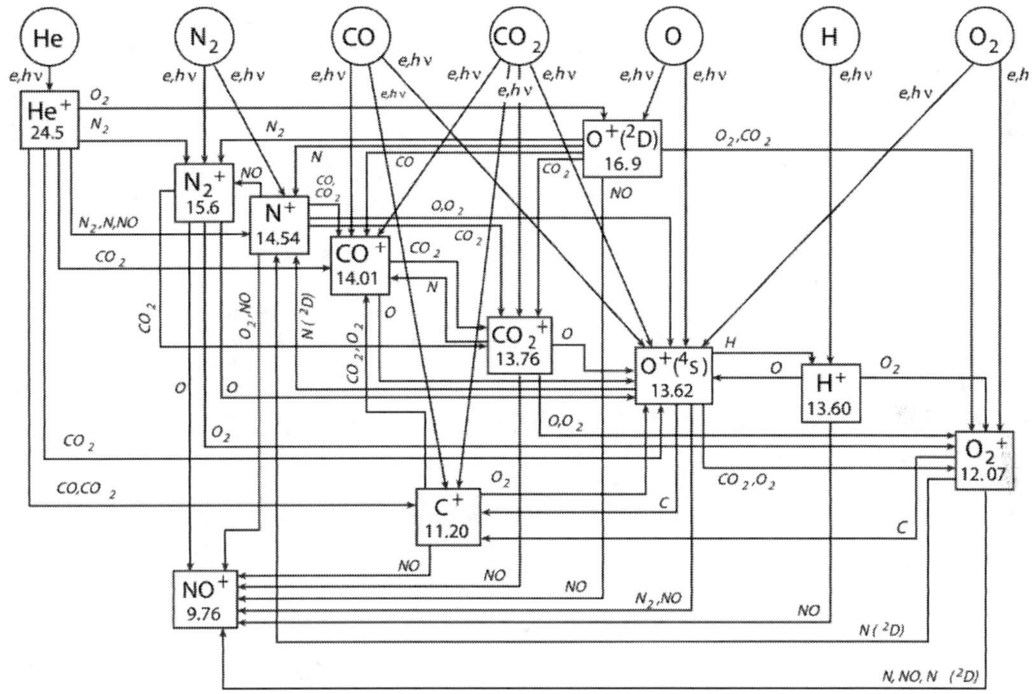

Figure 2. Summary of part of the ion chemistry on the terrestrial planets. The species in the top row enclosed by circles are the neutral species from which ions in the squares are formed. Each ion is labeled by the IP of its parent neutral, and its vertical position indicates the relative values of its IP. In the presence of sufficient neutral densities the ions are transformed from those whose parent neutrals have high IP's (near the top) to those whose parent neutrals have low IP's (near the bottom). Therefore the ionization tends to flow downward.

Thus, although there are many production mechanisms for NO^+ (IP_{NO}=9.76 eV) and O_2^+ (IP_{O_2}=12.07 eV) in the ionospheres of Venus, Earth and Mars (hereafter the terrestrial planets), there are few loss processes. Where collisions are sufficient, the terminal molecular ions are expected to be O_2^+ and NO^+, which are then destroyed mostly by DR. In fact, loss of NO^+ in the E and F regions is *only* by DR, although in the lower D regions, where the neutral number densities are large enough, NO^+ may form cluster ions, for which the DR rate coefficients may be necessary, by three-body recombination, or charge transfer to metal atoms (e.g., Mg, Fe, Ca, Na) and Si that are produced by ablation of meteors. The latter process usually occurs in a very narrow region, about 10 km in width, at pressures greater than 1μbar for the terrestrial planets. Table 1 shows that these species have very low IP's.

Although O_2 is abundant in the terrestrial lower thermosphere, we note here that the presence of O_2^+ is not evidence for the presence of large abundances of O_2; nor is the presence of NO^+ evidence for large abundances of NO. Both neutral species are present in the thermospheres of Mars and Venus in only trace amounts; the major neutral species are CO_2 at low altitudes and O at high altitudes. High solar activity models of the Martian thermosphere and ionosphere are shown in Figure 3.

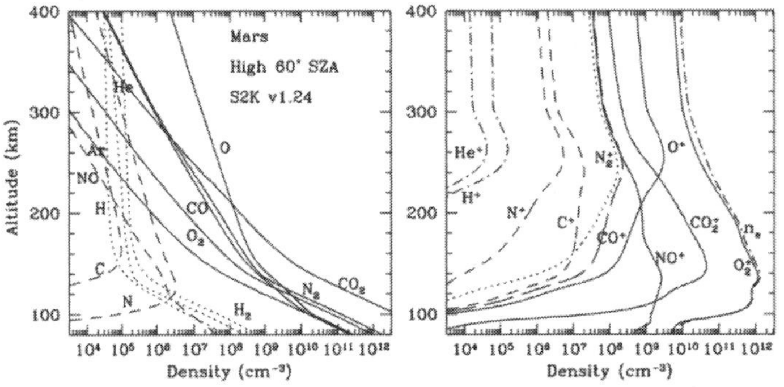

Figure 3. High solar activity models of the Martian thermosphere/ionosphere for 60° solar zenith angle. (left) Density profiles of the neutral species as a function of altitude included in the model. (right) Density profiles of important ions over the same altitude range.

The major sources of O_2^+ in the ionospheres of Venus and Mars are the reactions

$$CO_2^+ + O \rightarrow O_2^+ + CO \tag{2}$$

and

$$O^+ + CO_2 \rightarrow O_2^+ + CO. \tag{3}$$

Charge transfer reactions of ions to O_2 are negligible as sources of O_2^+. The major source of NO^+ on Venus and Mars is

$$O_2^+ + N \rightarrow NO^+ + O, \tag{4}$$

whereas the major source on Earth is

$$N_2^+ + O \rightarrow NO^+ + N \tag{5}$$

along with a few other ion-molecule reactions. Charge transfer reactions of ions to NO make up less than 10% of the production of NO^+ on Mars. Thus, even if we neglected the ambient NO and O_2 densities and charge transfer reactions for these two species, we would still predict fundamentally the same density profiles for O_2^+ and NO^+.

As shown in Table 1, the IP's of N_2 and H_2 are large, 15.58 and 15.43 eV, respectively. This means that there are few production mechanisms for N_2^+, but many loss mechanisms. DR is not important as a loss mechanism for N_2^+ or H_2^+ in the major regions of ionospheres, although it may be important at high altitudes. For example, DR becomes important as a loss process in the terrestrial ionosphere for N_2^+ only near and above an altitude of about 240 km, which is well above the F_1 peak, which appears at ~170 km (e.g., Dalgarno and Fox [6]). For the Mars model shown in Figure 3, the loss of N_2^+ by DR becomes important as a chemical loss mechanism in the region above ~ 300 km, but it is never dominant.

Table 1. Ionization Potentials of Some Species Relevant to Planetary Atmospheres[a]

High I_P Species	I_P (eV)	Medium I_P Species	I_P(eV)	Ionized by Ly α Species	I_P(eV)
He	24.59	H_2O	12.61	C_4H_2	10.18
Ne	21.56	CH_4	12.51	CH_3	9.84
Ar	15.76	SO_2	12.32	C_3H_6	9.73
N_2	15.58	CH_3CN	12.194	NO	9.264
H_2	15.43	C_2	12.11	C_6H_6	9.246
N	14.53	O_2	12.07	C_2H_3	8.9
CO	14.01	C_2H	11.7	Si	8.152
CO_2	13.77	HC_3N	11.64	C_2H_5	8.13
O	13.618	C_2H_6	11.52	HCO	8.10
H	13.598	C_2H_2	11.40	N_2H_4	8.10
HCN	13.60	C	11.26	C_3H_7	8.09
OH	13.00	C_3H_8	10.95	Fe	7.87
		CH	10.64	Mg	7.65
		C_2H_4	10.51	trans-HCNH	7.0[b]
		H_2S	10.45	cis-HCNH	6.8[b]
		CH_2	10.396	c-C_3H_3	6.6
		S	10.35	Ca	6.11
				Na	5.139

[a] Computed with data from Lias et al. [8], except as noted.
[b] From Nesbitt et al. [9]

Titan has an atmosphere of intermediate oxidation state, and the outer planets, Jupiter, Saturn, Uranus, and Neptune all have reducing atmospheres. In these atmospheres, the terminal ions are generally those whose parent neutrals have large proton affinities (PA's). Some PA's are shown in Table 2, and a large compendium of PA's can be found in Hunter and Lias [7]; see also Lias et al. [8]). Proton transfer reactions tend to be fast, near gas-kinetic (e.g., Su and Bowers [10], and many (but not all) of the ion-neutral reactions in reducing atmospheres are proton transfer reactions from a species whose parent neutral has a lower PA to another species whose parent neutral has a higher PA. Just as for ionization, the route to the protonated species need not be a proton transfer reaction. Therefore, the existence of a protonated neutral does not imply the existence of a large abundance of the parent neutral in the planetary thermosphere.

In the thermosphere of Titan (above about 1000 km), the major species is N_2 with a small admixture of ~ 2.2 % CH_4 and 0.4 %H_2 (e.g., Cravens et al. [11] and the references therein). The formation of ions in the ionosphere of Titan is initiated by photoionization and/or energetic electron-impact ionization of the major species. The energetic electrons may either be photoelectrons, or they may originate from the magnetosphere of Saturn. The dominant ions produced are N_2^+ and N^+, with small amounts of CH_4^+ and ion fragments produced from dissociative ionization of methane. In regions where the neutral densities are large, near the F_1 ion peak, the major ions produced are transformed by dissociative charge transfer and H atom transfer as well as proton transfer. For example, the following reactions have been found to have large rate coefficients:

$$N_2^+ + CH_4 \rightarrow CH_2^+ + N_2 + H_2, \quad (6a)$$
$$\rightarrow CH_3^+ + N_2 + H, \quad (6b)$$
$$\rightarrow N_2H^+ + CH_3, \quad (6c)$$

where (6*b*) is the dominant channel (McEwan et al. [12]). Because the proton affinity of CH_2 is about 8.56 eV, and that of CH_4 is 5.63 eV, the proton transfer reaction from CH_3^+ to CH_4, is highly endothermic. The reaction:

$$CH_3^+ + CH_4 \rightarrow C_2H_5^+ + H_2 \quad (7)$$

is, however, fast and exothermic and results in the formation of a higher hydrocarbon ion [13]. Proton transfer reactions transform ions within groups but will not significantly increase the size of an ion. Such reactions are, of course, also limited by the availability of the neutrals to which the proton can transfer. The resulting mix of ions is, however, dominated by ions with high proton affinities.

Figure 4 shows a mass spectrum from near the peak of the Titan ionosphere recorded by the ion and neutral mass spectrometer (INMS) on the Cassini Huygens probe for three altitudes (Cravens et al. [14]). The mass/charge ratios range from 1 (H^+) to 99.5 (probably $C_8H_3^+$). The limit of the INMS instrument is 99.5 Daltons, and larger ions are present, which have been measured up to 200 Daltons by the Cassini Plasma Spectrometer (Crary et al. [15]). The ions appear to be clustered around masses that are separated by 12–14 Daltons, which represents the addition of a C or N to species in the previous cluster. In general PA's increase with the size of the parent neutral, although the increase is not monotonic. The addition of an N atom seems to increase the PA of a hydrocarbon/nitrile molecule. The most important ions for which DR is important as a loss mechanism are the "terminal" ions in each mass group. The terminal $C_iN_jH_k^+$ species, where i, j, and k are integers, that are observed appear to be those in which the parent species is a moderately unsaturated stable molecule. The DR of these large ions could be the source of large neutral hydrocarbons and nitriles that eventually form the ever-present hazes and "soot" in the lower atmosphere.

In the atmospheres of Titan and the outer planets, and at high altitudes on the terrestrial planets, H_2^+ may be formed by ionization of H_2 or, theoretically, by protonation of H. H, however, has a very small PA (2.69 eV), and there are few species with smaller PA's (Table 2). Moreover, H_2 has a large IP. H_2^+ therefore reacts with almost any species with which it collides, including H_2 and H in the upper thermosphere, and with many other species present at lower altitudes. It is therefore never expected to be a terminal ion in planetary atmospheres, although Lundin et al. [16] have identified the mass-2 ion in outflow from Mars with the European Mars Express Ion Mass Analyzer as H_2^+.

The data that are needed to understand the loss of ions and the production of minor neutrals in the E and F_1 regions of ionospheres are the DR coefficients for the predicted terminal ions as a function of electron temperature, where T_e ranges from $\sim$ 100 to 5000 K. The ion temperature (T_i) is usually also elevated above that of the neutral temperature in the region above the peak, but its value is generally less than T_e. There is, of course an exception to this rule: on the nightside of Venus, anti-sunward flowing ions are accelerated by the plasma gradient force to supersonic speeds and form a shock near 150° solar zenith angle, which elevates T_i above that of T_e (e.g., Miller and Whitten [17]).

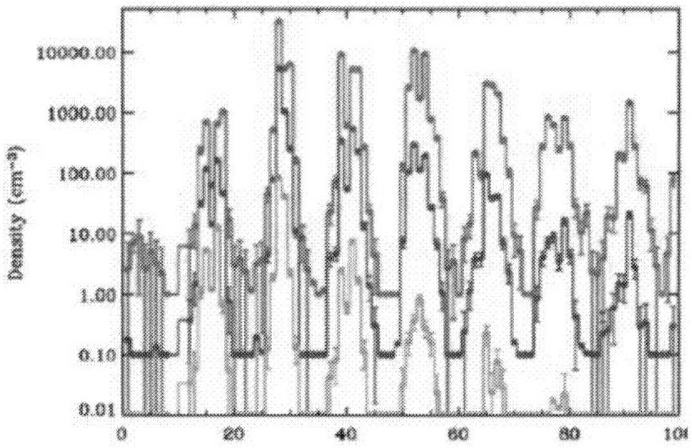

Figure 4. Measured mass spectra for 3 altitude intervals of Titan. Red: 1027–1200 km, multiplied by a factor of 100. Blue: 1200–1400 km, multiplied by a factor of 10. Green:1400–1600 km, multiplied by a factor of 1. From Cravens et al. [14].

In order to model the formation of minor neutrals in ionospheric DR reactions, a different branching ratio for each channel is required. This is because each channel is inserted into a model as a separate reaction. Total yields of the atomic or fragment species produced in DR of a molecular ion are interesting quantities, but are useless for modeling.

Electrons are replaced by negative ions in the lower ionospheres of Earth and Titan. On the Earth, the transition from electrons to negative ions appears at about 75–80 km. (e.g., Narcisi et al., [18]). On Titan, near 1000 km, the masses of the negative ions are found to range from 6 to 5000 Daltons (e.g., Vuitton et al. [19]). In order to model these regions, recombination mechanisms for negative ions, such as mutual neutralization, or associative detachment are required. The terminal ions are, of course, those with high electron affinities.

3. Neutral heating

Since we have discussed thermospheric heating due to DR reactions in a previous DR monograph (Fox [4]), we will confine our remarks to those here. DR reactions tend to be highly exothermic, especially those of diatomic ions. The exothermicity is deposited in the neutral atomic products as electronic and translational energy. For polyatomic molecular ions, the molecular fragments produced may also be vibrationally excited. Although these fragments may be excited rotationally as well, where there are sufficient collisions, rotational excitation is more readily transformed into heat than is vibrational excitation. The DR coefficients for the largest of these ions at room temperature are of the order of a few $\times 10^{-6}\,\mathrm{cm^3\,s^{-1}}$, about an order of magnitude larger than those of smaller ions (e.g., McLain et al. [20]; see also the review by Florescu-Mitchell and Mitchell [21]). In order to determine the heating rate due to DR reactions, in addition to DR coefficients for the various product channels as a function of T_e, we need measurements (or even estimates!) of the energy that is deposited as electronic, vibrational, rotational and translational excitation of molecular fragments.

4. The production of metastable and radiating states of fragments in DR

Electronically excited states of the atoms or fragments produced in DR may radiate to lower energy levels producing airglow emissions, or survive to participate in reactions that are not available to ground state species. A familiar example of atomic products for which both phenomena are important is the DR of O_2^+, which may proceed via 4 different channels, with

Eighth International Conference on Dissociative Recombination (DR2010) IOP Publishing
Journal of Physics: Conference Series **300** (2011) 012025 doi:10.1088/1742-6596/300/1/012025

Table 2. Selected Proton Affinities (eV)

Neutral Species	Ion Produced	Proton Affinity	Neutral Species	Ion Produced	Proton Affinity
He	HeH^+	1.84	H	H_2^+	2.69
N	NH^+	3.37	Ar	ArH^+	3.82
H_2	H_3^+	4.39	O_2	HO_2^+	4.40
CO	COH^+	4.74	O	OH^+	5.04
N_2	N_2H^+	5.13	CN	HCN^+	5.36
CH_3	CH_4^+	5.62	CH_4	CH_5^+	5.63
CO_2	CO_2H^+	5.67	NO	HNO^+	5.69
C_2N	C_2NH^+	5.75	CN	HNC^+	5.80
OH	H_2O^+	6.16	CO	HCO^+	6.16
C_2H_6	$C_2H_7^+$	6.23	C_2H_5	$C_2H_6^+$	6.42
C	CH^+	6.46	C_3H_8	$C_3H_9^+$	6.50
NH	NH_2^+	6.62	C_2H_2	$C_2H_3^+$	6.69
C_2	C_2H^+	6.92	C_3H_7	$C_3H_8^+$	7.03
C_2H_4	$C_2H_5^+$	7.05	H_2O	H_3O^+	7.24
HCN	$HCNH^+$	7.46	C_2H_3	$C_2H_4^+$	7.56
HN_3	$H_2N_3^+$	7.76	CH_3CCH	$CH_3CCH_2^+$	7.78
C_4H_2	$C_4H_3^+$	7.79	c-C_3H_6	(c-C_3H_6)H^+	7.80
C_3N_2	$C_3N_2H^+$	7.81	C_2H	$C_2H_2^+$	7.95
NH_2	NH_3^+	8.13	HNC	$HCNH^+$	8.14
CH_3CN	CH_3CNH^+	8.18	CH_2CHCN	CH_2CHCNH^+	8.23
CH_2	CH_3^+	8.56	NH_3	NH_4^+	8.85
N_2H_4	$N_2H_5^+$	8.87	CH_3NNCH_3	$CH_3NHNCH_3{}^+$	8.97
CH_3CHNH	$CH_3CHNH_2^+$	9.15	CH_2CHNH_2	$CH_3CHNH_2^+$	9.37
CH_2NCH_3	$CH_2NHCH_3^+$	9.41			

[a] Computed from data taken from Hunter and Lias [7].

exothermicities in the center of mass frame as shown below:

$$O_2^+(v,J) + e \rightarrow O(^3P) + O(^3P) + 6.99\,\text{eV} \quad (8a)$$
$$\rightarrow O(^1D) + O(^3P) + 5.02\,\text{eV} \quad (8b)$$
$$\rightarrow O(^1D) + O(^1D) + 3.06\,\text{eV} \quad (8c)$$
$$\rightarrow O(^1D) + O(^1S) + 0.835\,\text{eV} \quad (8d)$$

The exothermicities shown are for O_2^+ in the ground electronic, vibrational and rotational states, at zero collision energies, and they do not include the fine structure levels of the product $O(^3P)$ atoms. The DR of O_2^+ thus produces $O(^1D)$, which may radiate, giving rise to the visible O ($^1D \rightarrow ^3P$) red doublet at 6300, 6364 Å with a lifetime of $\sim$ 120 s. Since the species is metastable, it may survive to participate in reactions that are not available to ground state $O(^3P)$, and it is also easily quenched. The $O(^1S)$ state has a radiative lifetime of less than 1 s, and is more likely to radiate to either the $O(^1D)$ state, producing the 5577 Å green line, or to the ground $O(^3P)$ state, producing the 2972 Å emission line. The ratio of these lines is determined by the ratio of the transition probabilities, which has been calculated to be about 16 (e.g., Froese-Fischer and Tachiev [22]). This ratio, however, has been questioned by several investigators who have measured the emissions in the airglow and aurorae, and a ratio of $9-10$ has been proposed (e.g., Slanger et al. [23] and references therein). The green line has been detected in the nightglow of Venus (Slanger et al. [24]), but the identification of the mechanism for its production is not yet secure. Possible sources of $O(^1S)$ in the nightglows of the terrestrial planets include not only DR of O_2^+ (the ionospheric source) but also a mesospheric source that begins with the three body recombination of O atoms to produce O_2 in an excited state (denoted by an asterisk)

$$O + O + M \rightarrow O_2^* + M, \tag{9}$$

followed by excitation transfer to O

$$O_2^* + O(^3P) \rightarrow O_2 + O(^1S) \tag{10}$$

in the so-called Barth mechanism [25].

The branching ratios for the various channels of O_2^+ DR have been measured by Kella et al. [26] for a relaxed vibrational distribution and for an extended vibrational distribution. Peverall et al. [27] measured the DR coefficient and branching ratios for DR of O_2^+ in the ground vibrational and electronic state at several electron energies from 0 to 300 meV. The results shows that the branching ratios are somewhat oscillatory, and therefore difficult to convert to an electron temperature dependence. More recently, Petrignani et al. [28] measured the branching fractions as a function of electron energy from 1 to 281 meV for O_2^+ in the vibrational ground state. The results do not appear to agree with those of Peverall, except for channel (8*b*), which usually has the largest yield. Those for the production of channel (8*d*) are the smallest, and the agreement is fair, but the yield for a collision energy of 10 meV is 2%, whereas the measurements of Peverall et al. show the yield at 11.4 and 22 meV as zero. Petrignani et al. [28] did not measure the yields in this range.

The differences between the various measurements are difficult for modelers to ascertain or interpret; it is also difficult to determine the branching ratios as a function of T_e, especially when the largest collision energy is of the order of 300 meV. Of course, the measurements may be interpolated, extrapolated and then convolved with a Maxwell-Boltzmann distribution to give an electron temperature dependence, but doing so requires making the assumption that it is proper to do so. It would be better for the original investigators to carry out this exercise, in order to convert their results into a form that could be used by atmospheric modelers. Moreover, the applicability of the data to atmospheres is questionable when the measurements are confined to $O_2^+(v=0)$. Models have shown that O_2^+ is vibrationally excited in the thermospheres of Venus, Mars, and Earth (e.g., Fox and Hać [29,5] and references therein).

Petrignani et al. [30] reported the branching fractions and cross sections for DR of O_2^+ in vibrational levels $v = 0$, 1, and 2 at zero collision energy. They showed that the branching ratio for production of $O(^1S)$ (channel (8*d*)) is greatly enhanced for vibrationally excited O_2^+, although the cross section decreases as a function of vibrational level. Thus, there is clearly an abundance of data relating to DR of O_2^+, but in practice the available data are difficult to incorporate into models in which there are simultaneous variations in both T_e and in the vibrational distributions. Again, modelers require DR coefficients for each channel as a function of T_e and vibrational level simultaneously. Only when these data are available will it be possible to reliably predict the production rates of $O(^1D)$ and $O(^1S)$, the heating due to the DR reaction, the emission rates of the red, green and 2972 Å lines, and the escape rates or coronal populations of O atoms.

5. Excited electronic states of CO in CO_2^+ DR

DR of CO_2^+:

$$CO_2^+ + e \rightarrow CO(X^1\Sigma^+) + O(^3P) + 8.32\,\text{eV} \tag{11}$$

is sufficiently exothermic to produce CO in excited electronic states, including $A(^1\Pi)$ state, which is the upper level of the CO Fourth Positive bands, and the $a^3\Pi$ state, which radiates to the ground state in the CO Cameron band system. The Cameron bands are the brightest features in the UV dayglows of Mars and Venus. Jongma et al. [31] found that the radiative lifetime of the $\mathrm{CO(a^3\Pi}, v = 0, J = 1)$ state is about 3 ms. DR of CO_2^+ may also produce CO in the excited triplet states $a'^3\Sigma^+$, $d^3\Delta$, and $e^3\Sigma^-$, which may cascade into the $CO(a^3\Pi)$ state in the Asundi, Triplet, and Herman band systems, respectively. The branching ratios for production of excited states of CO are not well known, but it is clear that a significant fraction of the CO fragments are formed in excited states. Experimental identification of the branching ratios for the various excited states has not even begun to converge. Gutcheck and Zipf [32] found that 5% of the dissociative recombinations of CO_2^+ produce the $CO(A^1\Pi)$ state, while Tsuji et al. [33] suggested that this state is populated only 0.33% of the time. Wauchop and Broida [34] reported that 55% of the recombinations of CO_2^+ form CO in the a state, 0.8% in the d state, and 0.04 in the a' state. Skrzypkowski et al. [35] found that the yield of $CO(a^3\Pi)$ was 29%, including cascading from the higher triplet states. Further studies showed that the yields of the a', d and e states were approximately 18, 20, and 0.3%, respectively, and therefore about 38% of the total $CO(a^3\Pi)$ yield is due to cascading (Rosati et al. [36]). Tsuji et al. [37] suggested that the yields of the a', d and e excited triplet states in DR of CO_2^+ are 60%, 40% and 0.83%, thus producing Cameron band emission 100% of the time. In order to compute the emission rates of the Cameron bands and Fourth Positive bands in the dayglows of Mars and Venus, we need some consensus for the branching ratios for these excited states. Because of an accidental resonance of the solar Lyman alpha line with the (14,0) band of the CO Fourth Positive system, fluorescent scattering of the solar Lyman alpha line leads to emission in the $(14,v'')$ progression (e.g., Durrance et al. [38]). Although this progression dominates the Fourth Positive emission in the airglow of Venus, and to a lesser extent in the airglow of Mars, other bands in the fourth positive spectrum may arise from DR of CO_2^+, if the yield is large. Therefore, studies of the aeronomy of Venus and Mars would benefit from accurate determinations of the absolute yields of excited states of CO from CO_2^+ DR, the measurements of which currently conflict.

6. DR and escape of species from light bodies

Bodies that have substantial atmospheres and small gravitational wells, such as Mars, Titan, Triton, and Pluto, are subject to escape of heavy atoms and molecules. The escape may be thermal (Jeans escape and hydrodynamic escape), or non-thermal. Non-thermal escape may be physical, such as pickup ion escape, sputtering, or ion outflow (cf., Fox [39] and references therein). Photochemical escape includes photo- and electron-impact processes, such as dissociation and dissociative ionization, and exothermic reactions that occur at high altitudes, including DR. For example, on Mars, even though DR of N_2^+ is not important for determining the ion density, it is important as an escape mechanism for N. We have discussed this in detail in previous DR proceedings (e.g., Fox [4], and references therein).

For estimates of the escape due to DR, aeronomers need not only the rate coefficients, but the branching ratios of the various channels, and the variation of these parameters with T_e. For DR of homonuclear diatomics, such as O_2^+ and N_2^+, vibrationally resolved rate coefficients and branching ratios are required. DR coefficients for vibrationally excited heteronuclear ions such as NO^+ and CO^+ are interesting, but are not relevant to planetary atmospheres, since these molecular ions relax radiatively before they recombine. For example, the radiative lifetime of vibrational levels of NO^+ $v = 1$ to $v = 5$ have been measured by Wyttenbach et al. [40], and the values have been found to decrease from 90 ms to 19 ms. The lifetime of NO^+ against DR near the peak electron density at $\sim$ 135 km for the Martian high solar activity model shown in Figure 3 is $\sim$ 13 s, that near 100 km is $\sim$ 38 s, and that near 200 km is $\sim$ 104 s. Therefore there is ample time for vibrationally excited NO^+ to radiate to the ground state over the entire thermosphere before it can undergo DR. NO^+ may also be vibrationally quenched, and an estimate of the lifetime against this process is about 0.2 s at 100 km, where quenchers are abundant. Models of the NO^+ density and the production of translationally and electronically excited $N(^2D)$ atoms require DR coefficients that pertain to the ground vibrational and electronic states of the ions.

7. Isotope effects

Because DR processes may be isotope dependent, data for the analogous processes for the common isotopomers are also needed. Kella et al. [41, 26] measured the branching ratios for DR of $^{15}N^{14}N^+$ and $^{18}O^{16}O^+$ in ion storage ring experiments. In some of these studies, the heteronuclear isotopic species are used rather than the homonuclear species $^{14}N^{14}N^+$ and $^{16}O^{16}O^+$, because they are more likely to relax radiatively before undergoing DR. Guberman [42], however, carried out calculations of the DR coefficients for production of $O(^1S)$ (channel (8*d*)) from $v = 0$ of the $X^2\Pi_{1/2,g}$ state for $^{18}O^{16}O^+$ and $^{16}O^{16}O^+$. He reported that the DR coefficient of the isotopomer exceeds that of the more common species homonuclear ion over the T_e range from about 100 to 3000 K, by a factor which ranges from ~ 1.4 to ~ 2.6. Thus the use of isotopomers as proxies for the homonuclear species is apparently questionable to an extent that is unknown, but should be measured.

The relative escape rates of ^{18}O and ^{16}O (as well as ^{15}N and ^{14}N, and ^{13}C and ^{12}C) are important to the evolution of the Martian atmosphere. Dissociative recombination of O_2^+ is probably the most important escape mechanism for O in the Martian atmosphere at the current epoch. DR of O_2^+ is sensitive to vibrational excitation, both through changes in the rate coefficients, and in the branching ratios of the channels, for which there is conflicting data, as we have discussed above. Fox and Hać [5] have carried out a computation of the vibrational distribution on Mars as a function of altitude for both $^{32}O_2^+$ and $^{34}O_2^+$, and the results are shown in Figure 5. It is clear that $^{32}O_2^+$ is vibrationally excited for altitudes above about 180 km, but that $^{34}O_2^+$ is found to be overwhelmingly in the $v = 0$ ground state over the range from about 80 to 400 km. Because collisions are frequent at the lower altitudes, the vibrational levels of both isotopes are in equilibrium with the local thermospheric temperature, which ranges from about 130 to 250 K from 80 to 160 km. At higher altitudes radiative decay of the isotopic species limits the ion to mostly $v'' = 0$. The radiative lifetimes of levels $v'' = 1 - 10$ have been computed to range from 25.06 to 3.29 s (Amitay et al. [43]).

In addition to the different vibrational distributions, there are small differences in the energies of the O atoms produced in DR in the center-of-mass frame. In DR of $^{18}O^{16}O^+$, the ^{18}O carries away a fraction 16/34 of the total energy released, whereas in DR of $^{16}O^{16}O^+$, the ^{16}O carries away half the energy. The exothermicities of the $^{34}O_2^+$ channels (8) differ from those of $^{32}O_2^+$ by only ~ 0.02 eV, which is much smaller than the other uncertainties in the model. The computed nascent kinetic energy distributions and the escape energies of ^{18}O and ^{16}O due to DR of O_2^+ for the two isotopes at four altitudes are shown in Figure 6.

There is an isotope effect of a factor of 2 in DR reactions simply because there is one ^{18}O atom produced in DR of $^{18}O^{16}O^+$, but there are two ^{16}O atoms produced in DR of $^{16}O^{16}O^+$. In addition, the escape energy of an ^{18}O is larger than that of a ^{16}O atom. We have carried out Monte Carlo calculations of the escape probabilities of both isotopes, in which we follow the energetic O atoms from their altitude of production from collision to collision until they either escape or their velocity falls below the escape velocity. We modeled the escape processes for two assumptions about the differential cross sections in the center-of-mass frame. In the first model, the scattering is assumed to be isotropic, and in the second model, the scattering is strongly forward peaked. We have also compared our results to the exobase approximation. The exobase is the boundary of the "collisional" and the "collisionless" regions of the atmosphere. In the exobase approximation all atoms produced above the exobase, with velocity vectors oriented in the upward hemisphere, and with magnitudes greater than the escape speed, are

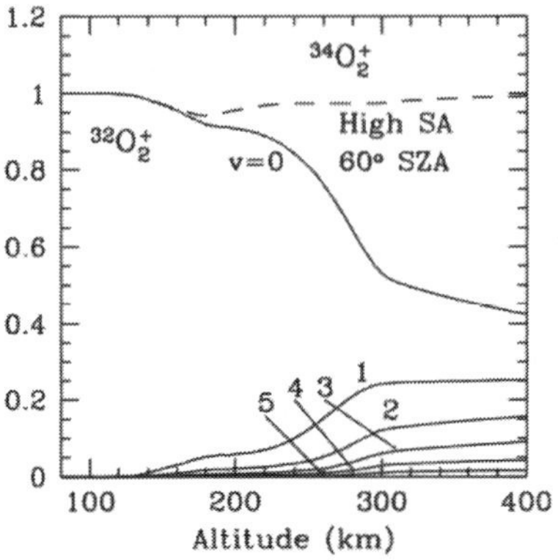

Figure 5. Computed fractional population of $^{32}O_2^+$ vibrational levels, $v = 0$, 1, 2, 3, 4, and 5 as a function of altitude from 80 to 400 km (solid curves). The fraction of $^{34}O_2^+$ in $v = 0$ is shown with a dashed curve. From Fox and Hać [5].

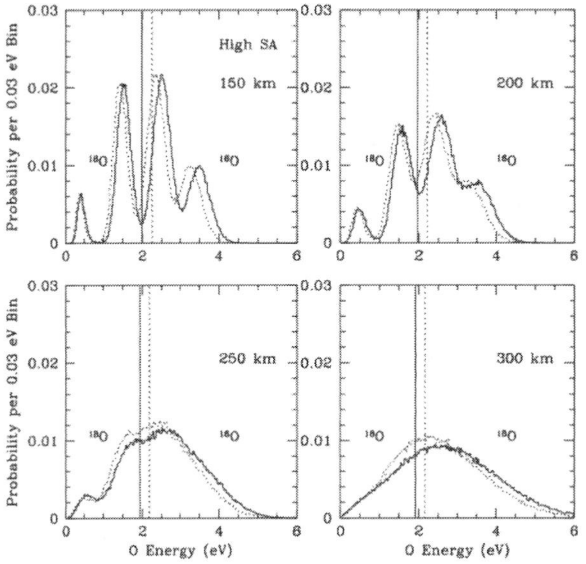

Figure 6. Examples of nascent energy spectra of ^{16}O (solid curves), and ^{18}O (dotted curves) due to DR of $^{32}O_2^+$ and $^{34}O_2^+$ for the high solar activity model at four altitudes. The panels are labeled by the altitudes. Also shown as vertical solid and dotted lines are the escape energies of the two isotopes. From Fox and Hać [5].

assumed to escape. The exobase approximation has been used extensively in the past, but is now outdated; the transition from collisional to collisionless conditions occurs gradually. We consider the forward scattering model to be superior to either the isotropic model or the exobase approximation.

In Figure 7 we show the resulting escape probabilities for ^{18}O and ^{16}O as a function of altitude of production for the three models, and for low and high solar activity. The ratio of escape probabilities of ^{18}O to ^{16}O is the isotope fractionation factor inherent in the DR mechanism, and it varies from about 0.20 to 0.37 as a function of altitude for the forward scattering model. This large fractionation factor operates in addition to that which arises from the different altitude distributions of $^{18}O^{16}O^+$, and $^{16}O^{16}O^+$. Since no fractionation (within about 5%) of the Martian O isotopes with respect to terrestrial isotope ratio has been found (e.g., Owen [44] and references therein), there must be effects operating in the Martian atmosphere that act over time to reverse or dilute the isotope fractionation that arises from preferential escape of the lighter isotope. This result has implications for processes that preferentially remove ^{18}O from the atmosphere; those due to exchange of the atmospheric isotopes of O with (large) unfractionated reservoirs of O, such as the polar caps or the regolith; the interaction of atomic O with the crust, or its incorporation into rocks.

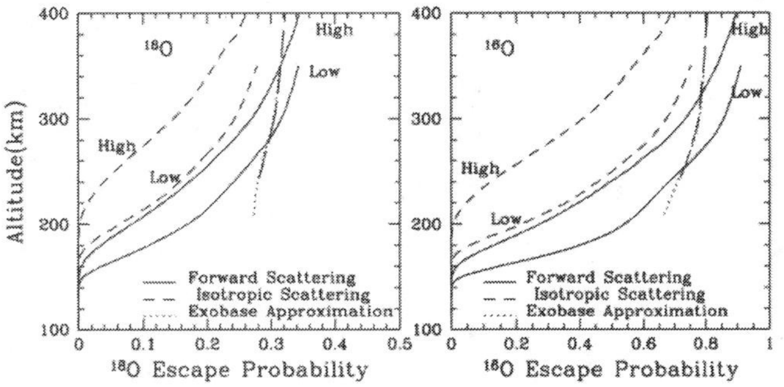

Figure 7. Computed O escape probabilities as a function of altitude for isotropic, forward scattering and exobase approximations. (left) ^{18}O escape probabilities. (right) ^{16}O escape probabilities. Note the differences in the scales for the left and right panels. From Fox and Hać [5]

Presumably, we will find similar isotope effects for escape of N in DR of N_2^+. The Martian atmospheric $^{15}N/^{14}N$ ratio is observed to be fractionated by a factor of about 1.62 with respect to the terrestrial value, due probably to preferential escape of ^{14}N operating over billions of years. Unlike O, however, there are several photochemical escape processes operating for N in addition to N_2^+ DR, such as photodissociation of N_2. The latter process cannot be computed with low resolution cross sections, and its role in escape of N from Mars has not yet been quantified.

8. Summary

Ionospheric modelers require DR coefficients for determining the loss rate of terminal ions, and for production rates of atoms or molecular fragments. Branching ratios of the allowed channels are required, including those which produce metastable or emitting fragments. For homonuclear diatomics, there is a need for vibrationally resolved DR coefficients and branching ratios. For heteronuclear diatomic ions, the DR coefficients and branching ratios must pertain to the ion in the ground vibrational level. For isotope evolution calculations of Mars (or other bodies), modelers need DR coefficients and branching ratios for the homonuclear molecules, such as O_2^+ and N_2^+, and their isotopomers, as well as lifetimes of the vibrationally excited ions and quenching coefficients. Above all, modelers need the expertise of the DR community in resolving discrepancies in the results of different measurements and computations, providing recommendations relative to which data are most appropriate and how best to incorporate the data into models. Rate coefficients are preferable to cross sections, and T_e dependences are preferable to collision energies.

Dave Barry [45] tells a story about an incident that occurred in 2009 on the Metromover, the public transportation system in Miami. Apparently one evening two men carried a live six-foot shark onto the Metromover, with the idea of selling it to a wholesaler in downtown Miami. When the men got to their intended stop, they carried the now-dead shark onto the street, but the wholesaler declined to purchase the fish. They then left the shark on the street, where it remained for several hours before it was removed. Bystanders were apparently unconcerned about the situation, partly because they were relieved that it was a shark and not a dead body, and partly because weirder things had happened in Miami. Barry points out, however, that for a while there was a live shark on the Metromover, and "had it been a little *more* alive, there is a very real possibility that it could actually have bitten somebody", which would "have been tragic". Therefore Barry recommends that the public "exercise caution when boarding public transportation", and further explains what he means when he says "exercise caution": "carry a speargun".

One might ask what this has to do with this chapter. I cannot count the number of times that experimentalists and theoreticians have admonished modelers to "use caution" in adopting DR data, but do not further explain what they mean, or make suggestions about what modelers should actually do with the apparently questionable data. The next time I see this phrase, you might see me with a spear-gun at my side as I am incorporating DR information into a model (Figure 8). Please, give me something I can use!

Figure 8. Using caution while inputting DR data into a model.

9. Acknowledgments
This work has been supported by grants NNX09AB70G and NNG084N46G from the National Aeronautics and Space Administration to Wright State University. NASA and NSF are acknowledged for providing travel support. William Wagner kindly provided Figure 8.

10. References
[1] Plane J M C 2003 *Chem. Rev.* **103** 4963
[2] Grebowsky J M, Moses J I and Pesnell W D 2002 *Atmospheres in the Solar System: Comparative Aeronomy, Geophysical Monograph 130*, ed Mendillo M, Nagy A, Waite J H (Washington DC: AGU) doi:10.1029/130GM15

[3] Bauer S J 1973 *Physics of Planetary Ionospheres, Physics and Chemistry in Space* **6** p 120
[4] Fox J L 1993 *Dissociative Recombination: Theory Experiment, and Applications*, ed Rowe B R, Mitchell J B A and Canosa A (New York: Plenum Press) p 219
[5] Fox J L and Hać A B 2010 *Icarus* **208** doi:10.1016/j.icarus.2010.01.0192010, 176
[6] Dalgarno A and Fox J L 1994 *Unimolecular and Bimolecular Reaction Dynamics*, ed Ng C Y, Baer T and Powis I (New York:Wiley) p 1
[7] Hunter E P L and Lias S 1998 *J. Phys. Chem. Ref. Data* **27** 413
[8] Lias S, Liebman J F and Levin R D 1984 *J. Phys. Chem. Ref. Data* **13** 696
[9] Nesbitt F L, Marston G, Stief L J, Wickramaaratchi M A, Tao W and Klemm R B 1991 *J. Phys. Chem.* **95** 7613
[10] Su T and Bowers M T 1979 *Gas Phase Ion Chemistry I.*, ed Bowers M T (New York: Academic Press) p 83
[11] Cravens T E, Yelle R V, Wahlund J-E, Shemansky D E and Nagy A F 2009 *Titan from Cassini-Huygens*, ed Brown R H, Lebreton J-P and Waite J H (New York: Springer) p 259
[12] McEwan M J, Scott G B I and Anicich V G 1998 *Int. J. Mass. Spectrom. Ion Proc.* **172** 209
[13] Anicich V G 1993 *J. Phys. Chem. Ref. Data* **22**, 1469
[14] Cravens T E, and 15 colleagues 2006 *Geophys. Res. Lett.* **33** L07105
[15] Crary F J, Magee B A, Mandt K, Waite Jr J H, Westlake J, and Young D T 2009 *Planet. Space Sci.* **57** 1847
[16] Lundin R, Barabash S, Holmström M, Nilsson H, Yamauchi M, Dubinin E M and Fraenz M 2009 *Geophys. Res. Lett.* **36** L17202, doi:10.1029/2009GL039341
[17] Miller K L and Whitten R C 1991 *Space Sci. Rev.* **55** 165
[18] Narcisi R S, Bailey A D, Wlodyka L E and Phylbrick C R, 1972 *J. Atmos. Terr. Phys.* **34** 647
[19] Vuitton V, Lavvas P, Yelle R V, Galand M, Wellbrock A, Lewis G R, Coates A J and Wahlund J-E 2009 *Planet. Space Sci.* **57** 1558
[20] McLain J L, Poterya V, Molek C D, Babcock L M and Adams N G 2004 *J. Phys. Chem. A* **108** 6704
[21] Florescu-Mitchell A I and Mitchell, J B A 2006 *Phys. Rep.* **430** 277
[22] Froese-Fischer C and Tachiev G 2004 *Atomic Data Nucl. Data Tables* **87** 1
[23] Slanger T G, Cosby P C, Sharpee B D, Minschwaner K R and Siskind D E 2006 *J. Geophys. Res.* **111** A12318, doi:10.1029/2006JA011972
[24] Slanger T G, Cosby P C, Huestis D L and Bida T A 2001 *Science* **291**, 463
[25] Barth C A 1961 *Science* **134** 1426
[26] Kella D, Vejby-Christenson L, Johnson J P, Pedersen H B and Andersen L H 1997 *Science* **276**, 1530; ibid **277**, 167
[27] Peverall R, Rosen S, Peterson J R, Larsson M, Al-Kahlili A, Vikor L, Semaniak J, Bobbenkamp R, Le Padellec A, Maurellis A N, and van der Zande W J 2001 *J. Chem. Phys.* **114** 6679
[28] Petrignani A, Hellberg F, Thomas R D, Larsson M, Cosby P C and van der Zande W 2005 *J. Chem. Phys.* **122** 23411
[29] Fox J L and Hać A B 1997 *J. Geophys. Res.* **102** 24,005
[30] Petrignani A, van der Zande W, Cosby P C, Hellberg F, Thomas R D and Larsson M 2005 *J. Chem. Phys.* **122** 014302
[31] Jongma R T, Berden G and Meijer G 1997 *J. Chem. Phys.* **107** 7034
[32] Gutcheck R A and Zipf E C 1973 *J. Geophys. Res.* **78** 5429
[33] Tsuji M, Nakamura M and Nishimura Y 1995 *J. Chem. Phys.* **103** 1413
[34] Durrance S T 1981 *J. Geophys. Res.* **86** 9115
[35] Wauchop T S and Broida H P 1972 *J. Chem. Phys.* **56** 330
[36] Skrzypkowski M P, Gougousi T, Johnsen R and Golde M F 1998 *J. Chem. Phys.* **108** 8400
[37] Rosati R E, Johnsen R and M F Golde 2003 *J. Chem. Phys.* **119** 11630
[38] Tsuji M, Nakamura M, Nishimura Y and Obase H 1998 *J. Chem. Phys.* **108** 8031
[39] Fox J L 2009 *J. Geophys. Res.* **114** E12005 doi:10.1029/2009JE003432
[40] Wyttenbach T, Beggs C G and Bowers, M T 1991 *Chem. Phys. Lett.* **177** 239
[41] Kella D, Johnson, P J, Pederson H B, Vejby-Christensen L and Andersen L H 1996 *Phys. Rev. Lett.* **77** 2432
[42] Guberman, S L 2000 *Dissociative Recombination: Theory, Experiment, and Applications IV*, ed Larrson M, Mitchell J B A, and I. F. Schneider I F 2000 (Singapore: World Scientific) p 111
[43] Amitay Z, Zajfman D and Forck P 1994 *Phys. Rev. A* **50** 2304
[44] Owen T 1992 *Mars* ed Kieffer H H, Jakosky B M, Snyder C W and Matthews M S (Tucson: University of Arizona Press) p 818
[45] Barry D 2010 *I'll Mature When I'm Dead: Dave Barry's Amazing Tales of Adulthood* (G. P. Putnam's Sons: New York) pp 82–84

Journal of Physics: Conference Series **300** (2011) 012026 doi:10.1088/1742-6596/300/1/012026

Dissociative recombination and electron attachment in regions of star formation

Eric Herbst[1], **Evelyne Roueff**[2]

[1] Departments of Physics, Chemistry, and Astronomy, The Ohio State University, Columbus, OH, USA

[2] Observatoire de Paris, 5 Place J. Janssen, 92195 Meudon Cedex, France

E-mail: herbst.6@osu.edu

Abstract. Dissociative recombination and electron attachment are important in interstellar chemistry, which is heavily dominated by ions and ionic processes. Here we consider how the competition between dissociative recombination and other reactions, such as H-atom transfer, can explain the unusually high observed abundances of the reactive cations OH^+ and H_2O^+ in the dense outflow source in front of the Orion Nebula. We also show how dissociative recombination and other processes might block the achievement of an equilibrium ortho-to-para abundance ratio for H_2O^+ in diffuse interstellar clouds. Finally, we consider the formation and destruction rates of molecular anions observed in the interstellar and circumstellar media, especially the formation mechanism of radiative attachment, the rate of which has only been estimated by a simple phase space theory, which is surprisingly successful in most instances.

1. Introduction

Peering at the sky in the THz region of the electromagnetic spectrum at frequencies up to 1.91 THz, the *Herschel Space Observatory* has changed our perception of some aspects of our Galaxy and many others [1]. In the Milky Way, Herschel observations with the *HIFI* heterodyne receiver have looked in a preliminary manner at emission from giant molecular clouds containing star-forming regions, and absorption from the outer and cooler environments surrounding these regions as well as from diffuse clouds in spiral arms along the line of sight from the more distant continuum sources, such as the Galactic Center. Among the regions studied in emission is the well-known bright extended object known as Orion KL, which is located in the larger Orion Molecular Cloud. This complex source contains several star-forming regions, known as "hot cores," which are of higher density ($n \approx 10^6$ cm^{-3}) and temperature ($100 - 300$ K) than the standard conditions of dense interstellar clouds. Previous studies at lower frequencies of these hot cores and similar sources in other giant molecular clouds have shown a high density of rotational emission lines, many of which have been assigned to common terrestrial organic molecules such as methanol, ethanol, methyl formate, dimethyl ether, and ethyl cyanide. A preliminary study of this source with Herschel [2] shows a somewhat lower density of rotational emission lines, most of which can be assigned to the same species, often known as "weeds" because their high spectral densities obscure the search for more exotic species with sparser spectra, known as "flowers." Further studies of these and other hot cores in the galaxy with *Herschel* will be undertaken shortly in a search for additional organic species.

Somewhat more unusual are the results of rotational absorption studies towards objects such as Orion KL, which emit a continuum in the THz region partially contaminated by emission spectra. The sources seen in absorption, whether they surround the star-forming regions or lie in galactic spiral arms between them and the earth, contain strong spectral features assigned to the light molecular cations OH^+ and H_2O^+ [3, 4]. This result is highly unusual, especially in the outer regions of the giant molecular clouds, because it is well known from laboratory studies that these cations are rapidly destroyed by reactions with H_2, the dominant neutral species in the dense interstellar medium, which lead to the saturated molecular ion H_3O^+; i. e.,

$$OH^+ + H_2 \longrightarrow H_2O^+ + H, \tag{1}$$

$$H_2O^+ + H_2 \longrightarrow H_3O^+ + H. \tag{2}$$

The hydronium ion subsequently undergoes a dissociative recombination reaction to produce the well-known neutral species H_2O and OH [5]:

$$H_3O^+ + e^- \longrightarrow H_2O + H; OH + 2H, \text{etc.} \tag{3}$$

One thus expects to see a strong signal from H_3O^+ and weaker signals from the simpler ions in dense regions. Although the hydronium ion is indeed detected, albeit weakly, in the diffuse spiral arm clouds along with OH^+ and H_2O^+, where much of the hydrogen is in the atomic form, it is not at all seen in the outer region surrounding Orion KL.

Another mystery concerns the H_2O^+ cation, which is detected in both its ortho and para forms [6]. The ratio of ortho-H_2O^+ to para H_2O^+ (OPR) is $\approx$ 4.8:1 in the diffuse spiral arm clouds, a ratio that can be converted to a spin temperature of $\approx$ 20 K but that must eventually be explained by a complex competition involving formation, equilibration, and destruction of the two spin modifications. In the body of this paper, tentative explanations for the surprising strength of the OH^+ and H_2O^+ spectral features, and the unusual OPR of H_2O^+ will be suggested.

As unusual as these detections are, the detection of molecular anions was equally unexpected by most astrochemists, although some very old papers had discussed their possible detection in cold interstellar clouds [7, 8]. Starting a few years ago, laboratory studies of the rotational spectra of a number of strongly-bound molecular anions were performed by Thaddeus, McCarthy and co-workers [9, 10]. Almost immediately, some of these anions were found in cold regions of interstellar clouds and one circumstellar source (IRC+10216) either from previously observed but unidentified lines or from new detections in the millimeter region of the electromagnetic spectrum [9, 11]. The anions detected, discussed below, come from precursor radical neutrals also known to be relatively abundant in cold dense clouds, especially the so-called cold core TMC-1 and the circumstellar source IRC+10216. The chemistry of formation and depletion of these anions is characterized by processes that, for the most part, have not been studied in the laboratory or by detailed theoretical methods, so that the reasonable agreement between simulations and observations obtained must be regarded as at least partially serendipitous.

2. The Orion Outflow Source

The absorption spectrum from material in front of Orion KL is dominated by a broad spectral feature which, when translated from frequencies to radial velocities, ranges from 0 to -50 km s^{-1} [3]. The negative velocities mean that the source is moving towards us, and so is known as an outflow. At 1 THz, the 50 km s^{-1} range corresponds to a line width of 170 MHz! Such line widths are often associated with turbulence and/or shocks, although it is also possible that there are a number of different outflows along the line of sight and that we are not resolving them. As a preliminary model, we first adopted what is known as a PDR, an acronym for "photon-dominated region." [12]. Such a region lies close to a strong source of vacuum ultraviolet radiation, which can strongly affect layers of the source closest to it. In our case, we assumed the radiation field to be 4×10^4 as strong as the standard radiation field in interstellar space. To try to explain the observations of OH^+ and H_2O^+, it is necessary to not only assume a strong VUV field, but also a high flux of either cosmic rays or X-rays to ionize a significant portion of the gas. The standard ionization rate coefficient used for cosmic rays (mainly high-energy protons), labelled by the letter ζ, is in the vicinity of 10^{-17} s^{-1}. This rate of ionization results in a fractional ionization in a dense region of approximately 10^{-7}, a very weak plasma, in which ions prefer to react with H_2 than with electrons. To break the cycle of reactions with H_2, one must raise the ionization rate by a factor of 1000 – 10000, a very dramatic increase but one in agreement with what another analysis indicates [13]. With a resulting greatly increased fractional ionization of perhaps 10^{-3}, dissociative recombination can compete with reaction with hydrogen. If we assume that the rate coefficient for dissociative recombination of an ion is 10^{-6} cm^3 s^{-1} and the rate coefficient for its reaction with H_2 is 10^{-9} cm^3 s^{-1}, the rate of reaction of the ion with electrons and with H_2 is the same. Thus, although reaction with H_2 still occurs, it must compete with dissociative recombination. Nevertheless we still need additional formation mechanisms to produce the OH^+ and H_2O^+ ions.

Let us look at what can happen in more detail. We start with the OH^+ ion, which can be formed in a number of ways, such as the reaction of O and H_3^+:

$$O + H_3^+ \longrightarrow OH^+ + H_2, \tag{4}$$

and the charge exchange between O and H^+ followed by reaction of O^+ with H_2:

$$O^+ + H_2 \longrightarrow OH^+ + H. \tag{5}$$

If OH^+ reacts with H_2 to form H_2O^+, the resulting water ion can undergo dissociative recombination to form OH and H. The strong VUV radiation field can then photoionize OH into its cation. Reaction of H_2O^+ with H_2 instead leads to the H_3O^+ cation, which is destroyed relatively quickly by dissociative recombination, so that it is hard to detect. Remember that, unlike the smaller ions, H_3O^+ cannot be produced by any photoionization process. The resulting neutral species H_2O and OH, however, are subsequently photoionized by the strong radiation field to produce OH^+ and H_2O^+. The high ionization rate from cosmic rays also leads to the production of a large abundance of protons, which in turn react with neutral water to produce more H_2O^+:

$$H^+ + H_2O \longrightarrow H_2O^+ + H. \tag{6}$$

It turns out that if one assumes that neutral water can flow into the outflow source or can be photodesorbed from icy mantles, the necessary cosmic ray ionization rate can be lowered by an order of magnitude, because of the enhanced photoionization rate of water. With a range of parameters, we can reproduce the observed column densities of OH^+ and H_2O^+ as well as the upper limit to the H_3O^+ column density. (Note that a column density is a concentration multiplied by a length; this quantity is what is directly measured along the line of sight.) The cloud is no longer a cold region; however, since the temperature at its edge facing the source of UV radiation is $\approx$ 400 K. Moreover, the atomic and molecular hydrogen abundances near the inner edge of the outflow are approximately equal, and the optimum density is perhaps 10^3 cm^{-3}.

3. The Ortho-to-Para Ratio of H_2O^+

Ortho-para abundance ratios for assorted species in interstellar sources can be the result of many processes, including those that occur both in the gas-phase and on granular surfaces. The case of

Journal of Physics: Conference Series **300** (2011) 012026 doi:10.1088/1742-6596/300/1/012026

H_2O^+ should be a *relatively* simple one, since it is likely to occur entirely in the gas phase. Using observations with Herschel, Schilke et al. [6] have just studied the ortho-to-para ratio (OPR) of the water ion in diffuse spiral arm sources, and determined that it is 4.8:1. Since the ground electronic state of H_2O^+ possesses an asymmetric ^{2}B symmetry, the rotational levels associated with the ortho and para forms are reversed from those of neutral water, and it is the ortho form that contains the lowest (0_{00}) rotational level while the lowest para rotational level is 1_{01}, which lies 30.1 K above the ground ortho level. Thus, as the rotational temperature approaches 0 K and dynamic processes interconvert the ortho and para forms, the ortho modification becomes increasingly dominant, while at high rotational temperatures, the OPR approaches its standard value of 3:1. The simplest way of interpreting the observed OPR of 4.8:1 is to assume that it is an equilibrium value, obtained by gas-phase processes in which protons can exchange from one species to another, such as:

$$\mathrm{p-H_2O^+ + H \rightleftharpoons o-H_2O^+ + H}, \quad (7)$$

where we have assumed that atomic hydrogen is the major equilibrant since the spiral arm clouds are diffuse, and because the water ion reacts with H_2 rather than be thermalized by it. Since the measured OPR is really a measurement of the abundance ratio between the lowest ortho and para rotational states, it can be expressed in simple terms via a so-called spin temperature T_s:

$$\mathrm{OPR} = \frac{g(0_{00})}{g(1_{01})}\exp(30.1/T_s), \quad (8)$$

where the g are combined nuclear-spin-rotational degeneracy factors and 30.1 K is the energy separation between the two rotational states in K, the astronomer's favorite unit of energy. Using this equation, an OPR of 4.8 translates into a spin temperature of $\approx$ 20 K since the degeneracies are both 6. But what can this spin temperature possibly mean? If the system equilibrates to the kinetic temperature of a diffuse cloud, then a temperature of 50-100 K is more likely. Such a temperature would correspond to an OPR for H_2O^+ closer to 3:1, or 1:1 in the absence of rotational excitation, as expressed in eq. (8), where the high-temperature limit goes to unity rather than to 3:1. Given the low density of a diffuse cloud, the 1:1 value seems more reasonable theoretically although it is far from the observed value of 4.8:1.

Let us now consider the complications caused by chemistry. If the destruction of H_2O^+ is faster or at least competitive with equilibration, then the ortho/para ratio obtained in the formation of H_2O^+ can be the main cause of the OPR. Consider the formation and destruction of ortho- and para- H_2O^+ via a small chemical network:

$$\mathrm{OH^+ + p-H_2 \longrightarrow p-H_2O^+, o-H_2O^+ + H} \quad (9)$$

$$\mathrm{OH^+ + o-H_2 \longrightarrow p-H_2O^+, o-H_2O^+ + H} \quad (10)$$

$$\mathrm{H_2O^+ + e^- \longrightarrow Products} \quad (11)$$

$$\mathrm{H_2O^+ + H_2 \longrightarrow H_3O^+ + H.} \quad (12)$$

Let the fraction of p-H_2O^+ and o-H_2O^+ produced in the formation steps be labelled f_{pp}, f_{po}, f_{op} and f_{oo}, where the first two parameters refer to the reaction with p-H_2, and the second two refer to the reaction with o-H_2. Although the steady-state solution to this series of equations and equilibration is rather complex, the result in the limit that equilibration is slow is given by

$$\mathrm{OPR} = \frac{f_{po} + f_{oo}\mathrm{OPR(H_2)}}{f_{pp} + f_{op}\mathrm{OPR(H_2)}}, \quad (13)$$

where OPR(H_2) represents the ortho-to-para ratio of molecular hydrogen. The f factors can be calculated using the angular-momentum approach of Oka [14] but an additional complication immediately ensues because it turns out that the f factors depend upon the mechanism of reactions (9) and (10); in particular, whether they occur via a long-range hopping of a hydrogen atom or via an intimate and long-lived complex. As for OPR(H_2), we can regard it as a variable parameter or we can use a value of near unity, determined by older studies for nearby diffuse clouds. With this value, we estimate that OPR $\approx$ 2:1 with the complex mechanism. In fact, to reproduce the observed value for OPR of 4.8:1, we appear to need almost pure o-H_2 for the complex mechanism, which makes little sense. Even worse, the hopping mechanism always leads to an OPR of 3:1. So, our simple chemistry-dominant approach fails.

One additional possibility is to consider ortho/para variations in the destruction reactions, as has been detected for the dissociative recombination reaction of H_3^+ [15]. Yet another possibility is that the small network used is incomplete. In particular, the charge exchange reaction between H^+ and H_2O (see eq. (6)) is a competitive source for both spin modifications of H_2O^+. Once H_2O becomes involved in the analysis, it becomes necessary to calculate its ortho-to-para ratio, which may well be governed by processes occurring on the surfaces of dust grains. Moreover, it is unlikely that the equilibration reactions involving atomic H are much slower than the destruction reaction involving H_2 in a diffuse cloud environment, since atomic hydrogen probably has a larger abundance than H_2. In at least some of the diffuse cloud spiral arm sources, an H/H_2 abundance ratio greater than unity can be determined based on the observation that OH^+ has a greater abundance than does H_2O^+ [4]. Given all of these complications, it is clear that much more work is needed to understand the ortho-to-para ratio of H_2O^+ and indeed other species exhibiting these spin modifications. It may well be that the OPR will not be reproduced until detailed models are found that reproduce the coupled abundances of all of the species, including their spin modifications, detected in these sources.

4. Interstellar Anions and their Chemistry

As of the present, six different negative molecular anions have been discovered in assorted sources, as shown in Table 1. These ions are all radical ions; that is, their neutral forms are all chemical radicals, possessing both an odd number of electrons and a large electron binding energy ("electron affinity") of $\approx$ 3-4 eV. All astronomical detections are based on firm laboratory studies with the exception of C_5N^-. The only source containing all of the anions is the outer envelope of the old carbon-rich star IRC+10216, in which the elemental abundance of carbon exceeds that of oxygen. The other sources are all interstellar ones: TMC-1 is the standard cold core studied by radio astronomers for its rich organic chemistry; the heading "other cold cores" includes L1544 and L1521F and a newly discovered source, known as Lupus-1A [16]. These objects, typically at temperatures of 10 K, are denser regions of larger clouds known as dark clouds for their obscuration of background star light. Some of those listed are pre-stellar cores, while one, Lupus-1A, like TMC-1 shows no evidence of collapsing to form stars. The final source, L1527, is a region outside of a protostellar source, and has a somewhat higher temperature. Despite individual differences, the chemistry of all of these objects is dominated by ion-molecule, dissociative recombination, and radical-neutral reactions, which produce large numbers of carbonaceous unsaturated radicals. The family of interstellar radicals with structure C_nH is particularly well studied, and consists of molecules with n=1 to n=8. None of the anions is abundant in an absolute sense, but their existence has strong implications for the chemistry of complex molecule formation [17].

The production of anions in cold interstellar clouds and in carbon-rich stellar envelopes is very different from their production in the laboratory, typically at much higher densities. In the low-density interstellar and circumstellar media, the most likely process of formation is thought to be radiative attachment, a two-body process in which electrons attach themselves

Table 1. Molecular anions and their sources

Ion	IRC+10216	TMC-1	Other Cold Cores	L1527
CN^-	X	—	—	—
C_3N^-	X	—	—	—
C_5N^-	X	—	—	—
C_4H^-	X	—	X	X
C_6H^-	X	X	X	X
C_8H^-	X	X	X	—

to neutral species with a large electron affinity to form an unstable complex, which then either auto-detaches or radiates sufficient energy to stabilize the nascent negative ion. The radiation can be produced by relaxation of vibrational levels above the attachment continuum to those below it or by relaxation of a complex in an excited electronic state to stable vibrational levels of the ground electronic state. The former process is thought to be the more common. The first estimate of the rate of radiative attachment was made by Herbst [7], who also suggested the likely anions to be formed. The theory used, updated by Terzieva & Herbst [18] and Herbst & Osamura [19], is a very simple one, and more detailed quantum calculations are urgently needed. Herbst also estimated that the anion-to-neutral ratio achievable via this process could be as high as 1%, and that this ratio could only be achieved for anions with at least 4-5 atoms.

The method used by Herbst [7] is known as phase-space theory, which is a statistical approach with rigorous conservation of angular momentum. The additional assumption was made that only s-wave electrons are involved. The overall mechanism consists of three different processes: complex formation, complex detachment, and radiative stabilization, and can be written as:

$$A + e^- \rightleftharpoons A^{*-} \rightarrow A^- + h\nu, \tag{14}$$

where the complex is indicated by an asterisk. The rates of complex formation and detachment are governed by microscopic reversibility [20], while the rate of radiative stabilization via purely vibrational emission can be obtained using quantum chemical or experimental information of transition strengths for the various modes [21]. As the size of the complex and its binding energy increase, the rate of detachment lessens, to the point that eventually the complex is stabilized by emission of radiation on 100% of the collisions that form it. Some calculated radiative attachment rate coefficients, $k_{\rm radatt}$, are listed in Table 2 [19].

Another formation mechanism is dissociative attachment, in which an electron attaches itself to a neutral molecule with the concomitant breaking of a chemical bond. This process is normally endothermic, but there are odd radicals in the interstellar medium for which the process can indeed be exothermic. The most important example is the reaction between electrons and the high-energy cumulene isomer H_2C_6 [19, 22]:

$$e^- + H_2C_6 \longrightarrow C_6H^- + H, \tag{15}$$

which is calculated to be exothermic by 16 kJ mol^{-1}. This reaction appears to be more important than radiative attachment in denser sources where there is little neutral C_6H. Finally, negative ions can be formed via normal ion-molecule reactions, especially when the anions are too small to be formed efficiently by radiative attachment. An example is the anion CN^-, recently detected in IRC+10216 [23], which is most likely formed from a large anion of the form C_n^- by the reaction with nitrogen atoms [24]:

$$N + C_n^- \longrightarrow CN^- + C_{n-1}, \tag{16}$$

Table 2. Estimated Attachment Rate Coefficients for C_nH^{-a}

No. of C Atoms	k_{radatt} (cm^3 s^{-1}, 10 K)
2	1×10^{-14}
4	6×10^{-8}
6	3×10^{-7}
8	3×10^{-7}

[a] Herbst & Osamura [19]

and possibly by other ion-molecule reactions involving H^- [25].

Once formed, anions can be destroyed by a number of processes including photodetachment, associative detachment with hydrogen atoms, normal ion-molecule reactions, and positive ion-negative ion recombination. The last process is currently poorly understood, although new experiments are planned in several laboratories [26].

In Table 3, we compare the results of observations and current models for the chemistry of the carbon-rich envelope IRC+2016 in the form of anion-to-neutral ratios [23, 25, 27, 28]. It can be seen that agreement is reasonable (order-of-magnitude) for most of the anions. The largest disagreement concerns the abundance ratio C_4H^-/C_4H, which is calculated to be roughly two orders of magnitude too large. As can be seen in Table 2, the calculated radiative association rate coefficient for C_4H lies in between the very small rate coefficient for C_2H radiative attachment, and the large ones for C_6H and C_8H. It is likely that this rate coefficient is calculated incorrectly by the statistical theory, which is typically accurate only for large systems. Detailed calculations of radiative attachment are clearly needed.

5. Radiative recombination??

The very recent detection of C_{60} in the interstellar medium [29], currently limited to so-called planetary nebulae and reflection nebulae, raises the question of its synthesis. Bettens & Herbst [30] suggested a synthesis of this species many years ago, based on higher-density laboratory work by Bowers and co-workers [31] In this synthesis, ion-molecule reactions first produce linear carbon chains, which, as they grow, eventually isomerize into single rings. These rings, as they continue to grow, isomerize into triple rings, which finally convert to fullerenes. But only fullerene ions; e.g., C_{60}^+, are produced, unless the process of radiative recombination; viz., :

$$C_{60}^+ + e^- \longrightarrow C_{60} + h\nu \tag{17}$$

occurs efficiently. Bettens & Herbst used a very simple statistical theory, known as RRKM theory, to estimate that the reaction is quite efficient for unsaturated hydrocarbon cations of all shapes with more than $\approx$ 20-30 carbon atoms. To the best of our knowledge, there is currently no information, either experimental or theoretical, on this process. Once again, some detailed work is needed.

Acknowledgments

E. H. acknowledges the support of the National Science Foundation for his astrochemistry program through grant AST-0702876, and his program in chemical kinetics through the Center for the Chemistry of the Universe. He also acknowledges support from NASA NAI for studies in the evolution of pre-planetary matter. E. R. acknowledges support from PCMI, a national Program on Interstellar Molecules of CNRS-INSU.

Journal of Physics: Conference Series **300** (2011) 012026 doi:10.1088/1742-6596/300/1/012026

Table 3. Molecular anionic abundances in IRC+10216

Ratio	Model[a]	Observation[b]
CN^-/CN	0.001	0.0025
C_2H^-/C_2H	1.3×10^{-5}	$< 1 \times 10^{-5}$
C_3N^-/C_3N	0.013	5.0×10^{-3}
C_4H^-/C_4H	0.015	2.4×10^{-4}
C_5N^-/C_5N	0.059	0.12
C_6H^-/C_6H	0.069	0.062 - 0.086
C_8H^-/C_8H	0.049	0.26

[a] Agúndez et al. [23]; Cordiner & Millar [25]; Agúndez et al. [27]; Cernicharo et al. [28]
[b] See [25] for a list of observational references.

6. References

[1] Walmsley C M, Bertout C, Combes F, Forveille T, Guillot T, Jones A and Shore S 2010 *Astron. Astrophys.* **521** E1
[2] Crockett N R et al. 2010 *Astron. Astrophys.* **521** L21
[3] Gupta H et al. 2010 *Astron. Astrophys.* **521** L47
[4] Neufeld D et al. 2010 *Astron. Astrophys.* **521** L10
[5] Buhr H et al. 2010 *Phys. Rev. Lett.* **105** 103202
[6] Schilke P et al. 2010 *Astron. Astrophys.* **521** L11
[7] Herbst E 1981 *Nature* **289** 656
[8] Lepp S and Dalgarno A 1988 *Astrophys. J.* **324** 553
[9] McCarthy M C, Gottlieb C A, Gupta H and Thaddeus P 2006 *Astrophys. J.* **625** L141
[10] Gupta H, Brünken S, Tamassia F, Gottlieb C A, McCarthy M C and Thaddeus P 2007 *Astrophys. J.* **655** L33
[11] Brünken S, Gupta H, Gottlieb C A, McCarthy M C and Thaddeus P 2008 *Astrophys. J.* **664** L43
[12] Rimmer P, Herbst E. and Roueff E 2010, in preparation
[13] Shaw G, Ferland G J, Henney W J, Stancil P C, Abel N P, Pellegrini E W, Baldwin J A and van Hoof P A M 2009 *Astrophys. J.* **701** 677
[14] Oka, T 2004 *J. Molec. Spectros.* **228** 635
[15] Plasil R, Varju J, Hejduk M, Dohnal P, Kotrik T and Glosik J 2010 *J. Phys. Conf. Ser.* this volume
[16] Sakai N, Shiino T, Hirota T, Sakai T and Yamamoto S 2010 *Astrophys. J.* **718** L49
[17] Walsh C, Harada N, Herbst E and Millar T J 2009 *Astrophys. J.* **700** 752
[18] Terzieva R and Herbst E 2000 *Int. J. Mass Spectrom.* **201** 135
[19] Herbst E and Osamura Y 2008 *Astrophys. J.* **679** 1670
[20] Herbst E 1985 *Astron. Astrophys.* **153** 151
[21] Herbst E 1982 *Chemical Physics* **65** 185
[22] Sakai N, Sakai T, Osamura Y and Yamamoto S 2007 *Astrophys. J.* **667** L65
[23] Agúndez M et al. 2010 *Astron. Astrophys.* **517** L2
[24] Eichelberger B, Snow T P, Barckholtz C and Bierbaum V M 2007 *Astrophys. J.* **667** 1283
[25] Cordiner M A and Millar T J 2009 *Astrophys. J.* **697** 68
[26] Thomas R D et al. 2010 *J. Phys. Conf. Series*, this volume
[27] Agúndez M, Cernicharo J, Guélin M, Gerin M, McCarthy M C and Thaddeus P 2008 *Astron. Astrophys.* **478** L19
[28] Cernicharo J, Guéllin M, Agúndez M, McCarthy M C and Thaddeus P 2008 *Astrophys. J.* **688** L83
[29] Cami J, Bernard-Salas J, Peeters E and Malek S E 2010 *Science* **329** 1180
[30] Bettens R P A and Herbst E 1995 *Int. J. Mass Spectrom.* **149/150** 321
[31] von Helden G, Gotts N G and Bowers M T 1993 *Nature* **363** 60

Journal of Physics: Conference Series **300** (2011) 012027 doi:10.1088/1742-6596/300/1/012027

Probabilistic representations of partial branching ratios: bridging the gap between experiments and chemical models

Pascal Pernot[1,2], Sylvain Plessis[1] and Nathalie Carrasco[3]
[1]Univ Paris-Sud, Laboratoire de Chimie Physique, UMR 8000, Orsay, F-91405
[2]CNRS, Orsay, F-91405
[3]Laboratoire Atmosphères, Milieux, Observations Spatiales, Université de Versailles Saint-Quentin, UMR 8190, F-78280 Guyancourt

E-mail: `pascal.pernot@u-psud.fr`

Abstract. Dissociative recombination (DR) of polyatomic positive ions with electrons is a very complex process, for which the breakup patterns into neutral fragments are presently unpredictable. Measured branching ratios to the various products channels are thus the only data source for modelers of chemical plasmas. Because of limited detector resolution, data for many ions are incomplete, and implementation of DR processes in chemical models is problematic. In a recent paper [Plessis S *et al.* 2010 *J. Chem. Phys.* **133** 134110], we proposed a novel approach, based on probabilistic trees, enabling modelers to fully account for the hierarchy of available information. For the first time, we were able to build a chemical model for Titan's ionosphere including all available data about DR branching ratios. The improvement on previous models is considerable, and the data needs for Titan are redefined accordingly.

1. Dissociative recombination in chemical models

Despite the accumulation of reference data in the past years [1, 2], there are still major difficulties in the modeling of dissociative recombination (DR) of polyatomic ions in natural or laboratory chemical plasmas [3–11]. The chief problem is the lack of knowledge about branching ratios, mainly due to the difficulty of detection and quantification of neutral fragments, and their isomeric and/or electronic configurations. In the absence of tractable and predictive theories, experimental data are at the moment the only source of information available to modelers. Even for smaller species (3-4 heavy atoms), information is often incomplete: for hydrocarbon ions, the distribution of H atoms between fragments is presently not accessible (see Table 1 and Fig. 1). Also, the isomeric form of the parent ion and the isomeric form and electronic state of the neutral products are generally not known [12]. This lack of information about the chemical identity, and therefore the reactivity, of the products of DR can introduce substantial biases in the predictions of chemical models [13].

Through the production of energetic radicals and heavy neutral species, DR is an important pathway of molecular complexification and diversification. It is therefore crucial to incorporate all the available knowledge about branching ratios into chemical models. This can be difficult, notably when the chemical identity of the fragments are not fully elucidated (see Table 1). If we

Journal of Physics: Conference Series **300** (2011) 012027 doi:10.1088/1742-6596/300/1/012027

Products	Measured probability	Exoergic channels	Selected channels	(b_i)
$C_3NH_x + y\,H + wH_2$	0.50±0.04	6	$CH_2CHCN + H$	(0.50)
$C_2H_x + CNH_y + zH + wH_2$	0.49±0.04	7	$\{ C_2H_2 + HCN + H$ $C_2H_2 + CN + H_2$	(0.25) (0.25)
$\{ C_3H_x + NH_y$ $C_2NH_x + CH_y + zH$	0.01±0.01	9	–	
with $x + y + z + 2w = 4$		$\Sigma = 22$	$\Sigma = 3$	

Table 1. Typical experimental results and their interpretation for the branching ratios of the dissociative recombination of a middle-sized ion: CH_2CHCNH^+ [14].

consider recent results about the DR of CH_2CHCNH^+ [14] (Table1), it is striking that from 22 exoergic channels, Vigren *et al.* have selected only three channels for astrochemistry databases, one from the first group of 6, and two from the second group of 7. The latter ones have been given identical branching ratios. This oversimplified description, ignoring all minor channels and plausible isomers (*e.g.* HNC *vs.* HCN), is dictated by the lack of an appropriate representation of such complex data in chemical databases. It can be feared that the systematic use of such simplifications would lead to an artefactual impoverished role of DR in chemical models.

In a recent paper [15], we proposed an original solution to this problem by treating partial knowledge about branching ratios as an uncertainty source. Accordingly, probabilistic representations were designed to account for the various uncertainty patterns identified through an extensive review of the literature [15]. In order to deal with partial branching ratios data and their hierarchical structure, we used probabilistic trees. This enabled us to design and run the first chemical model of Titan's ionosphere implementing all available data about DR.

In the following, we present the main steps of the method, and through our main results for Titan's ionosphere, we demonstrate how they contribute to redefine the data needs for the modeling of chemical plasmas.

2. Probabilistic representation of partial sets of uncertain branching ratios

This work extends the previous study by Carrasco and Pernot [18] about the representation of uncertain branching ratios in chemical models, where the necessity of a good description of correlations between these parameters was underlined in the context of uncertainty propagation and sensitivity analysis [13]. Our new developments acknowledge the fact that information on branching ratios, or lack thereof, has often a hierarchical structure which can be translated into probabilistic trees incorporating all available data.

The first step in our approach is to consider that reaction rates and branching ratios are typically measured by different experiments/techniques [1, 2]. Good practice in uncertainty management is to keep an explicit separation of uncertainty sources. Therefore, we will not consider here the problem of the rate coefficients [15].

For a reaction with n products channels, the branching ratios b_i have two constraints ($b_i \geq 0$ and $\sum_{i=1}^{n} b_i = 1$), which define them as probabilities. The sum-to-one is essential to ensure that, when estimating uncertainties of the outputs of a chemical model, uncertainties on branching ratios of a given process do not propagate to products not directly related to this process [15, 18, 19].

In the following, we call *probabilistic tree* a hierarchy of branching ratios, such that, at each node (represented in the schemes by an opening brace '{'), all contributions sum to 1.

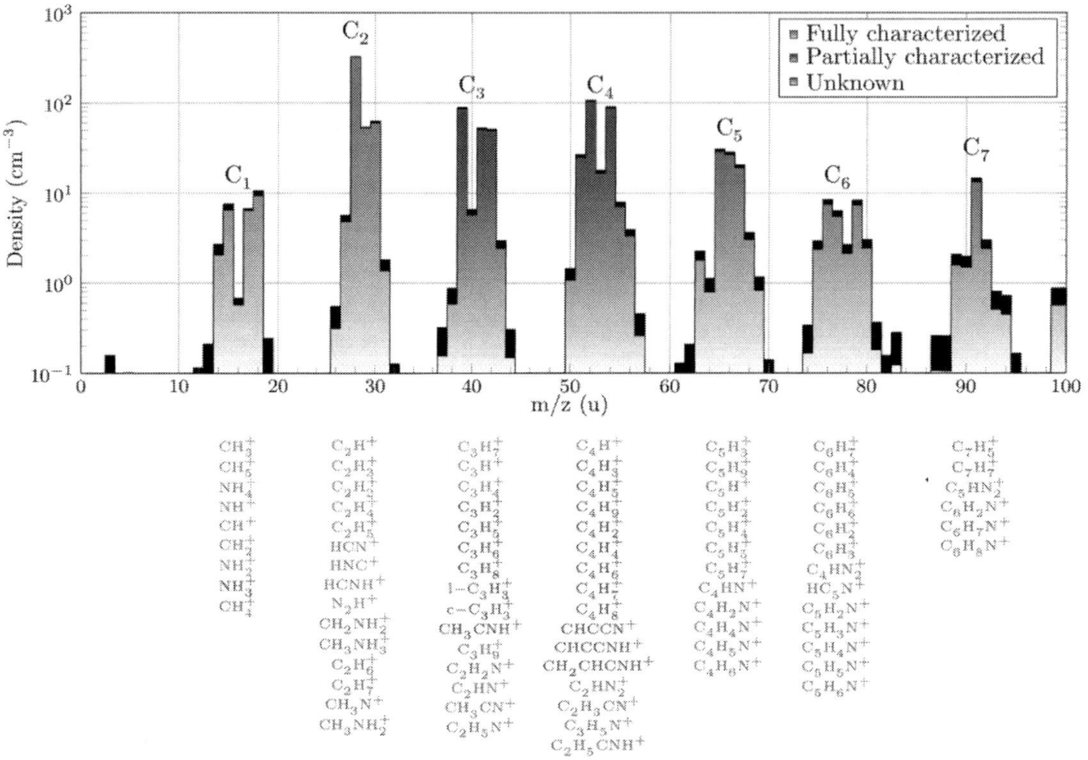

Figure 1. Typical mass spectrum of ions in Titan's ionosphere, as recorded by Cassini INMS at an altitude of ca. 1100 km [16]. The ionic species, identified by chemical modeling [17], are specified below each C_x block of peaks, where x is the number of heavy atoms in the ions. Ions are colored according to the available information on their DR products. Blocks of peaks are colored according to the status of the dominant ions.

2.1. Representation of branching ratios by probabilistic trees

We consider a few typical cases extracted from the literature and see how they can be incorporated into our representation scheme.

2.1.1. Fully characterized products. A few ions (up to class C_4, Fig. 1) have fully characterized branching ratios. For instance, the branching ratios for seven exoergic products pathways for $C_2H_4^+$ have been measured[20]

$$C_2H_4^+ + e^- \longrightarrow \left\{ \begin{array}{l} \xrightarrow{b_1=0.11\pm0.07} C_2H_3 + H \\ \xrightarrow{b_2=0.66\pm0.06} C_2H_2 + 2H \\ \xrightarrow{b_3=0.06\pm0.03} C_2H_2 + H_2 \\ \xrightarrow{b_4=0.10\pm0.04} C_2H + H_2 + H \\ \xrightarrow{b_5=0.01\pm0.01} CH_4 + C \\ \xrightarrow{b_6=0.02\pm0.02} CH_3 + CH \\ \xrightarrow{b_7=0.04\pm0.02} 2CH_2 \end{array} \right.$$

In this case, the representation is simple and does not require a hierarchy.

2.1.2. Combining complementary experiments. NH_2^+ is an interesting case, where we have to combine two sets of branching ratios coming from different experiments. The main channels have been measured first: $B_1(N + H_2) = 0.04 \pm 0.03$, $B_2(N + 2H) = 0.58 \pm 0.09$ and $B_3(NH + H) = 0.38 \pm 0.06$ [21, 22]. In a second stage, the spin state of N through the second channel was elucidated: $B_{21}((^4S)N + 2H) = 0.53 \pm 0.04$, $B_{22}((^2D)N + 2H) = 0.45 \pm 0.05$ and $B_{23}((^2P)N + 2H) = 0.02 \pm 0.02$ [22].

Preserving the measurement uncertainties and the norm constraint for each experiment leads to a natural embedding of the data into a probabilistic tree

$$\mathrm{NH_2^+ + e^-} \longrightarrow \begin{cases} \xrightarrow{B_1=0.04\pm0.03} \mathrm{N + H_2}(b_1) \\ \xrightarrow{B_2=0.58\pm0.09} \begin{cases} \xrightarrow{B_{21}=0.53\pm0.04} (^4\mathrm{S})\mathrm{N + 2H}(b_2) \\ \xrightarrow{B_{22}=0.45\pm0.05} (^2\mathrm{D})\mathrm{N + 2H}(b_3) \\ \xrightarrow{B_{23}=0.02\pm0.02} (^2\mathrm{P})\mathrm{N + 2H}(b_4) \end{cases} \\ \xrightarrow{B_3=0.38\pm0.06} \mathrm{NH + H}(b_5) \end{cases}$$

For a given pathway, the branching ratio is obtained by multiplication of the probabilities from the root to the terminal leave, *e.g.* $b_1 = B_1$, or $b_4 = B_2 * B_{23}$. Note that the spin state is left unspecified for the first pathway, an inconsistency that can be resolved by the method exposed in the next example.

2.1.3. Dealing with partial measurements. When one has partial measurements, as in Table 1, one can build a hierarchy similar to the one in the previous case, with measured values of branching ratios at the first level. This defines a partition of the products in a number of classes, and within each class, on can use unspecified (totally uncertain) branching ratios.

The example of $C_3H_2^+$ is developed in steps (1,2) of Fig. 2. One has measurements for two heavy atoms configurations $B_1(C_3)$ and $B_2(C_2 + C)$ [23, 24]. For the C_3 pathway, three exoergic channels are considered with unknown branching ratios: $(C_3 + H_2)$, $(C_3 + 2H)$ and $(C_3H + H)$, and for the $C_2 + C$ pathway, two channels are invoked: $(C_2H_2 + C)$ and $(C_2 + CH_2)$, also with unknown proportions.

In the resulting representation, the branching ratio to channel $(C_3 + H_2)$, $b_1 = B_1 * B_{11}$, lies between 0 and 0.88 ± 0.02. The same is valid for the two other channels in its class, b_2 and b_3, with the constraint that $b_1 + b_2 + b_3 = 0.88 \pm 0.02$. By construction, this representation incorporates all available information and respects the experimental data.

2.1.4. Empirical rules for increased precision. An important property of the probabilistic trees is that one can implement constraints locally (within a branch) without disturbing other parts of the tree. In particular, any constraint at the lower levels will not affect the fact that the experimental data at the first level are respected.

It is therefore possible to implement *empirical rules*, notably through order constraints ($b_1 \geq ... \geq b_n$). We emphasize that this is a strong point of this approach that non-numerical information can be incorporated in the representation to increase its predictivity.

Unfortunately, empirical rule correlating the values of branching ratios to easily estimated properties, are still mostly nonexistent. There is clearly no straight correlation between the exothermicity of products channels and their branching ratios. Using more complex indicators, weak correlations between thermodynamics and preferred breakup patterns have been detected by Vigianno *et al.* [25] and Roueff [26]. Implementation of these weak correlations in our framework is under study.

The only empirical rule we have used so far is the so-called "2H $\geq H_2$" rule [15], which states that the production of two hydrogen atoms is always more probable than one hydrogen molecule.

Main steps in generating a probabilistic tree for branching ratios

(1) Prior information: open channels

$$C_3H_2^+ + e^- \longrightarrow \begin{cases} \xrightarrow{b_1\in[0,1]} C_3 + 2H \\ \xrightarrow{b_2\in[0,1]} C_3 + H_2 \\ \xrightarrow{b_3\in[0,1]} C_3H + H \\ \xrightarrow{b_4\in[0,1]} C_2H + CH \\ \xrightarrow{b_6\in[0,1]} C_2H + C + H \end{cases}$$

(2) Experimental information introduces a hierarchy

$$C_3H_2^+ + e^- \longrightarrow \begin{cases} \xrightarrow{B_1=0.88\pm0.02} \begin{cases} \xrightarrow{B_{11}\in[0,1]} C_3 + 2H \\ \xrightarrow{B_{12}\in[0,1]} C_3 + H_2 \\ \xrightarrow{B_{13}\in[0,1]} C_3H + H \end{cases} \\ \xrightarrow{B_2=0.12\pm0.02} \begin{cases} \xrightarrow{B_{21}\in[0,1]} C_2H + CH \\ \xrightarrow{B_{22}\in[0,1]} C_2H + C + H \end{cases} \end{cases}$$

(3) Using empirical rule(s)

$$C_3H_2^+ + e^- \longrightarrow \begin{cases} \xrightarrow{B_1=0.88\pm0.02} \begin{cases} \xrightarrow{B_{11}\in[0,1]} \begin{cases} \xrightarrow{B_{111}\in[0,1]} C_3 + 2H \\ \xrightarrow{B_{112}\leq B_{111}} C_3 + H_2 \end{cases} \\ \xrightarrow{B_{12}\in[0,1]} C_3H + H \end{cases} \\ \xrightarrow{B_2=0.12\pm0.02} \begin{cases} \xrightarrow{B_{21}\in[0,1]} C_2H + CH \\ \xrightarrow{B_{22}\in[0,1]} C_2H + C + H \end{cases} \end{cases}$$

Figure 2. Probabilistic tree structure of the available information on branching ratios for the dissociative recombination of $C_3H_2^+$.

This rule is validated by all experimental data for the ions considered in Titan's ionosphere, at the exception of NH_4^+. Implementation of this rule for $C_3H_2^+$ is presented in Fig. 2(3).

However modest in its scope, each instance of this rule in a tree contracts the space of accessible branching ratios by a factor two. Ordering of more than two channels would result in still better uncertainty reduction. Considering their practical renewed interest, the search for empirical rules should be actively pursued.

2.2. Managing probabilistic trees in chemical models

For numerical applications, the treatment of probabilistic trees requires to generate representative samples to be used in a Monte Carlo Uncertainty Propagation framework [18, 19, 27]. We designed and use a toolbox of *knowledge-adapted* Dirichlet-based distributions [15]. As an illustration, the nested Dirichlet representation of the tree in Fig. 2(3) is translated in the database to

$$\{b_1, \ldots, b_5\} \sim \mathrm{Dirg}(0.88 \otimes \mathrm{Diri}(0.5 * \mathrm{Dior}(2), 0.5\,;\, 2), 0.12 \otimes \mathrm{Diun}(2)\,;\, 0.02, 0.02),$$

where Dirg, Diri, Dior and Diun are different "flavors" of the Dirichlet distribution: Dirg, the generalized Dirichlet distribution, is used when we have a set of branching ratios and their uncertainties; Diun is the uniform Dirichlet distribution, to be used in absence of information; Dior is the ordered version of Diun; and Diri is the standard Dirichlet distribution. This expression is parsed by a code, which generates random numbers at each node and makes the

Journal of Physics: Conference Series **300** (2011) 012027 doi:10.1088/1742-6596/300/1/012027

adequate products to compute the b_i's. This code is available upon request to the corresponding author.

3. Application to Titan's ionosphere

The mass spectrometers (INMS and CAPS) aboard the Cassini probe have revealed the presence in Titan's ionosphere of unexpectedly high densities of positive ions with m/z as large as 300 u [16, 35]. Typical INMS data are shown in Fig. 1, along with the tentative identification of the measured ions. In ionospheric pressure and temperature conditions, DR is a major pathway for the destruction of ions to produce heavy neutral species and/or (smaller) highly reactive radicals. The observations demonstrate also the importance of representing minor channels in the DR products, because the density range of INMS mass spectra covers at least 2 orders of magnitude: a one percent channel of a major ion might contribute as much to a given product as a 99 percent channel of a minor ion.

In fact, the extreme complexity of DR products formation is in sharp contrast with the fact that none of the ion-neutral coupled models for Titan's ionosphere integrates the multi–pathway nature of this process [28–34]. With only few exceptions, the present paradigm is to consider the H-loss channel as the only pathway for all H-bearing ions: $MH^+ + e^- \rightarrow M + H$, which is at best a severe approximation. Recent experimental studies have clearly shown that DR could not only break bonds between heavy atoms efficiently, but also break more than one bond (Fig. 3) [1, 2]. As our probabilistic approach enables to implement a DR scheme accounting for all plausible pathways, we tested it in a ionospheric chemistry model of Titan [19, 27], comparing the production rates of neutral species to those of the H-loss approximation. This is the first model implementing all available DR data for Titan's ions (58 species) [15].

Our "full scheme" includes 448 partial DR reactions producing 62 neutral species. From it, we derived an "H-loss" scheme with 63 partial reactions and 48 neutral products. This comparison showed unambiguously that the "H-loss" scheme dampens the production of radicals and favors the production of heavy stable neutral species. By contrast, the full model produces much more reactive radicals which could boost the neutral chemistry in the ionosphere. Preliminary tests in a coupled ion-neutrals models show indeed that the densities of N-bearing species are very sensitive to this flow of radicals [36]. The effect is less marked for hydrocarbons. In addition, DR is globally much more efficient than bimolecular ion-molecule reactions to produce radicals.

We also observe a "cascade" effect, where heavier ions contribute significantly to the formation of smaller neutral species. In absence of data about their DR branching ratios, most heavy species are not presently included in the model. Their breakup patterns by DR and their role in the chemistry of Titan's ionosphere is therefore a major open issue.

4. Conclusions and data needs

Probabilistic tree representation of branching ratios is a powerful method to deal with partial measurements and to implement empirical rules. In combination with Monte Carlo methods, this is a very promising tool for consistent kinetic modeling of chemical plasmas.

A major result of this study is that we are able to deal with partial measurements, which is a reason to redefine data needs for Titan's ionosphere.

To understand the major fluxes in the ionospheric chemistry, notably the impact of heavy ions on the neutral chemistry, it is *not* a priority to get exhaustive measurements of branching ratios for individual ions. If one considers the groups of ions in Fig. 1, it is much more important to get partial branching ratios for heavy ions $C_{x-y}N_yH_z^+$ ($x \geq 5$) or smaller unmeasured ions, than to get fully characterized branching ratios for already partially characterized ions (*i.e.* green-to-blue rather than green/blue-to-red, in terms of group colors).

Several strategies can be invoked to complement the experimental data. Analogies with other dissociative processes [37] and isotopomers [38] can provide precious information. Identification

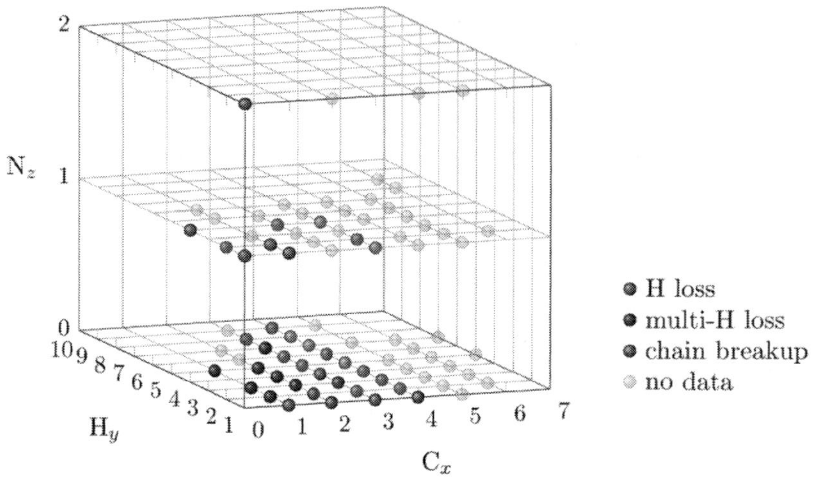

Figure 3. Major breakup patterns for $C_xH_yN_z^+$ ($x + z \neq 0$ and $y \neq 0$) ions assigned in INMS mass spectra.

of tendencies in breakup patterns would enable us to propose a first level of nesting, and thus constrain the role of heavier ions.

Empirical rules constraining the relative behavior of several branches within an inner level of nesting have also the potential to increase dramatically the model's predictivity (*e.g.* the "2H $\geq H_2$" rule). Production of "isomers" (*e.g.* HCN/HNC, $^1CH_2/^3CH_2$, spin states of N) is an ubiquitous consequence of DR, and data on their relative production are strongly needed.

Acknowledgments
The first author is grateful to NSF and NASA for travel support to DR2010.

References

[1] Florescu-Mitchell A and Mitchell J 2006 *Phys. Reports* **430** 277–374
[2] Larsson M and Orel A 2008 *Dissociative Recombination of Molecular Ions* (Cambridge University Press)
[3] Masuzaki M and Tsuzuki T 1972 *Phys. Lett. A* **38** 441–442
[4] Millar T, DeFrees D, McLean A and Herbst E 1988 *Astron. Astrophys.* **194** 250–256
[5] Herbst E and Lee H H 1997 *Astrophys. J.* **485** 689–696
[6] Schram D 2002 *Pure Appl. Chem.* **74** 369–380
[7] Geppert W, Thomas W, Ehlerding A, Semaniak J, Österdahl F, af Ugglas M, Djurić N, Paál A and Larsson M 2004 *Faraday Discuss.* **127** 425–437
[8] Geppert W *et al.* 2005 *J. Phys.: Conf. Series* **4** 26–31
[9] Fox J 2005 *J. Phys.: Conf. Series* **4** 32–37
[10] Huestis D, Bougher S, Fox J, Galand M, Johnson R, Moses J and Pickering J 2008 *Space Sci. Rev.* **139** 63–105
[11] Schram D, Zijlmans R, Gabriel O and Engeln R 2009 *J. Phys.: Conf. Series* **192** 012012
[12] Green S and Herbst E 1979 *Astrophys. J.* **229** 121–131

[13] Carrasco N, Alcaraz C, Dutuit O, Plessis S, Thissen R, Vuitton V, Yelle R and Pernot P 2008 *Planet. Space Sci.* **56** 1644–1657

[14] Vigren E, Hamberg M, Zhaunerchyk V, Kaminska M, Thomas R D, Larsson M, Millar T J, Walsh C and Geppert W D 2009 *Astrophys. J.* **695** 317–324

[15] Plessis S, Carrasco N and Pernot P 2010 *J. Chem. Phys.* **133** 134110

[16] Waite J H., Young D T, Cravens T E, Coates A J, Crary F J, Magee B, and Westlake J 2007 *Science* **316** 870–875

[17] Vuitton V, Yelle R V and McEwan M J 2007 *Icarus* **191** 722–742

[18] Carrasco N and Pernot P 2007 *J. Phys. Chem. A* **11** 3507–3512

[19] Carrasco N, Alcaraz C, Dutuit O, Plessis S, Thissen R, Vuitton V, Yelle R and Pernot P 2008 *Planet. Space Sci.* **56** 1644–1657

[20] Ehlerding A, Hellberg F, Thomas R, Kalhori S, Viggiano A, Arnold S, Larsson M and af Ugglas M 2004 *Phys. Chem. Chem. Phys.* **6** 949–954

[21] Vikor L, Al-Khalili A, Danared H, Djurić N, Dunn G H, Larsson M, Le Padellec A, Rosén S and af Ugglas M 1999 *Astron. Astrophys.* **344** 1027–1033

[22] Thomas R, Hellberg F, Neau A, Rosén S, Larsson M, Vane C, Bannister M, Datz S, Petrignani A and van der Zande W 2005 *Phys. Rev. A* **71** 032711–16

[23] Angelova G, Novotny O, Mitchell J, Rebrion-Rowe C, Le Garrec J, Bluhme H, Svendsen A and Andersen L 2004 *Int. J. Mass Spec.* **235** 7–13

[24] Mitchell J, Angelova G, Rebrion-Rowe C, Novotny O, Le Garrec J, Bluhme H, Seiersen K, Svendsen A and Andersen L 2005 *J. Phys.: Conf. Series* **4** 198–204

[25] Viggiano A A, Ehlerding A, Arnold S T and Larsson M 2005 *J. Phys.: Conf. Series* **4** 191–197

[26] Roueff E 2005 *J. Phys.: Conf. Series* **4** 1–9

[27] Carrasco N, Dutuit O, Thissen R, Banaszkiewicz M and Pernot P 2007 *Planet. Space Sci.* **55** 141–157

[28] Keller C, Cravens T and Gan L 1992 *J. Geophys. Res.* **97** 12117

[29] Keller N, Anicich V and Cravens T 1998 *Planet. Space Sci.* **46** 1157

[30] Banaszkiewicz M, Lara L, Rodrigo R, Lopez-Moreno J and Molina-Cuberos G 2000 *Icarus* **147** 386–404

[31] Wilson E and Atreya S 2004 *J. Geophys. Res.* **109** E06002

[32] De La Haye V 2005 *Corona Formation and Heating Efficiencies in Titan's Upper Atmosphere: Construction of a coupled Ion, Neutral and Thermal Structure Model To Interpret the First INMS Cassini Data* Ph.D. Thesis (The University of Michigan)

[33] De La Haye V, Waite J, Cravens T, Robertson I and Lebonnois S 2008 *Icarus* **197** 110–136

[34] Krasnopolsky V 2009 *Icarus* **201** 226–256

[35] Wahlund J E *et al.* 2009 *Planet. Space Sci.* **57** 1857–1865

[36] Plessis S 2010 *Modélisation probabiliste pour l'étude de l'influence des ions sur la composition des espèces neutres dans l'atmosphère de Titan.* Ph.D. Thesis (Univ Paris-Sud: Orsay, France).

[37] Chabot M *et al.* 2010 *Astron. Astrophys.* in press

[38] Geppert W. D. *et al.* 2004 *Astrophys. J.* **613** 1302–1309

Eighth International Conference on Dissociative Recombination (DR2010) IOP Publishing
Journal of Physics: Conference Series **300** (2011) 012028 doi:10.1088/1742-6596/300/1/012028

Dissociative Recombination: building and destroying aspects in interstellar chemistry

Evelyne Roueff [1] **and Eric Herbst** [2]

[1] Observatoire de Paris, 5 Place J. Janssen, 92195 Meudon cedex, France
[2] Departments of Physics, Chemistry, & Astronomy, The Ohio State University, Columbus, OH, USA

E-mail: `evelyne.roueff@obspm.fr`

Abstract. We discuss the ionic composition of the interstellar molecular gas. Particular emphasis is put on the halogenic ions such as H_2Cl^+ and H_2F^+ for which no DR studies have been published up to now.

1. Introduction

The molecular inventory of interstellar and circumstellar environments is now close to 150 different species without counting isotopic substitutions. Most of these molecules are neutral. Amongst the 25 molecular ions detected, 6 molecular anions have now been identified thanks to the combined efforts of spectroscopists and observers. Even though their relative number is small, atomic and molecular ions trace the ionization content of the interstellar medium and play a considerable role in the stability of interstellar clouds via their coupling to the magnetic field. Table 1 reports all the presently detected molecular ions in interstellar clouds, including various isotopologues with D, ^{13}C and ^{15}N nuclei. A number of new ionic detections, reported in italics, have taken place since the last DR meeting in 2007 thanks to the opening of the submillimeter domain with HIFI (Heterodyne Instrument for the Far Infrared) on board the *Herschel* satellite, launched in July 2009, and the ground-based APEX telescope on the Atacama plateau.

In this paper, we describe briefly the basic steps of gas phase interstellar chemistry in Section 2. Section 3 focuses on halogen chemistry in the context of new detections as well as existing theoretical and experimental studies in dissociative recombination (DR). The continuing study of isomers, such as HNCO and two higher energy isomers, may be a signature of specific branching ratios, a topic we address in Section 4. Finally, Section 5 focuses on the newly detected SH^+ molecular ion, for which no DR studies have been performed.

2. General features of positive ion chemistry

We focus the discussion on positive ions as the chemistry of negative molecular ions is discussed in detail in [1]. The ionization source in molecular clouds is provided both by ultraviolet radiation with photon energies less than 13.6 eV, corresponding to the ionization energy of atomic hydrogen, and by cosmic rays. External UV radiation is most important in diffuse clouds, where the radiation can penetrate. In such clouds, C^+ and Cl^+ are formed via this radiation. Of these, only the latter is able to react rapidly with H_2. Indeed, it is remarkable that very

Table 1. List of detected interstellar molecular ions.

2-atoms	3-atoms	4-atoms	5-atoms	6-atoms	7-atoms	9-atoms
CH^+	H_3^+	H_3O^+	H_2CO^+	HC_3NH^+	C_6H^-	C_8H^-
$^{13}CH^+$	H_2D^+	$HCNH^+$	C_4H^-	HC_4H^+		
OH^+	D_2H^+	$HOCO^+$		C_5N^-		
SH^+	*H_2O^+*	C_3N^-				
CO^+	HCO^+					
SO^+	DCO^+					
CF^+	$H^{13}CO^+$					
CN^-	HOC^+					
	HCS^+					
	N_2H^+					
	N_2D^+					
	$^{14}N^{15}NH^+$					
	H_2Cl^+					

few atomic ions (Cl^+, O^+, F^+) react directly and rapidly with H_2. Ionization via cosmic rays is critical for hydrogen atoms and molecules. In a molecular environment, H_3^+ , resulting from the $H_2^+ + H_2$ reaction, is then able to transfer its proton to other neutral molecules such as CO and N_2, eventually forming polyatomic molecular ions. The detection of these species; e.g., HCO^+, confirmed the first sketches of interstellar chemistry [2]. In dark clouds where hydrogen is the molecular reservoir, dissociative recombination reactions compete with possible exothermic ion-neutral reactions with neutrals. In diffuse environments, photodissociation needs also to be considered although detailed corresponding cross sections are not always available. In dense molecular clouds the fractional ionization fraction is about 10^{-8} whereas in diffuse gas the main ionic reservoir is ionized carbon and on the order of 10^{-4}. From complex molecular ions, DR can lead to various neutral products. In this way, DR may have a profound impact on the formation of specific neutral molecules and isomers. We consider both of these aspects in Sections 3 and 4.

3. Halogen chemistry

Halogen chemistry has received recent attention [3, 5] with the detections of HF [4], CF^+ [6], HCl [7], and H_2Cl^+ [8]. As quoted before, Cl^+ is photoionized in diffuse regions and HCl^+ is formed by the $Cl^+ + H_2$ reaction as soon as some H_2 is present. H_2Cl^+ is then obtained from the reaction of HCl^+ with H_2. Figure 1 displays the chemical network for both chlorine (left) and fluorine (right) containing species. We see the main differences between these chemical networks: whereas HCl is mainly formed from dissociative recombination of H_2Cl^+ and reactions of H_2Cl^+ with neutrals, HF results from a neutral reaction between F and H_2. This highly exothermic reaction has been studied in great detail experimentally and theoretically, and is found to have a small barrier of about 1.5 kcal mol^{-1}. One of the latest studies [9] reports the reaction rate coefficient as a function of the rotational level of H_2 for $J = 0$ and 1, which are relatively close. The theoretical limit of the reaction rate coefficient at 0 K has been determined to be 1.3 10^{-12} cm^3 s^{-1} as a result of resonance effects [10]. The DR of H_2Cl^+ and H_2F^+ has not yet been studied and corresponding product branching ratios are not known. We suggest that these branching ratios can be obtained from the general trend of the branching ratios of dihydrides as a function of energy release in the three-body channel. Such a trend was first established by [11] and a

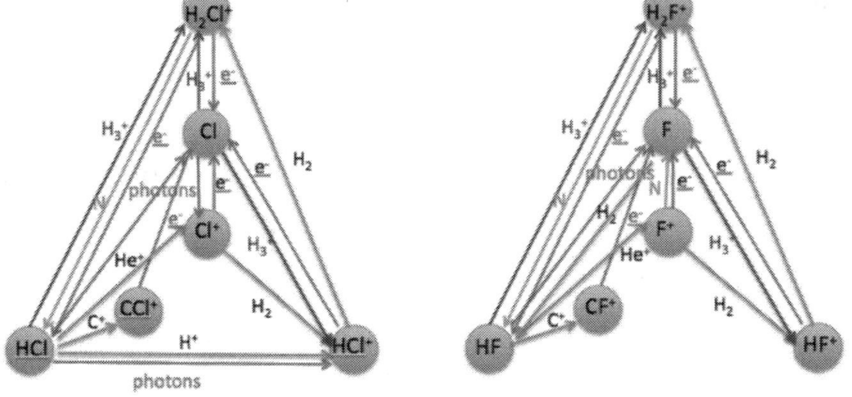

Figure 1. Chemical network for halogen chemistry. Photon induced reactions are displayed in red and recombination reactions are represented by green arrows. Reactions with neutrals (N) involve orange arrows.

clear correlation is obtained by plotting the branching ratio of the triatomic ions for 3-body dissociation as a function of the corresponding energy release, as shown in Fig. 2. From this correlation, we obtain that the 3-body dissociation fraction of H_2Cl^+ and H_2F^+ is respectively 0.56 and 0.40. Although the total DR rates are not available yet from experiments, a provisional value of 1.2 10^{-7} $(T/300)^{-0.85}$ is reported in [3] for H_2Cl^+. The rather large column density of H_2Cl^+ detected towards NGC 6334I and Sg B2(S), on the order of 10^{13} cm^{-2} [8], remains to be understood. The observations allowed also to detect HCl at an amount of about 10^{14} cm^{-2}, so that the HCl over H_2Cl^+ ratio is about 10. It would also be interesting to determine the HF / H_2F^+ ratio. However, the H_2F^+ molecular ion has not yet been detected, due to the missing information on the microwave spectroscopy of this stable molecular ion.

4. Building the isomers

Different isomers are found in interstellar clouds, despite the very low ambient temperature in all but some star-forming regions. Table 2 displays their present observational status. The HCN/HNC problem has been extensively discussed and the similar amount of both isomers at low temperatures (e.g. 10 K) is explained in part by equivalent product branching fractions for HNC + H and HCN + H in the dissociative recombination reaction of the linear $HCNH^+$ ion. We focus instead on the 3 isomers of HNCO which have been recently detected in various interstellar clouds ranging from cold to warm (50-100 K) [12]. It is remarkable that HOCN

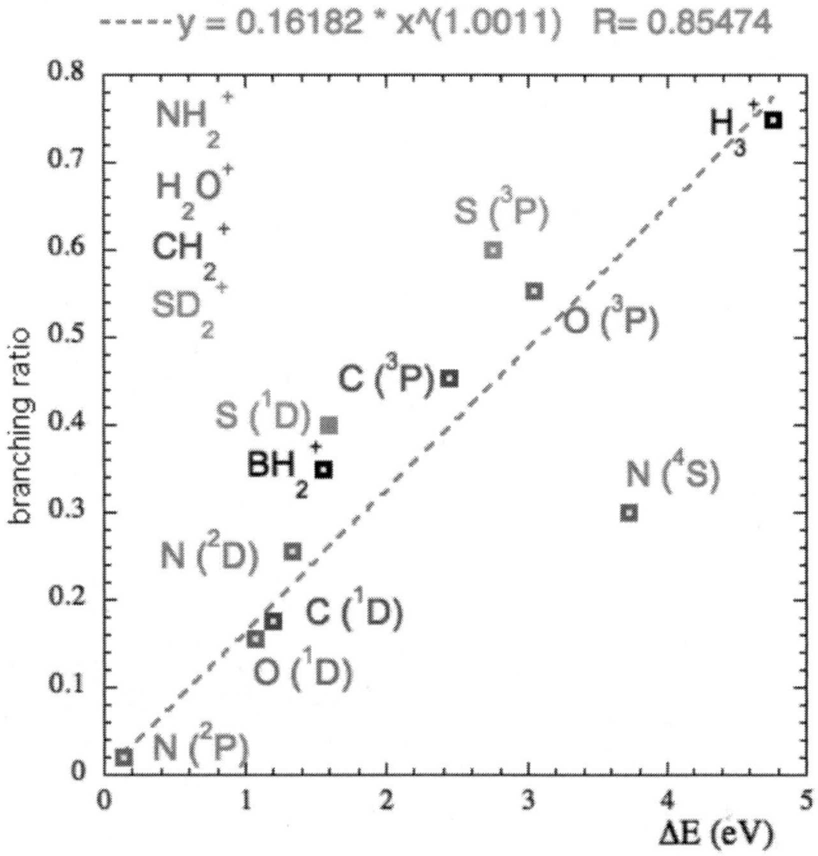

Figure 2. Branching ratio towards 3 body dissociation of dihydrides as a function of the released energy. A power law fitting function is reported on the top of the Figure where y and x are respectively the branching ratio and ΔE in eV and displayed as a dashed light blue curve. R is the corresponding correlation factor.

and HCNO, lying respectively above 24.7 and 70.7 kcal mol^{-1} above the stable HNCO isomer, are found in regions where a temperature of 10 - 20 K pertains. It is equally remarkable that in warmer regions, the HOCN isomer is detected but the HCNO isomer is not. A gas-phase chemical scenario to explain the observations is discussed in [12] where results of the chemical models rely heavily on the assumed DR branching ratios of isomeric molecular ions. Table 3 displays these reactions and the assumed DR reaction rate coefficients. However, it seems relatively well established that HCNO formation occurs via a specific gas phase mechanism, involving the neutral-neutral reaction between NO and CH_2. Grain-surface chemistry may also be at work, especially in the warmer sources, and a definite conclusion is difficult to derive in view of all the underlying hypotheses and uncertainties, as discussed in [13].

Journal of Physics: Conference Series **300** (2011) 012028 doi:10.1088/1742-6596/300/1/012028

Table 2. List of detected isomers.

HCN	HNC	
MgCN	MgNC	
SiCN	SiNC	
c-C_3H	l-C_3H	
HNCO	HOCN	HCNO
HNCS	HSCN	
c-C_3H_2	l-C_3H_2	
HC_2CN	HC_2NC	HNCCC
CH_3CN	CH_3NC	
CH_3CHO	c-CH_2OCH_2	H_2CCHOH
$HCOOCH_3$	$HOCH_2CHO$	CH_3COOH
$H_2CCCHCN$	CH_3C_2CN	
$(CH_3)_2O$	CH_3CH_2OH	
$(CH_3)_2CO$	CH_3CH_2CHO	

Table 3. Assumed DR reaction rate coefficients of H_2NCO^++ isomers at 300K.

							eV	cm^3s^{-1}
H_2NCO^+	+	e^-	→	HNCO	+	H	6.21	$1.5 \cdot 10^{-7}$
H_2NCO^+	+	e^-	→	NCO	+	H_2	5.93	$1.5 \cdot 10^{-7}$
$HNCOH^+$	+	e^-	→	HNCO	+	H	7.03	$1.5 \cdot 10^{-7}$
$HNCOH^+$	+	e^-	→	HOCN	+	H	5.96	$1.5 \cdot 10^{-7}$
$HCNOH^+$	+	e^-	→	HCNO	+	H	6.66	$1.5 \cdot 10^{-7}$
$HCNOH^+$	+	e^-	→	HONC	+	H	6.04	$1.5 \cdot 10^{-7}$
H_2OCN^+	+	e^-	→	HOCN	+	H	8.82	$1.5 \cdot 10^{-7}$
H_2OCN^+	+	e^-	→	NCO	+	H_2	9.61	$1.5 \cdot 10^{-7}$
H_2CNO^+	+	e^-	→	HCNO	+	H	7.00	$1.5 \cdot 10^{-7}$

5. The SH^+ molecular ion

Amongst the new hydrides detected recently with high frequency heterodyne receivers (Herschel, APEX), some of which are discussed in the contribution Herbst & Roueff [1], we now focus on SH^+, which has been detected towards the galactic center in both dense components near the galactic center and diffuse clouds in spiral arms surrounding it [14]. As already pointed out in [15], the reaction between S^+ and H_2 giving SH^+ + H is endothermic by a significant value of about 0.84eV. In diffuse regions, where S is ionized and where the kinetic temperature is about 40-80 K, additional energy input is required to allow the reaction to proceed. MHD shocks or turbulence driven processes may provide such a chemical impetus. In dense regions, the S + H_3^+ reaction is supposed to form SH^+, in a reaction analogous to O + H_3^+. The specificity of SH^+ is that the reaction of SH^+ with H_2 giving H_2S^+ + H is also endothermic, with an endothermicity is 0.73eV, so that dissociative recombination may become the dominant destruction channel in such environments (see below). This is different from CH^+, another important astrophysical ion, which is produced by an endothermic reaction between C^+ and

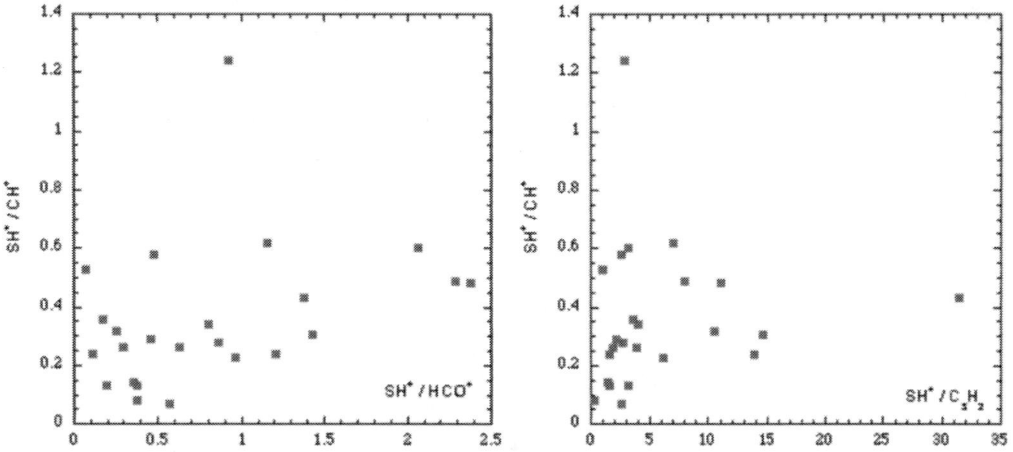

Figure 3. Comparison of column density ratios observed in various cloud components towards Sgr B2, as reported in [14].

H_2 but destroyed straightforwardly by H_2. Another channel of formation of SH^+, suggested by Abel et al. [16], involves the doubly charged sulfur ion, which reacts with H_2 to produce SH^+ and H^+ amongst other possible exothermic channels. Again, some energy input is required to form S^{++}, which can be due to X rays but cannot be provided by UV photons, as the ionization energy of S^+ is larger than 23eV.

The dissociative recombination rate coefficient of SH^+ has not been studied either theoretically nor experimentally. The usually assumed reaction rate coefficient of 2×10^{-7} $(T/300)^{-0.5}$ cm^3 s^{-1} needs to be checked. We may point out that this value is significantly larger than the DR rate coefficient of OH^+ (0.375×10^{-7} $(T/300)^{-0.5}$ cm^3 s^{-1}) and NH^+ (0.43 $\times$ 10^{-7} $(T/300)^{-0.5}$ cm^3 s^{-1}), as given in [17]. An even smaller value, $(6.3 \pm 0.7) \times 10^{-9}$ $(T/300)^{-0.48}$, of the OH^+ DR rate has been calculated theoretically by Guberman [18]. We have displayed in Fig 3 the column density ratios of SH^+/CH^+ as a function of SH^+/HCO^+ and SH^+ /c-C_3H_2 given in [14]. We see that SH^+/CH^+ is always smaller than 1 except in one case, which occurs, however, for an average value of the other ratios but is somewhat larger than the elemental sulfur to carbon ratio. Values of SH^+/HCO^+ span a range between 0.1 and 2.5, whereas SH^+ is almost always larger than c-C_3H_2. The chemical significance of these data is not clear yet.

5.1. Acknowledgments
ER wishes to thank for the support of the ANR SUMOSTAI. EH acknowledges the support of the National Science Foundation for his program in astrochemistry.

6. References
[1] Herbst E and Roueff E 2011 *J. Phys. Conf* 2010, this volume
[2] Herbst E and Klemperer W 1973 *Astrophys. J.* B **185** 505
[3] Neufeld D and Wolfire M 2010, *Astrophys. J.* **628**, 260
[4] Sonnentrucker P, Neufeld D A et al. 2010, *A&A* **521**, L12
[5] Neufeld D, Wolfire M and Schilke P 2005, *Astrophys. J.* **628**, 260
[6] Neufeld D A, Schilke P et al. 2006, *A&A* **454**, L37
[7] Schilke P, Phillips T G and Wang N, 1995, *Astrophys. J.* **441**, 334

[8] Lis D, Pearson J C, Neufeld D et al. 2010, *A&A* **521**, L9
[9] Zhu C, Krems R, Dalgarno A and Balakrishnan N, 2002 *Astrophys. J.* **577**, 795 , 2009, *Astrophys. J.* **703**, 1176
[10] Balakrishnan N and Dalgarno A, 2001, *Chem. Phys. Lett* **341**, 652
[11] Le Petit F and Roueff E, 2001, *Dissociative Recombinationof Molecular Ions with Electrons* Ed. Guberman S L, 373, Kluwer Acad. Press
[12] Marcelino N, Cernicharo J, Tercero B and Roueff E 2009 *Astrophys. J.* **690**, L27; Marcelino N, Bruenken S, Cernicharo J, Quan D, Roueff E, Herbst E and Thaddeus P 2010, *A&A* **516**, A105
[13] Quang D, Herbst E, Osamura Y and Roueff E 2010, *ApJ*, in press
[14] Menten K, Wyrowski F., Belloche A., et al. 2011, *A&A* **525**, 77
[15] Horani M, Rostas J and Roueff E 1985, *Astrophys. J.* **142**, 346; Pineau des Forêts G, Roueff E. and Flower D R 1986, *MNRAS* **223**, 743
[16] Abel N P, Federman S R and Stancil P C 2008, *Astrophys. J.* **675**, L81
[17] Florescu-Mitchell A I and Mitchell J B A 2006, *Phys. Rep.* **430**, 277
[18] Guberman S 1995, *J. Chem. Phys.* **102**, 1699

KIDA : The new kinetic database for astrochemistry

D Talbi[1] and V. Wakelam[2] for the KIDA team
[1]Université de Montpellier II - GRAAL, CNRS - UMR 5024, place Eugène Bataillon, 34095 Montpellier, France
[2]Université de Bordeaux, Laboratoire d'Astrophysique de Bordeaux, CNRS/INSU, UMR 5804, BP 89, Floirac, F-33271.

E-mail: dahbia.talbi@univ-montp2.fr

Abstract. KIDA (KInetic Database for Astrochemistry) is a project initiated by astrochemists and physical-chemists (both theoreticians and experimentalists) in order to reinforce their mutual interaction and to simplify the work of modeling the chemistry of astrophysical environments, such as the interstellar medium and planetary atmospheres.

Keywords: astrochemistry - interstellar medium - planetary atmospheres - kinetic data

1.Introduction to KIDA

If astrochemical modeling helps our understanding of how the molecules that are observed in space are formed, it is also used as a tool to study the physical conditions prevailing in the astronomical objects where these molecules are observed. However, for astrochemical models to be successful, all the chemical processes that are important in determining the molecular abundances must be included in the model and the values of the associated rate coefficients need to be close to the correct values. Unfortunately, the fraction of chemical processes that have been studied experimentally for the physical conditions prevailing in space is small, and most of the experimentally studied reactions have their rate coefficient determined for room temperature and extrapolated to the temperatures of space, leading either to large uncertainties or to misleading values. The fraction of reactions that have been studied theoretically is even smaller and for all the processes that are unknown, rate coefficients are assigned on the basis of chemical intuition or by analogy to other reactions. In such circumstances it is obvious that a stronger interaction between modelers and physical chemists (both theoreticians and experimentalists) is the best approach to improve astrochemical models and to help prevent the use of inappropriate or obsolete data/extrapolations in chemical networks. Moreover such an interaction is the best way for modelers to publicize their most urgent data needs and for physical chemists to advertise new data. Let us point out here that since the chemical networks used to model the chemistry of the interstellar medium and of the planetary atmospheres are similar, the physical chemists who work on the determination of reaction rate coefficients for both types of environments tend to be the same individuals.

KIDA, whose logo is shown in Figure 1, is by design a collaborative project with an interface that allows users to add new data as well as information and comments. New data are validated by experts before publication. The innovative functionalities of the KIDA database are outlined in the next section.

Journal of Physics: Conference Series **300** (2011) 012029 doi:10.1088/1742-6596/300/1/012029

Figure 1. The logo of KIDA

2. Kinetic data

KIDA is designed to gather all the kinetic data that can be of interest for the chemical modeling of the interstellar medium and planetary atmospheres. The reactions that are included in KIDA are subdivided into nine groups: direct cosmic-ray processes; photo-processes induced by cosmic-rays; photo-processes; bimolecular reactions and dissociative neutral attachment; charge exchange reactions; radiative association; associative detachment; electronic (dissociative) recombination and attachment; and third-body assisted association. The database provides the user with extensive information about the data such as references, details on the methods used to obtain the data, validity, range of temperature, etc.

2.1 User inputs

Since KIDA is by design a collaborative project, a friendly interface is provided to enable users to attach comments or new information to data already stored in KIDA, or to enrich the database with new data. The reviewing of new data by a group of experts prior to publication in the database is a strong asset of KIDA (see Figure 2).

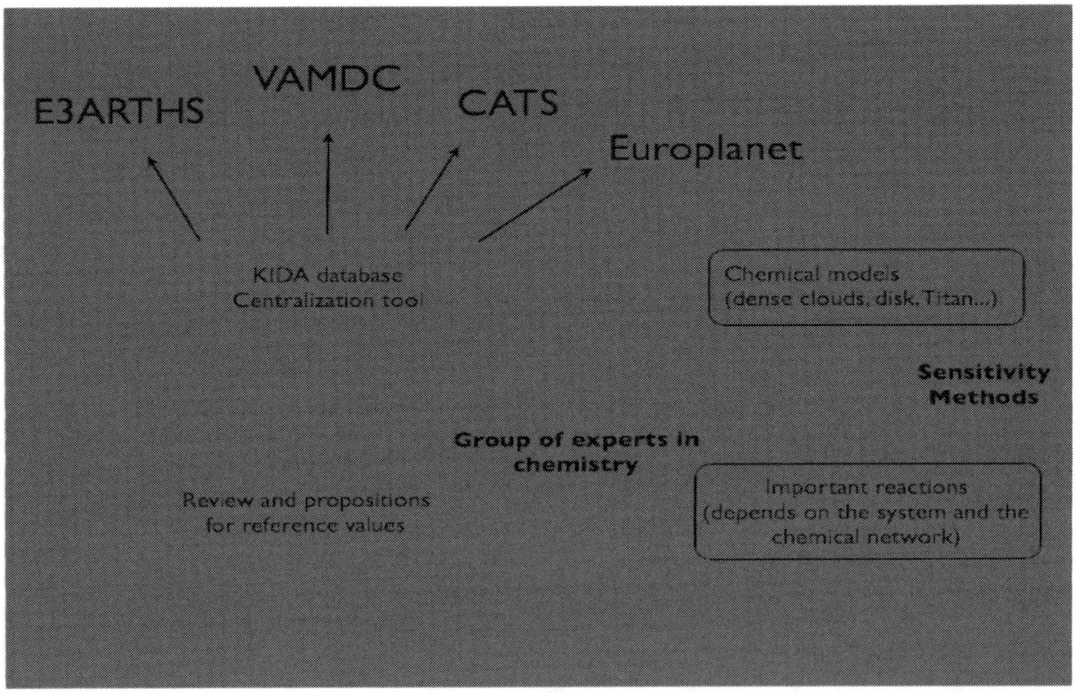

Figure 2. The specifics of KIDA: the improvement of data stored in KIDA is done by the identification of important reactions for specific objects and by the review by groups of experts on these reactions. KIDA is linked to four other projects, discussed below.

2.2 The experts
The role of the KIDA experts is to validate the addition of data from data providers, giving recommendations about the rate coefficients to use in specific physical conditions. In most cases, when a recommendation is given, the details of the expertise can be seen in a data sheet provided on the reaction page. Recommendations can be of four types: (1) not recommended; (2) not rated; (3) valid; and (4) recommended.

2.3 Data outputs
Modelers can extract lists of reactions from KIDA, based on different search criteria. The lists are automatically commented to alert the user about reactions that might be problematic. Output formats should enable insertion into various chemistry codes without further processing by the user.

2.4 Chemical networks
A second section of the database is an archive of chemical networks published by modelers. This feature should enable the sharing and intercomparison of models and ensure the traceability and reproducibility of models outputs.

3. Uncertainty management
Knowing the precision of input reaction rates is a basic requirement for the assessment of the precision of model outputs. These last are crucial for any comparison with observations or identification of key reactions [1, 2]. KIDA therefore puts a strong effort on uncertainty management and seeks to provide accuracy factors for all stored reaction rates, based on review by the experts. A major advantage of proper uncertainty management is to enable identification of key reactions as those having a major effect on the (im)precision of models outputs. On this basis, KIDA lists the key reactions to be studied in priority by physical chemists, as has been shone in a publication recently accepted by Space Science Reviews [3]. With such a procedure one can expect to improve optimally the accuracy of kinetic data for astrochemistry and thereby the precision of model predictions.

4. Links with other projects
KIDA is by essence an interdisciplinary project and, as shown in Figure 2, is linked to other projects, including:

Europlanet: a European network for planetary science, funded by the FP7. KIDA is part of the new databases constructed in this context.

VAMDC: a virtual Atomic and Molecular Data Center, aimed at interfacing several databases. Also funded by FP7, this project is led by M-L. Dubernet (Observatoire de Paris, France).

E3ARTHs: a project funded by the European Research Council (Starting Grant) and led by F. Selsis (Bordeaux University, France). Its goal is to build modeling tools for exo-planetary atmospheres. Simulations of the chemical composition of habitable planets will be done with KIDA.

CATS: a set of tools that are developed to optimize the analysis of future data from the giant radio telescope interferometer named ALMA. CATS is funded by Astronet and led by P. Schilke (Max Plank Institute, Bonn). KIDA will be directly interfaced with the chemical models of CATS.

5. Conclusions
KIDA was built by and for astrochemists and physical chemists. The KIDA website has been online since May 2010 (http://kida.obs.u-bordeaux1.fr). Anyone can register with KIDA, as well as submit

and download data. More details about the kinetic data and the spirit of KIDA can be found in the Space Science Review cited above [3], which can be downloaded from the following web site: http://kida.obs.u-bordeaux1.fr/uploads/documents/issi_database.pdf

Acknowledgment

We acknowledge the following sources of funding: the Agence Nationale de la Recherche from France, the European Research Council, Astronet, the Seven Framework Program of the European Commission FP7, the Centre National de la Recherche Scientique from France (French programs PCMI and PNP) and the University of Bordeaux. The KIDA team also thanks the International Space Science Institute from Switzerland for financing the meetings of some of the team members. D. Talbi thanks NSF and NASA for providing additional travel support.

References:

[1] Hébrard E, Dobrijevic M, Pernot P, Carrasco N, Bergeat A, Hickson K M, Canosa A, Le Picard S D and Sims I R 2009 *The Journal of Physical Chemistry A* **113** 11227

[2] Wakelam V, Herbst E, Le Bourlot J, Hersant F, Selsis F and Guilloteau S 2010 *Astronomy & Astrophysics* in press (arXiv:1004.1902).

[3] Wakelam V, Smith I W M, Herbst E, Troe J, Geppert W, Linnartz H, Oberg K, Rouff E, Aguùndez M, Pernot P, Cuppen H M, Loison J C, Talbi D 2011, accepted in Space Science Review, arXiv:1011.1184v1

AUTHOR INDEX

AUTHOR INDEX